2/08

BATS IN FORESTS

CONSERVATION AND MANAGEMENT

Edited by Michael J. Lacki, John P. Hayes, and Allen Kurta
Foreword by Merlin D. Tuttle

THE **JOHNS HOPKINS** UNIVERSITY PRESS

BALTIMORE

This book was brought to publication
with the generous assistance of the
National Congress of American Indians,
Bat Conservation International,
and the Bat Diversity Network.

The Johns Hopkins University Press
2715 North Charles Street
Baltimore, Maryland 21218-4363
www.press.jhu.edu

Library of Congress Cataloging-in-Publication Data
Bats in forests : conservation and management /
edited by Michael J. Lacki, John P. Hayes,
Allen Kurta.
p. cm.
Includes bibliographical references and index.
ISBN 0-8018-8499-3 (hardcover : alk. paper)
1. Bats—Ecology—North America. 2. Bats—
Effect of forest management on—
North America. 3. Bats—Conservation—
North America. I. Lacki, Michael J., 1956–
II. Hayes, John P. (John Parker)
III. Kurta, Allen, 1952–
QL737.C5B373 2007
639.97′94—dc22   2006013034

A catalog record for this book is available
from the British Library.

Photographs at chapter openers are by
Merlin D. Tuttle, Bat Conservation International

To Sonia, Ashley,
and Marty–
for always being there

To Joan, Heather,
and Forest

To my 'Little Red Wife,'
for being
my best friend

CONTENTS

FOREWORD

Today, many bat populations are only small remnants of former numbers. Prior to the arrival of European settlers in the United States, for example, millions of now endangered Indiana and gray myotis lived in single caves. Even species that are still considered common have, in fact, declined markedly. For example, in the 1870s great migratory flocks of eastern red bats were reported to pass over for days at a time during fall migration, a sight now vanished from American skies.

Despite their decline and an increase in conservation concern, bats still rank among the world's most neglected and misunderstood mammals. Few members of the public understand that the world's more than 1,100 species play essential ecological roles in all but the most extreme desert and polar regions. Fewer still know that most of America's 51 species rely on forests to varying degrees, either for food or shelter. In turn, forests nearly everywhere benefit from bats, probably much more than is realized. The most abundant bats inhabiting forests consume vast quantities of insects nightly, including costly "pests" such as moths and beetles. In short, the available evidence suggests that bats are essential to ecological balance and forest health, fulfilling the same roles by night as birds do by day.

A variety of factors related to forest conditions have contributed to bat decline, including disturbance of hibernation and nursery caves in forests, loss of snags and major cavities in ancient trees, careless use of pesticides in forest settings, and early neglect of forest management. Forest bats often require loose, exfoliating bark and other cavities in extra tall trees that are dying or newly dead, and these are now scarce in many places. Bats also require alternative roosts in close proximity, in part because exfoliating bark and old snags are ephemeral in nature, and in part because they must find varied temperatures and avoid predators.

Management practices that ensure age diversity of trees, retain snags and damaged trees, and aid plant diversity can greatly benefit bat populations. When forests become monocultures, bats and other insectivorous animals encounter unfavorable conditions such as prey cycles that produce fewer species of insects and longer gaps between hatch cycles. The eventual explosions of these insects then results in their designation as major forest pests. Both forests and bats are more secure when structural, floral, and faunal diversity are maintained. For these reasons, the growing collaboration between bat biologists and forestry agencies and industries is a wise investment.

In this book, America's leading experts on bats collaborated to produce an extraordinarily comprehensive review of current knowledge that is the best available anywhere. Both the volume editors and chapter authors are unsurpassed in their respective areas. Twenty-seven authors have contributed eleven chapters of up-to-date research and review that is a must read for anyone interested in either bats or forest management.

Merlin D. Tuttle

PREFACE

At least 45 species of bats occur in the United States and Canada, and 27 of these use trees for roosting at least some of the time (see table P.1), and presumably forage in or near forested areas as well. All forest-dwelling bats in North America are insectivorous, and all except one species, the pallid bat (*Antrozous pallidus;* Antrozoidae), belong to the family Vespertilionidae. Many species of bats that inhabit forests are species of concern, threatened, or endangered, because of low or unknown population size, apparent population declines, high vulnerability to disturbance, or limited geographic distribution. Although some species, such as the little brown myotis (*Myotis lucifugus*) and big brown bat (*Eptesicus fuscus*), have been studied for more than 50 years, there is surprisingly little information on how these or other bats use forested areas. In fact, some species of bats are so poorly studied that current approaches to conservation and management of these species within forests is reduced to an educated guess. The recent spate of lawsuits to halt logging on public lands in the eastern United States, due largely to the endangered Indiana myotis (*Myotis sodalis*), emphasizes the need for better information on bats in forests.

Data on forest-dwelling bats are mostly scattered throughout scientific journals and the immense gray literature of governmental agencies and private industry. To date, only two consolidated volumes of information on forest-dwelling bats in North America have been published (R. M. R. Barclay and R. M. Brigham, 1996, *Bats and Forests Symposium;* A. Kurta and J. Kennedy, 2002, *The Indiana Bat: Biology and Management of an Endangered Species*). In both books, however, individual chapters are limited in scope and, in general, present new data on a specific topic, rather than synthetic reviews of existing data or development of strategies for management of bats and their habitats.

The First Bats and Forests Symposium held in 1995 in Victoria, British Columbia, was an initial attempt to bring together individuals engaged in the study or management of forest-dwelling bats in the United States and Canada. Since then, advances in technologies and approaches to collecting data, including reduction in mass of radiotransmitters and improvement of acoustic-monitoring systems, have led to new insights into the biology of many species of forest-dwelling bats. Moreover, the number of biologists, consultants, and land managers involved with bats in forests has increased in recent decades, resulting in a large volume of both published and unpublished data and a growing need for synthesis of existing information. For example, of the papers presented at the North American Symposium on Bat Research from 1991 to 1993, 24 (or 9.3% of the total presentations) addressed topics on North American forest-dwelling bats, whereas from 2001 to 2003, 71 papers (19.1%) were associated with bats in forests. This increase in attention to forest-dwelling bats is evident in other scientific outlets as well. From 1986 to 1991, the *Journal of Wildlife Management* published only a single article addressing bats in forests, but from 1999 to 2004 the number of articles on bats in North American forests had increased to 12. Additionally, articles addressing management of forest-dwelling bats are now appearing in forestry journals, such as *Forest Science, Forest Ecology and Management,* and *Northern Journal of Applied Forestry;* outlets in which data on bats and their habitats had previously not been published.

To help synthesize new information on bats in forests and provide a foundation for future directions in the field, the Second Bats and Forests Symposium and Workshop was held in Hot Springs, Arkansas, in 2004.

Table P.1. Regional distribution of occurrence of North American tree-roosting bats

Species	Common name	Distribution in North America
Antrozous pallidus	pallid bat	NW, SW
Corynorhinus rafinesquii	Rafinesque's big-eared bat	SE
Corynorhinus townsendii	Townsend's big-eared bat	NW, SE, SW
Eptesicus fuscus	big brown bat	All
Euderma maculatum	spotted bat	NW, SW
Idionycteris phyllotis	Allen's big-eared bat	SW
Lasionycteris noctivagans	silver-haired bat	All
Lasiurus blossevillii	western red bat	NW, SW
Lasiurus borealis	eastern red bat	All
Lasiurus cinereus	hoary bat	All
Lasiurus ega	southern yellow bat	SW
Lasiurus intermedius	northern yellow bat	SE
Lasiurus seminolus	Seminole bat	SE
Lasiurus xanthinus	western yellow bat	SW
Myotis auriculus	southwestern myotis	SW
Myotis austroriparius	southeastern myotis	SE
Myotis californicus	California myotis	NW, SW
Myotis evotis	long-eared myotis	NW, SW
Myotis keenii	Keen's myotis	NW
Myotis lucifugus	little brown myotis	All
Myotis septentrionalis	northern myotis	NE, NW, SE
Myotis sodalis	Indiana myotis	NE, SE
Myotis thysanodes	fringed myotis	NW, SW
Myotis volans	long-legged myotis	NW, SW
Myotis yumanensis	Yuma myotis	NW, SW
Nycticeius humeralis	evening bat	NE, SE, SW
Pipistrellus subflavus	eastern pipistrelle	NE, SE, SW

This meeting was built around a series of invited presentations that reviewed our scientific understanding of key issues related to the ecology and management of bats in North America forests. The presentations were followed by workshops and field tours that brought together biologists and forest managers to help focus future research directions and improve strategies for management of bat habitat. The present book, Bats in Forests: *Conservation and Management,* grew from the seeds planted at that meeting and was written to meet the need for a synthesis of information on forest-dwelling bats in North America.

The book is structured around 11 chapters, beginning with an introduction to the topic of bats in North American forests that highlights key areas of progress and recommended avenues for future study of forest-dwelling bats. This chapter is followed by a series of papers that discuss use of roost trees, including cavity, bark, and foliage roosts, foraging habitat, and night roosts of bats. The final section of the book includes chapters that address silviculture and forest management, including management for bats on private industry lands, monitoring and censusing of bat populations, and response of bats to forest management at the level of the stand and landscape.

Authors of chapters were chosen for their expertise in particular areas, and, to ensure breadth and depth of coverage, most chapters have multiple authors representing different regions of the continent and/or areas of expertise. This range of backgrounds and knowledge provides the reader with differing perspectives regarding bats in forests. The book is intended to serve as a resource for students, academic biologists, governmental employees, environmental organizations, biological consultants, legal experts, and private land managers. The emphasis of this book is on bats in North American forests, so the literature cited and data sets evaluated primarily represent studies on North American species; but this volume should also serve as a comparative reference for anyone concerned with forest-dwelling bats in other temperate regions of the globe because many of the problems associated with management of bat habitat are the same.

Each manuscript was peer-reviewed to facilitate a more comprehensive coverage of material and to set a standard of quality in presentation of each chapter. We thank the following individuals for reviewing various chapters: Ed Arnett, Mike Baker, Burr Betts, Mark Brigham, Carol Chambers, John Cox, Paul Cryan, Mark Ford, Matina Kalcounis-Rüppell, Steve Langenstein, Jeff Larkin, Susan Loeb, Alice Chung-MacCoubrey, Chester Martin, Darren Miller, Susan Murray, Ralph Nyland, Tom O'Shea, Dan Taylor, Ted Weller, and Craig Willis. Financial assistance with both the Symposium and Workshop and with the preparation of this book was provided by Bat Conservation International, Inc., Weyerhaeuser Company, National Council for Air and Stream Improvement, Inc., Southeastern Bat Diversity Network, and the College of Agriculture at the University of Kentucky. We thank Dan Taylor and Merlin Tuttle for their assistance with development of the Symposium and Workshop and for support in completing this project. The editors also wish to recognize our home institutions, the University of Kentucky, University of Florida, and Eastern Michigan University for infrastructural and financial support of our research programs over the past two decades, and to all the sponsors, federal, state, and private, that have contributed technical and logistical support, and financial resources to assist us in the study of bats in forests.

CONTRIBUTORS

SYBILL K. AMELON, Research Wildlife Biologist, North Central Research Station, University of Missouri, Columbia, Missouri 65211

EDWARD B. ARNETT, Conservation Scientist, Bat Conservation International, Inc., Austin, Texas 78746

MICHAEL D. BAKER, Post-doctoral Scholar, Department of Forestry, University of Kentucky, Lexington, Kentucky 40546

ROBERT M. R. BARCLAY, Professor, Department of Biological Sciences, University of Calgary, Calgary, Alberta, Canada T2N 1N4

R. MARK BRIGHAM, Professor, Department of Biology, University of Regina, Regina, Saskatchewan, Canada S4S 0A2

S. ANDREW CARTER, Wildlife Biologist, Southern Research Station, USDA Forest Service, Nacogdoches, Texas 75965

TIMOTHY C. CARTER, Associate Scientist, Department of Zoology, Southern Illinois University, Carbondale, Illinois 62901

PAUL M. CRYAN, Wildlife Biologist, USGS, Fort Collins Science Center, Fort Collins, Colorado 80526

JOSEPH E. DUCHAMP, Doctoral Candidate, Department of Forestry and Natural Resources, Purdue University, West Lafayette, Indiana 47907

WILLIAM H. EMMINGHAM, Emeritus Professor, Department of Forest Science, Oregon State University, Corvallis, Oregon 97331

JAMES M. GULDIN, Supervisory Ecologist, Southern Research Station, USDA Forest Service, P.O. Box 1270, Hot Springs, Arkansas 71902

JOHN P. HAYES, Professor & Chair, Department of Wildlife Ecology and Conservation, University of Florida, Gainesville, Florida 32611

JAMES D. KISER, Wildlife Ecologist, Daniel Boone National Forest, Whitley City, Kentucky 42653

ALLEN KURTA, Professor, Department of Biology, Eastern Michigan University, Ypsilanti, Michigan 48197

MICHAEL J. LACKI, Professor, Department of Forestry, University of Kentucky, Lexington, Kentucky 40546

MICHAEL A. LARSON, Wildlife Research Scientist, Forest Wildlife Research Group, Minnesota Department of Natural Resources, Grand Rapids, Minnesota 55744

SUSAN C. LOEB, Wildlife Ecologist, Southern Research Station, Department of Forest Resources, Clemson University, Clemson, South Carolina 29634

JENNIFER M. MENZEL, Research Wildlife Biologist, Northeastern Research Station, USDA Forest Service, Parsons, West Virginia 26287

DARREN A. MILLER, Southern Wildlife Program Manager, Weyerhaeuser Company, Columbus, Mississippi 39704

PATRICIA C. ORMSBEE, Wildlife Ecologist, Willamette National Forest, Eugene, Oregon 97440

STUART I. PERLMETER, WELL Project Coordinator, Springfield Public Schools, Springfield, Oregon 97477

DAVID A. SAUGEY, Wildlife Biologist, Ouachita National Forest, Jessieville, Arkansas 71949

ROBERT K. SWIHART, Professor & Interim Head, Department of Forestry and Natural Resources, Purdue University, West Lafayette, Indiana 47907

JACQUES P. VEILLEUX, Assistant Professor, Department of Biology, Franklin Pierce College, Rindge, New Hampshire 03461

THEODORE J. WELLER, Research Wildlife Ecologist, Pacific Southern Research Station, USDA Forest Service, Arcata, California 95521

T. BENTLY WIGLEY, Manager of the Sustainable Forestry and Eastern Wildlife Program, National Council for Air and Stream Improvement, Inc., Clemson, South Carolina 29634

GREG K. YARROW, Professor, Department of Forestry and Natural Resources, Clemson University, Clemson, South Carolina 29634

BATS IN FORESTS: WHAT WE KNOW AND WHAT WE NEED TO LEARN

R. Mark Brigham

Ten years ago, reports about the ecology, management, and behavior of bats in North America were, with few exceptions, centered on aggregations of animals in caves and human-made structures. Over the past decade, however, the focus has changed toward trying to understand how many of these same species interact with forested environments and how we need to manage forests with a view to conserving the bats inhabiting them. Hayes (2003) recently pointed out that the degree of association between bats and forests varies considerably and that the degree of association represents a continuum rather than a set of discrete categories. Of the 45 species of bats recognized in North America (O'Shea and Bogan 2003), 25 species use forests for roosting or foraging at least some of the time. Most "forest-dwelling bats" are small (<30 g), and virtually all are aerial insectivores, although some glean invertebrates from the ground or the surface of vegetation (Faure et al. 1993; Faure and Barclay 1992; Lacki et al., chap. 4 in this volume). A significant portion of the bat community in North America is associated with wooded areas, whether for foraging, roosting, or both. Thus, it is important to consider the nature of the interaction between these animals and forested habitats.

This chapter provides a brief synthetic perspective of the general features of what biologists know about forest-dwelling bats in North America, most notably information published in the past decade, and it offers some opinion about the priorities for future studies. My review is by no means exhaustive, and thorough assessments of various aspects of the lives of forest-dwelling bats can be found in Hayes (2003), Miller et al. (2003), and subsequent chapters in this volume. This chapter is North American in outlook, but I encourage readers to examine recent studies of bats emanating from New Zealand (O'Donnell and Sedgeley 1999; Sedgeley and O'Donnell 1999), Australia (Law and Chidel 2002; Lumsden et al. 2002; Pavey and Burwell 2004; Turbill et al. 2003), and Europe (Kerth and König 1999; Russo et al. 2002; Russo and Jones 2003). Although there may be geographic differences, these studies provide a valuable ecological corollary to North American research, and it would be a grave mistake to adopt a continentally insular perspective.

ROOSTS

Roosts are critically important for bats because they provide shelter from the elements and a haven from potential predators, as well as a location to mate, raise young, hibernate, digest food, and socially interact with other individuals (Barclay and Kurta, chap. 2 in this volume; Kunz and Lumsden, 2003; Ormsbee et al., chap. 5 in this volume). In North American forests, most bats use some sort of a cavity or crevice—under exfoliating bark (California myotis, *Myotis californicus;* Brigham et al. 1997), in a furrow in the bark (silver-haired bat, *Lasionycteris noctivagans;* Barclay et al. 1988), or within the tree itself (Indiana myotis, *Myotis sodalis;* Kurta and Kennedy 2002). For the purpose of this review, I will simplistically label all such sites as "cavity roosts." A small but significant minority of North American bats roost in the foliage of trees (e.g., lasiurines, *Lasiurus* spp.; Hutchinson and Lacki 2001; Mager and Nelson 2001; Menzel et al. 1998; Willis and Brigham 2005; eastern pipistrelle, *Pipistrellus subflavus;* Veilleux and Veilleux 2004); however, comparatively little research has focused on the ecology and behavior of these animals (Carter and Menzel, chap. 3 in this volume).

Selection of cavity roosts and research on movement among sites has received more attention than all other questions concerning bats in forests combined (Hayes 2003; Kunz and Lumsden 2003; Miller et al. 2003). Although most studies focus on species living in coniferous forests of the Pacific Northwest, some general trends are evident. Kalcounis-Rüppell et al. (2005) quantitatively reviewed the available North American literature and showed that roost trees of bats consistently are tall trees with large diameters and are located in stands with an open canopy and a high density of dead trees (snags) relative to randomly selected sites. These authors noted that cavity roosts are surrounded by forest stands with more open canopies and are closer to sources of water than random sites, but foliage roosts are in stands with a more closed canopy, with no evidence of influence of distance to water on roost choice.

Forest managers need to identify general features necessary for bats that are relevant to the manager's particular jurisdiction (Marcot 1996). Thus, I expect that there will be a continued emphasis on characterizing roost-site choice by bats. It remains to be seen, however, whether further investigations will support the general patterns found to date or uncover new trends. I believe that the bats will continue to show us new things. Although there are many good published models for these types of studies and, in general, convergence in the protocols used, it is worth emphasizing that appropriate multivariate analysis of the data is essential. The recent trend toward an information-theoretic approach using the Akaike information criterion (AIC), as an alternative to null-hypothesis testing, is a good step forward (Broders and Forbes 2004). The null-hypothesis approach is prone to errors because there is high risk of eliminating biologically significant variables, even with analyses based on a high alpha level. Furthermore, although statistically comparing roosts used by bats with randomly chosen sites is a useful paradigm, recent work suggests that random sites actually are used by bats in some situations, which makes the comparisons invalid (Kalcounis and Brigham 1998). When designing new projects, biologists should remember that comparing roost trees with randomly selected sites is only an indirect means of assessing choices made by bats.

Additional facets of roost-site selection also require further study.
Studies directly addressing the motives for the choices that bats make are limited (Dechmann et al. 2004; Willis and Brigham 2005), and innovative approaches for evaluating choices in roosting structures are needed. Several recent reports have begun to address the hypothesis that the interplay of reproduction and thermoregulation promotes differences in site selection by males and females (Brack et al. 2002; Broders and Forbes 2004; Cryan and Wolf 2003; Veilleux et al. 2004). Cryan et al. (2000), for example, provide tantalizing evidence that males and females of several species select roosts, at least in part, as a function of elevation. We also must investigate the scale at which bats select roosts. Most evidence suggests that bats choose sites based on characteristics at the level of the stand (Crampton and Barclay 1998; Jung et al. 1999), likely as a result of the availability of potential roosts, but choices at the level of the landscape have not been rigorously addressed (Carter et al. 2002; Duchamp et al., chap. 9 in this volume). Finally, hardly any research has been published about the consequences of roost selection for fitness of individuals (Brigham and Fenton 1986). Obviously fitness consequences are biologically significant, but legal implications are also associated with targets for reproductive success that are written into recovery plans for endangered species. Although it is logistically difficult to collect data on reproductive success in forested situations, the incentive to try should be high.

Biologists have made great strides in understanding why bats commonly move among roosts. The emerging tenet of these studies (Barclay and Kurta, chap. 2 in this volume) is that groups of individuals generally exhibit interannual fidelity to an area of forest (Gumbert et al. 2002; Kurta et al. 2002; Veilleux and Veilleux 2004; Vonhof and Barclay 1996; Vonhof et al. 2004; Willis and Brigham 2004). There is also some evidence that bats remain loyal to specific structures and areas for several years (Barclay and Brigham 2001; Kurta and Murray 2002; Veilleux and Veilleux 2004; Willis et al. 2003; Winhold et al. 2005). In a seminal paper, Lewis (1995) reviewed hypotheses concerning roost switching (e.g., competition for space, microclimate advantages, gender differences, and social interactions), but Barclay and Kurta (chap. 2 in this volume) show that these reasons do not explain most cases of switching by cavity-roosting species. In contrast to regular movements by bats among cavity roosts, Willis and Brigham (2005) demonstrate that hoary bats (*Lasiurus cinereus*) use the same foliage roosts over days or even weeks in a study area that appears to offer an unlimited number of potential sites, and Veilleux and Veilleux (2004) provide similar results for eastern pipistrelles. Loyalty to specific sites by these foliage-roosting species discredits the hypothesis that roost fidelity is negatively correlated with ephemerality and abundance.

To further enhance understanding of fidelity, investigators would be well advised to mimic the experimental rigor employed by Gerald Kerth and his colleagues in their studies of Bechstein's myotis (*Myotis bechsteinii*) use of bat boxes in Germany (Kerth et al. 2001a; 2001b). Kerth and his colleagues were able to radiotrack repeatedly the same individuals within and between years, and they used molecular genetic techniques to assess genetic similarity between individuals. They were also able to test experimentally whether bats selected between warm and cool roosts. Admittedly, applying these approaches to bats using natural roosts in forested settings will be a challenge.

From a management perspective, roost switching has a significant implication that cannot be overemphasized. Managers must establish policies that concentrate on conserving entire forested areas for bats, because conserving specific sites alone will not suffice. In addition, management should focus on the landscape across different temporal scales. In the short term, we must manage for landscapes that provide bats with immediate roosting opportunities, but just as important, long-term management of landscapes is required to allow recruitment of older-aged stands that will generate roosts of the future. Unfortunately, management at these longer timescales often conflicts with harvesting schedules (rotation times) that are typically set by economic priorities.

FORAGING

The emphasis of studies on forest-dwelling bats to date has been on characterizing roosts and assessing roost-switching behavior, with studies on foraging occurring less often (Lacki et al., chap. 4 in this volume). Most studies addressing foraging have employed ultrasonic detectors to monitor echolocation calls in lieu of direct measures of activity by individual animals (Russo et al. 2002). Bat detectors have proven extraordinarily valuable in allowing scientists to eavesdrop on bats and, as a result, to learn about where and when they are active. Bats appear more active in some habitats than others, presumably reflecting foraging activity, and both vertical and horizontal edges of forests are important sites for commuting and foraging (Grindal and Brigham 1999; Hayes and Gruver 2000; Kalcounis et al. 1999; Menzel et al. 2005). Likewise, linear landscape elements (i.e., hedgerows and forest corridors) that connect habitats seem essential for some species (Murray and Kurta 2004; Verboom and Spoelstra 1999), although few North American biologists have addressed this concept. Stand age, structural composition, and the size of remnant patches of forest are also likely to affect patterns of foraging behavior (Guldin et al., chap. 7 in this volume; Thomas 1988; Swystun et al. 2001).

One physical attribute of the landscape that has received considerable focus in foraging studies is riparian corridors (Grindal et al. 1999; Hayes and Adam 1996; Zimmerman and Glanz 2000). Perhaps because of this emphasis, riparian buffer strips are now incorporated into many forest management plans that consider bats (Hayes and Adam 1996). Although there are some imaginative experimental studies that illustrate the importance of water surfaces that are calm and uncluttered for foraging by bats (Mackey and Barclay 1989; von Frenckell and Barclay 1987), more data are needed to evaluate the absolute significance of riparian areas to bats because much of our current knowledge of their role is anecdotal (Grindal et al. 1999).

Structure of habitat likely affects the composition of bat communities directly because of constraints imposed on individuals and species by their sensory systems (i.e., echolocation and morphology) and indirectly through the types of insect that are present. We have only limited data about the loyalty of bats to specific foraging sites and the spatial distribution of foraging areas on the landscape, both horizontally and vertically (Adam et al. 1994; Brigham 1991; Burford and Lacki 1995; Erickson and West 2003; Hurst and Lacki 1999; Kalcounis et al. 1999; Menzel et al. 2001, 2005; Murray and Kurta 2004; Waldien and Hayes 2001). In addition, explicit consideration of the relationship between habitats selected by bats for foraging and those selected for roosting is limited. Bats, for example, apparently prefer to forage in older-aged stands (Crampton and Barclay 1998; Jung et al. 1999), but this has typically been explained as a consequence of where bats roost rather than as an active choice for foraging locations. Likewise, there are some data suggesting that bats preferentially forage in stands dominated by certain tree species (Kalcounis et al. 1999) and that silvicultural thinning has minimal effects (Patriquin and Barclay 2003; Tibbels and Kurta 2003).

Kalcounis-Rüppell et al. (2005) attempted to distill general patterns in habitat use by foraging bats by conducting a meta-analysis on data from published reports, but they were unable to make any definitive conclusions. They attributed this to the small number of studies that have been conducted and to widely varying data-collection protocols. To address rigorously questions about bat foraging in space and time, experimental evaluations will be necessary.

To understand how bats might influence populations of insects, especially pests, we need to examine the basis for dietary selection by bats. One aspect of this question that has attracted considerable research interest is whether insectivorous bats forage opportunistically or selectively (Jones and Rydell 2003). The inherent difficulty of measuring prey that are actually available to bats, because of the limitations of sampling methods, represents a continuing problem for prey-selection studies (Lacki et al., chap. 4 in this volume; Whitaker 1994), and this difficulty is compounded by our lack of understanding about precisely what the sensory capabilities of bats are. These issues notwithstanding, diets of many species apparently vary both seasonally (Kurta and Whitaker 1998) and geographically (e.g., big brown bat, *Eptesicus fuscus;* Black 1972; Brigham 1990; Brigham and Saunders 1990; Whitaker 1972). Such dietary variation is not surprising when one considers the morphological variation that bats exhibit and the range of habitats available to them in North America. Burford et al. (1999) proposed that vegetative structure and plant species diversity in a particular habitat are correlated with production of prey (e.g., moths) for bats—a novel idea in the context of managing for foraging habitat and one that deserves more attention. The possibility that bats select foraging areas based on vegetative characteristics raises questions about habitat use and provides incentive to collect more detailed dietary information for a variety of species.

In the context of studies on use of roosts and foraging patterns, projects funded by forestry-related organizations have focused mostly on coniferous ecosystems (Wigley et al., chap. 11 in this volume). Jamieson et al. (2001), however, discuss the importance of hardwoods, especially cottonwood (*Populus trichocarpa*) stands, for insectivorous wildlife, arguing that these types of forests have fewer defenses against herbivory by insects and, therefore, support greater populations of insects than do coniferous stands. A large diversity of herbivorous insects use hardwoods, especially lepidopteran and coleopteran defoliators. The likely relationship between the nature and abundance of insect prey and different types of vegetation should provide incentive to manage habitats from a landscape perspective.

Although energy is typically the currency most commonly employed to test models of optimal foraging, diet may also be dictated by limiting nutrients. Barclay (1994), for example, proposed that the need for calcium might affect dietary choice within and between species of bats. If nutrients are more important than energy, at least in some situations, bats may hunt in restricted or unusual habitats, searching for specific insects that supply the necessary component.

From a forest-management perspective, the potential for active selection of prey by bats, combined with their voracious appetites, leads to the hypothesis that these mammals exert control on outbreaks of insect pests in forests (Aubrey et al. 2003). In one of the few tests of this hypothesis, Wilson (2004) showed that insect-eating bats did consume spruce budworm moths (*Choristoneura occidentalis*); however, the evidence was equivocal as to whether bats altered their foraging behavior to specialize on these moths during outbreaks. Although the notion that bats perform a valuable pest-control service is certainly logical and likely to be true, this hypothesis needs to be evaluated explicitly in a variety of forested ecosystems.

Another commonly cited view stemming from the high intake of insects by bats is that forest-roosting species are responsible for movement of significant amounts of nutrients into forested ecosystems (Gellman and Zielinski 1996; Pierson 1998; Rainey et al. 1992; Zielinski and Gellman 1999). Although numerous authors have proposed this concept, there are no substantive data to support it (Brigham et al. 2002). Unfortunately, the importance of bats as "fertilizers of the forest" has been used by some to justify efforts at conservation, even without corroborating data. Although a role for bats in the distribution of nutrients is eminently plausible and certainly deserving of research, we should not accept it without quantitative evidence.

METHODOLOGY

During the past decade, new technology has been both a boon and a bane to our understanding of bats and the forests in which they live. The advent of increasingly lightweight radiotransmitters has enhanced our ability to collect information on small-bodied bats, including various species of *Myotis*. In North America, *Myotis* has the highest diversity of species that live in forests, so the incentive to study these small mammals is high. Nevertheless, we have to resist the temptation to attach radiotags to bats just because we can, and we must be cognizant of the potential impact of added mass on behavior (Aldridge and Brigham 1988; Fenton 2003; Kalcounis and Brigham 1995). Due to the short range of transmitters and the fact that studies of forest-dwelling bats typically take place in areas with a limited number of roads, there have been few studies of foraging by bats that employed radiotelemetry (Lacki et al., chap. 4 in this volume), but the number of such studies is bound to increase. The mass of transmitters is an important issue in this context, because large transmitters are likely to have more impact on foraging than on roosting behavior.

Most studies of foraging by bats in forests involve monitoring echolocation calls using ultrasonic detectors. Although advances in hardware and software have been impressive, this has spawned debate about which detector system is most sensitive (Fenton 2000; Fenton et al. 2001; Jones 2004; Larson and Hayes 2000), the limitations to identifying species from calls (Barclay 1999; Barclay and Brigham 2004; Biscardi et al. 2004; Corben 2004; Fenton 1999; O'Farrell et al. 1999), the influence of habitat on detector function (Patriquin et al. 2003), and the design of studies and statistical interpretation of data (Gannon and Sherwin 2004; Hayes 1997, 2000; Sherwin et al. 2000). Perhaps most unfortunate is the apparent entrenched reliance on certain types of detectors depending on a researcher's continent of origin (Brigham et al. 2004); clearly the question being addressed should be the primary driver of detector choice (Jones 2004). In addition, correct interpretation of acoustic data is hampered by individual variation in the structure of echolocation calls, the influence of habitat and conspecifics on structure of calls, and variation among species in the intensity of calls (Barclay and Brigham 2004; Obrist 1995). Furthermore, an implicit bias to any detector-based study is that there is no means of associating acoustic measures of activity with actual numbers of bats (Fenton 2003).

Four emerging technologies—passive integrated transponders (PITs), thermal imaging, stable-isotope analysis, and molecular genetic analyses—will become important in future studies. PIT tags should enhance our knowledge about the interactions between bats and forests over the long term, although at present use of this system is limited by the small distance between bat and antenna that is needed for the system to work. Thermal imaging should become an important tool for enumerating bats and addressing questions about population sizes; however, the equipment currently is out of the reach of most researchers because of its cost. Stable isotopes have been used to identify the migratory routes of some forest-dwelling (Cryan et al. 2004) and nectar-feeding bats (Fleming 1995), and this technique potentially may be used to address whether bats play a significant role in movement of nutrients into and out of forested ecosystems. Although not really new, the application of molecular genetic techniques to forest bats eventually will clarify relationships among individuals and populations both at roosts (Vonhof et al. 2004) and at foraging sites (Rossiter et al. 2002). Further, molecular techniques should allow identification of prey to species, thus permitting an explicit evaluation of the impact of bats on specific insect pests (G. F. McCracken, pers. comm.).

Although not strictly a methodological issue, one of the most pressing constraints on how to enhance our understanding of bat–forest interactions is the paucity of truly experimental investigations (Barclay and Brigham 1996; Hayes 2003; Miller et al. 2003). Experimental approaches are especially needed to assess the motivations and processes that bats use to choose roosts. Large-scale, long-term, experimental investigations are required to evaluate the effects of management practices and human-caused disturbance on bats in forested ecosystems. In addition, we need to make greater use of experimental physiological data, obtained in the field or in the laboratory, to assess how global climatic change may impact forest-roosting bats in North America, akin to reports for bats in tropical cloud forests (LaVal 2004) and for hibernation sites (Humphries et al. 2002).

Finally, perhaps the greatest limitation on effective management of bats is our inability to measure population size (Bogan et al. 2003) and ultimately the effects of management efforts on fitness. Although there are many standard techniques for censusing bats that roost in caves or buildings, roost switching and the distribution of roosts across the landscape make these logistically impractical for tree-roosting species (Kunz 2003). Most studies of bats purporting to evaluate management practices and human-caused disturbance to forested ecosystems rely on indices of abundance and not actual data on populations of bats (Grindal et al. 1999).

NEW DIRECTIONS

Several recent studies suggest new directions for examining bat–forest interactions. Akin to ornithologists in the Northern Hemisphere emphasizing studies of breeding birds in summer, research on forest-dwelling bats to date also is almost entirely focused on the reproductive period of May through August. The documentation of eastern red bats (*Lasiurus borealis*) hibernating in leaf litter and roosting in trees throughout winter in Missouri (Mormann et al. 2004) illustrates the temporal bias of when most investigations have been conducted (Gellman and Zielinski 1996). To understand life histories of bats more fully, but also to manage forested ecosystems better, we must understand use of forested ecosystems by bats outside the reproductive season. Clearly work by Lynn Robbins and his students in Missouri illustrates that just because trees in temperate forests become dormant over winter, does not mean that forested habitats cease to be important for bats.

I return to my opening point about bats exploiting human-made structures to discuss another important theme. Although I heartily applaud the recent emphasis on investigating the ecology and behavior of bats in natural environments, we should not ignore bats in buildings. The logistics of conducting research on forest-dwelling bats will continue to be a constraint for the foreseeable future (Kunz and Reynolds 2003); therefore, studies of building-dwelling bats may prove particularly important. In some cases, the species are the same, and I would argue that insights gleaned from bats living in buildings could be used to formulate testable hypotheses about how bats interact with forests. An example is the work of Tom O'Shea and his colleagues in Colorado that focuses on big brown bats and the incidence of rabies (O'Shea et al. 2004). Forest-dwelling bats have not been considered at all in the context of the impact of disease, and work like O'Shea's is an important reminder of issues that we might otherwise forget.

9

What We Know

CONCLUSION

Many workers perceive an increasing recognition in both the public and private sectors of the importance of bats for sustaining the health of forested ecosystems (Fenton 1997; Medellin et al. 2000). Although there has been significant growth in our understanding of how bats in North America interact with forests, the idea that bats are critically important to those systems, as opposed to just dependent on them, is not yet supported by rigorous empirical research and may represent a degree of wishful thinking by those concerned with the conservation of these fascinating animals. Nevertheless, conservation of bats is becoming an increasingly high priority in North America. Most North American bats are "forest-dwelling bats," and many are designated as threatened or endangered by some jurisdiction due to low or unknown population sizes and apparent vulnerability to disturbance (Pierson 1998). In short, many species of wildlife need forests for their survival and a complete management plan should consider the needs of all residents, including bats. Unfortunately, some species of bat are so little studied that management is based on little more than educated guesses, because the data available are deemed insufficient from a management perspective (Miller et al. 2003; although see Kalcounis-Rüppell et al. 2005). Although I certainly applaud the efforts to understand, conserve, and manage for bats, an important focus should be to increase our understanding of the interactions between bats and the forests in which they live (Fenton 2003). We must provide land managers with the information needed to generate conservation plans and, perhaps even more compelling, we need to use our imaginations and resourcefulness to make bats divulge their many secrets. During the past decade, we have made excellent strides toward unraveling some mysteries and although we have much left to uncover, I am positive that the journey will be an exciting one.

ACKNOWLEDGMENTS

I am grateful to R. M. R. Barclay, S. Holroyd, A. Kurta, and three anonymous reviewers for thoughtfully commenting on an earlier draft of the manuscript and to various funding agencies, most notably the Natural Sciences and Engineering Research Council (NSERC), that have given my students and me the opportunity to learn a little bit about the lives of bats in forests.

LITERATURE CITED

Adam, M.D., M.J. Lacki, and T.G. Barnes. 1994. Foraging areas and habitat use of the Virginia big-eared bat in Kentucky. Journal of Wildlife Management 58:462–469.

Aldridge, H.D.J.N., and R.M. Brigham. 1988. Load carrying and maneuverability in an insectivorous bat: a test of the 5% "rule" of radio-telemetry. Journal of Mammalogy 69:379–382.

Aubrey, K.B., J.P. Hayes, B.L. Biswell, and B.G. Marcot. 2003. Ecological role of arboreal mammals in western coniferous forests, pp. 405–443, *in* Mammal community dynamics in coniferous forests of western North America: management and conservation (C.J. Zabel and R.G. Anthony, eds.). Cambridge University Press, Cambridge, MA.

Barclay, R.M.R. 1994. Constraints on reproduction by flying vertebrates: energy and calcium. American Naturalist 144:1021–1031.

———. 1999. Bats are not birds—a cautionary note on using echolocation calls to identify bats: a comment. Journal of Mammalogy 80:290–296.

Barclay, R.M.R., and R.M. Brigham, eds. 1996. Bats and forests symposium. Research Branch, British Columbia Ministry of Forests, Victoria, BC.

———. 2001. Year-to-year reuse of tree-roosts by California bats (*Myotis californicus*) in southern British Columbia. American Midland Naturalist 146:80–85.

———. 2004. Geographic variation in the echolocation calls of bats: a complication for identifying species by their calls, pp. 144–149, *in* Bat echolocation research: tools, techniques and analysis (R.M. Brigham, E.K.V. Kalko, G. Jones, S. Parsons, and H.J.G.A. Limpens, eds.). Bat Conservation International, Austin, TX.

Barclay, R.M.R., P.A. Faure, and D.R. Farr. 1988. Roosting behavior and roost selection by migrating silver-haired bats (*Lasionycteris noctivagans*). Journal of Mammalogy 69:821–825.

Biscardi, S, J. Orprecio, M.B. Fenton, A. Tsoar, and J.M. Ratcliffe. 2004. Data, sample sizes and statistics affect the recognition of species of bats by their echolocation calls. Acta Chiropterologica 6:347–363.

Black, H.L. 1972. Differential exploitation of moths by the bats, *Eptesicus fuscus* and *Lasiurus cinereus*. Journal of Mammalogy 53:598–601.

Bogan, M.A., P.M. Cryan, E.W. Valdez, L.E. Ellison, and T.J. O'Shea. 2003. Western crevice and cavity-roosting bats, pp. 69–77, *in* Monitoring trends in bat populations of the United States and Territories: problems and prospects (T.J. O'Shea and M.A. Bogan, eds.). U.S. Geological Survey Information and Technology Report ITR-2003-0003.

Brack, V., Jr., C.W. Stihler, R.J. Reynolds, C.M. Butchkoski, and C.S. Hobson. 2002. Effect of climate and elevation on distribution and abundance in the mid-eastern United States, pp. 21–28, *in* The Indiana bat: biology and management of an endangered species (A. Kurta and J. Kennedy, eds.). Bat Conservation International, Austin, TX.

Brigham, R.M. 1990. Prey selection by big brown bats (*Eptesicus fuscus*) and common nighthawks (*Chordeiles minor*). American Midland Naturalist 124:73–80.

———. 1991. Flexibility in foraging and roosting behaviour by the big brown bat (*Eptesicus fuscus*). Canadian Journal of Zoology 69:117–121.

Brigham, R.M., R.M.R. Barclay, J.M. Psyllakis, D.J.H. Sleep, and K.T. Lowrey. 2002. Guano traps as a means of assessing habitat use by foraging bats. Northwest Naturalist 85:15–18.

Brigham, R.M., and M.B. Fenton. 1986. The influence of roost closure on the roosting and foraging behaviour of *Eptesicus fuscus* (Chiroptera, Vespertilionidae). Canadian Journal of Zoology 64:1128–1133.

Brigham, R.M., E.K.V. Kalko, G. Jones, S. Parsons, and H.J.G.A. Limpens, eds. 2004. Bat echolocation research: tools, techniques and analysis. Bat Conservation International, Austin, TX.

Brigham, R.M., and M.B. Saunders. 1990. Diet of big brown bats (*Eptesicus fuscus*) in relation to insect availability in southern Alberta, Canada. Northwest Science 64:7–10.

Brigham, R.M., M.J. Vonhof, R.M.R. Barclay, and J.C. Gwilliam. 1997. Roosting behavior and roost-site preferences of forest-dwelling California bats (*Myotis californicus*). Journal of Mammalogy 78:1231–1239.

Broders, H.G., and G.J. Forbes. 2004. Interspecific and intersexual variation in roost-site selection of northern long-eared and little brown bats in the Greater Fundy National Park ecosystem. Journal of Wildlife Management 68:602–610.

Burford, L.S., and M.J. Lacki. 1995. Habitat use by *Corynorhinus townsendii virginianus* in the Daniel Boone National Forest. American Midland Naturalist 134:340–345.

Burford, L.S., M.J. Lacki, and C.V. Covell Jr. 1999. Occurrence of moths among habitats in a mixed mesophytic forest: implications for management of forest bats. Forest Science 45:323–329.

Carter, T.C., S.K. Carroll, J.E. Hofmann, J.E. Gardner, and G.A. Feldhamer. 2002. Landscape analysis of roosting habitat in Illinois, pp. 160–164, *in* The Indiana bat: biology and management of an endangered species (A. Kurta and J. Kennedy, eds.). Bat Conservation International, Austin, TX.

Corben, C. 2004. Zero-crossings analysis for bat identification: an overview, pp. 195–107, in Bat echolocation research: tools, techniques and analysis (R.M. Brigham, E.K.V. Kalko, G. Jones, S. Parsons, and H.J.G.A. Limpens, eds.). Bat Conservation International, Austin, TX.

Crampton, L.H., and R.M.R. Barclay. 1998. Selection of roosting and foraging habitat by bats in different-aged mixed-wood stands. Conservation Biology 12:1347–1358.

Cryan, P.M., M.A. Bogan, and J.S. Altenbach. 2000. Effect of elevation on distribution of female bats in the Black Hills, South Dakota. Journal of Mammalogy 81:719–725.

Cryan, P.M., M.A. Bogan, R.O. Rye, G.P. Landis, and C.L. Kester. 2004. Stable hydrogen isotope analysis of bat hair as evidence for seasonal molt and long-distance migration. Journal of Mammalogy 85:995–1001.

Cryan, P.M., and B.O. Wolf. 2003. Sex differences in the thermoregulation and evaporative water loss of a heterothermic bat, *Lasiurus cinereus,* during its spring migration. Journal of Experimental Biology 206:3381–3390.

Dechmann, D.K.N., E.K.V. Kalko, and G. Kerth. 2004. Ecology of an exceptional roost: energetic benefits could explain why the bat *Lophostoma silvicolum* roosts in active termite nests. Evolutionary Ecology Research 7:1037–1050.

Erickson, J. L., and S.D. West. 2003. Associations of bats with local structure and landscape features of forested stands in western Oregon and Washington. Biological Conservation 109:95–102.

Faure, P.A., and R.M.R. Barclay. 1992. The sensory basis of prey detection by the long-eared bat, *Myotis evotis,* and the consequences for prey selection. Animal Behaviour 44:31–39.

Faure, P.A., J.H. Fullard, and J.H. Dawson. 1993. The gleaning attacks of the northern long-eared bat, *Myotis septentrionalis,* are relatively inaudible to moths. Journal of Experimental Biology 178:173–189.

Fenton, M.B. 1997. Science and the conservation of bats. Journal of Mammalogy 78:1–14.

———. 1999. Describing the echolocation calls and behaviour of bats. Acta Chiropterologica 1:127–136.

———. 2000. Choosing the "correct" bat detector. Acta Chiropterologica 2:215–224.

———. 2003. Science and the conservation of bats: where to next? Wildlife Society Bulletin 31:6–15.

Fenton, M.B., S. Bouchard, M.J. Vonhof, and J. Zigouris. 2001. Time-expansion and zero-crossing period meter systems present significantly different views of echolocation calls of bats. Journal of Mammalogy 82:721–727.

Fleming, T.H. 1995. The use of stable isotopes to study the diets of plant-visiting bats, pp. 99–110, *in* Bats: ecology, behaviour, and evolution (P.A. Racey, U. McDonnell, and S. Swift, eds.). Oxford University Press, Oxford, U.K.

Gannon, W.L., and R.E. Sherwin 2004. Are acoustic detectors a "silver bullet" for assessing habitat use by bats? pp. 38–45, *in* Bat echolocation research: tools, techniques and analysis (R.M. Brigham, E.K.V. Kalko, G. Jones, S. Parsons, and H.J.G.A. Limpens, eds.). Bat Conservation International, Austin, TX.

Gellman, S.T., and W.J. Zielinski. 1996. Use by bats of old-growth redwood hollows on the northern coast of California. Journal of Mammalogy 77:255–265.

Grindal, S.D., and R.M. Brigham. 1999. Impacts of forest harvesting on habitat use by foraging insectivorous bats at different spatial scales. Ecoscience 6:25–34.

Grindal, S.D, J.L. Morissette, and R.M. Brigham. 1999. Concentration of bat activity in riparian habitats over an elevational gradient. Canadian Journal of Zoology 77:972–979.

Gumbert, M.W., J.M. O'Keefe, and J.R. MacGregor. 2002. Roost fidelity in Kentucky, pp. 143–152, *in* The Indiana bat: biology and management of an endangered species (A. Kurta and J. Kennedy, eds.). Bat Conservation International, Austin, TX.

Hayes, J.P. 1997. Temporal variation in activity of bats and the design of echolocation-monitoring studies. Journal of Mammalogy 78:514–524.

———. 2000. Assumptions and practical considerations in the design and interpretation of echolocation-monitoring studies. Acta Chiropterologica 2:225–236.

———. 2003. Habitat ecology and conservation of bats in western coniferous forests, pp. 81–119, *in* Mammal community dynamics in coniferous forests of western North America: management and conservation (C.J. Zabel and R.G. Anthony, eds.). Cambridge University Press, Cambridge, MA.

Hayes, J.P., and M.D. Adam 1996. The influence of logging riparian areas on habitat utilization by bats in western Oregon, pp. 228–237, *in* Bats and forests symposium (R.M.R. Barclay and R.M. Brigham, eds). Research Branch, British Columbia Ministry of Forests, Victoria, BC.

Hayes, J.P., and J.C. Gruver 2000. Vertical stratification of bat activity in an old-growth forest in western Washington. Northwest Science 74:102–108.

Humphries, M.M., D.W. Thomas, and J.R. Speakman. 2002. Climate-mediated energetic constraints on the distribution of hibernating mammals. Nature 418:313–316.

Hurst, T.E., and M.J. Lacki. 1999. Roost selection, population size and habitat use by a colony of Rafinesque's big-eared bats (*Corynorhinus rafinesquii*). American Midland Naturalist 142:363–371.

Hutchinson, J.T., and M.J. Lacki. 2001. Selection of day roosts by red bats in mixed mesophytic forests. Journal of Wildlife Management 64:87–94.

Jamieson, B., E. Peterson, M. Peterson, and I. Parfitt. 2001. The conservation of hardwoods and associated wildlife in the CBFWCP area in southeastern British Columbia. Unpublished report. Columbia Basin Fish and Wildlife Compensation Program, Nelson, BC.

Jones, G. 2004. Conference wrap-up: Where do we go from here? pp. 166–167, *in* Bat echolocation research: tools, techniques and analysis (R.M. Brigham, E.K.V. Kalko, G. Jones, S. Parsons, and H.J.G.A. Limpens, eds.). Bat Conservation International, Austin, TX.

Jones, G., and J. Rydell. 2003. Attack and defense: interactions between echolocation bats and their insect prey, pp. 301–345, *in* Bat ecology (T.H. Kunz and M.B. Fenton, eds.). University of Chicago Press, Chicago, IL.

Jung, T.S., I.D. Thompson, R.D. Titman, and A.P. Applejohn. 1999. Habitat selection by forest bats in relation to mixed-wood stand types and structure in central Ontario. Journal of Wildlife Management 63:1306–1319.

Kalcounis, M.C., and R.M. Brigham. 1995. Intraspecific variation in wing loading affects habitat use by little brown bats (*Myotis lucifugus*). Canadian Journal of Zoology 73:89–95.

———. 1998. Secondary use of aspen cavities by tree-roosting big brown bats. Journal of Wildlife Management 62:603–611.

Kalcounis, M.C., K.A. Hobson, R.M. Brigham, and K.R. Hecker. 1999. Bat activity in the boreal forest: importance of stand type and vertical strata. Journal of Mammalogy 80:673–682.

Kalcounis-Rüppell, M.C., J.M. Psyllakis, and R.M. Brigham. 2005. Tree roost selection by bats: an empirical synthesis using meta-analysis. Wildlife Society Bulletin 33:1123–1132.

Kerth, G., and B. König. 1999. Fission, fusion and nonrandom associations in female Bechstein's bats (*Myotis bechsteini*). Behaviour 136:1187–1202.

Kerth, G., M. Wagner, and B. König. 2001a. Roosting together, foraging apart: information transfer about food is unlikely to explain sociality in female Bechstein's bats (*Myotis bechsteinii*). Behavioural Ecology and Sociobiology 50:283–291.

Kerth, G., K. Weissmann, and B. König. 2001b. Day roost selection in female Bechstein's bats (*Myotis bechsteinii*): a field experiment to determine the influence of roost temperature. Oecologia 126:1–9.

Kunz, T.H. 2003. Censusing bats: challenges, solutions, and sampling biases, pp. 9–19, *in* Monitoring trends in bat populations of the United States and Territories: problems and prospects (T.J. O'Shea and M.A. Bogan, eds.). U.S. Geological Survey Information and Technology Report ITR-2003-0003.

Kunz, T.H., and L.F. Lumsden. 2003. Ecology of cavity and foliage roosting bats, pp. 3–89, *in* Bat ecology (T.H. Kunz and M.B. Fenton, eds.). University of Chicago Press, Chicago, IL.

Kunz, T.H., and D.S. Reynolds. 2003. Bat colonies in buildings, pp. 91–102, *in* Monitoring trends in bat populations of the U.S. and territories: problems and prospects (T.J. O'Shea and M.A. Bogan, eds.). U.S. Geological Survey Information and Technology Report ITR-2003-0003.

Kurta, A., and J. Kennedy, eds. 2002. The Indiana bat: biology and management of an endangered species. Bat Conservation International, Austin, TX.

Kurta, A., and S.W. Murray. 2002. Philopatry and migration of banded Indiana bats (*Myotis sodalis*) and effects of radio transmitters. Journal of Mammalogy 83:585–589.

Kurta, A., S.W. Murray, and D.H. Miller. 2002. Roost selection and movements across the summer landscape, pp. 118–129, *in* The Indiana bat: biology and management of an endangered species (A. Kurta and J. Kennedy, eds.). Bat Conservation International, Austin, TX.

Kurta, A., and J.O. Whitaker Jr. 1998. Diet of the endangered Indiana bat (*Myotis sodalis*) on the northern edge of its range. American Midland Naturalist 140:280–286.

Larson, D.J., and J.P. Hayes. 2000. Variability and sensitivity of Anabat II bat detectors and a method of calibration. Acta Chiropterologica 2:209–213.

LaVal, R.K. 2004. Impact of global warming and locally changing climate on tropical cloud forest bats. Journal of Mammalogy 85:237–244.

Law, B., and M. Chidel. 2002. Tracks and riparian zones facilitate use of Australia regrowth forests by insectivorous bats. Journal of Applied Ecology 39:605–617.

Lewis, S.E. 1995. Roost fidelity of bats: a review. Journal of Mammalogy 76:481–496.

Lumsden, L.F., A.F. Bennett, and J.E. Silinsa. 2002. Location of roosts of the lesser long-eared bat *Nyctophilus geofroyi* and Gould's wattled bat *Chalinolobus gouldii* in a fragmented landscape in south-eastern Australia. Biological Conservation 106:237–249.

Mackey, R.L., and R.M.R. Barclay. 1989. The influence of physical clutter and noise on the activity of bats over water. Canadian Journal of Zoology 67:1167–1170.

Mager, K.J., and T.A. Nelson. 2001. Roost-site selection by eastern red bats (*Lasiurus borealis*). American Midland Naturalist 145:120–126.

Marcot, B.G. 1996. An ecosystem context for bat management: a case study of the interior Columbia River basin, USA, pp. 19–36, *in* Bats and forests symposium (R.M.R. Barclay and R.M. Brigham, eds.). Research Branch, British Columbia Ministry of Forests, Victoria, BC.

Medellin, R.A., M. Equihua, and M.A. Amin. 2000. Bat diversity and abundance as indicators of disturbance in Neotropical rainforests. Biological Conservation 14:1666–1675.

Menzel, M.A., T.C. Carter, B.R. Chapman, and J. Laerm. 1998. Quantitative comparison of tree roosts used by red bats (*Lasiurus borealis*) and Seminole bats (*L. seminolus*). Canadian Journal of Zoology 76:630–634.

Menzel, M.A., J.M. Menzel, W.M. Ford, J.W. Edwards, T.C. Carter, J.B. Churchill, and J.C. Kilgo. 2001. Home range and habitat use of male Rafinesque's big-eared bats (*Corynorhinus rafinesquii*). American Midland Naturalist 145:402–408.

Menzel, J.M., M.A. Menzel, J.C. Kilgo Jr., W.M. Ford, J.W. Edwards, and G.F. McCracken. 2005. Effect of habitat and foraging height on bat activity in the coastal plain of South Carolina. Journal of Wildlife Management 69:235–245.

Miller, D.A., E.B. Arnett, and M.J. Lacki. 2003. Habitat management for forest-roosting bats of North America: a critical review of habitat studies. Wildlife Society Bulletin 31:30–44.

Mormann, B., M. Milam, and L. Robbins. 2004. Hibernation: red bats do it in the dirt. Bats 22:6–10.

Murray, S.W., and A. Kurta. 2004. Nocturnal activity of the endangered Indiana bat (*Myotis sodalis*). Journal of Zoology (London) 262:197–206.

Obrist, M.K. 1995. Flexible bat echolocation: the influence of individual, habitat and conspecifics on sonar signal design. Behavioral Ecology and Sociobiology 36:207–219.

O'Donnell, C.F.J., and J.A. Sedgeley. 1999. Use of roosts in the long-tailed bat, *Chalinolobus tuberculatus,* in temperate rainforest in New Zealand. Journal of Mammalogy 80:913–923.

O'Farrell, M.J., B.W. Miller, and W.L. Gannon. 1999. Qualitative identification of free-flying bats using the Anabat detector. Journal of Mammalogy 80:1–23.

O'Shea, T.J., and M.A. Bogan. 2003. Introduction, pp. 69–77, *in* Monitoring trends in bat populations of the United States and Territories: problems and prospects (T.J. O'Shea and M.A. Bogan, eds.). U.S. Geological Survey Information and Technology Report ITR-2003-0003.

O'Shea, T.J., V. Shankar, R.A. Bowen, C.E. Rupprecht, and J.H. Wimsatt. 2004. Serological status of bats in relation to rabies: What does the presence of anti-rabies virus neutralizing antibodies mean? Bat Research News 45:249–250.

Patriquin, K.J., and R.M.R. Barclay. 2003. Foraging by bats in cleared, thinned and unharvested boreal forest. Journal of Applied Ecology 40:646–657.

Patriquin, K.J., L.K. Hogberg, B.J. Chruszcz, and R.M.R Barclay. 2003. The influence of habitat structure on the ability to detect ultrasound using bat detectors. Wildlife Society Bulletin 31:475–481.

Pavey, C.R., and C.J. Burwell. 2004. Foraging ecology of the Horseshoe bat, *Rhinolophus megaphyllus* (Rhinolophidae), in eastern Australia. Wildlife Research 31: 403–413.

Pierson, E.D. 1998. Tall trees, deep holes, and scarred landscapes: conservation biology of North American bats, pp. 309–325, *in* Bat biology and conservation (T.H. Kunz and P.A. Racey, eds.). Smithsonian Institution Press, Washington, DC.

Rainey, W.E., E.D. Pierson, M. Colberg, and J.H. Barclay. 1992. Bats in hollow redwoods: seasonal use and role in nutrient transfer into old growth communities. Bat Research News 33:71.

Rossiter, S.J., G. Jones, R.D. Ransome, and E.M. Barratt. 2002. Relatedness structure and kin-biased foraging in the greater horseshoe bat (*Rhinolophus ferrumequinum*). Behavioral Ecology and Sociobiology 51:510–518.

Russo, D., and G. Jones. 2003. Use of foraging habitats by bats in a Mediterranean area determined by acoustic surveys: conservation implications. Ecography 26:197–209.

Russo, D., G. Jones, and A. Migliozzi. 2002. Habitat selection by the Mediterranean horseshoe bat, *Rhinolophus euryale* (Chiroptera: Rhinolophidae) in a rural area of southern Italy and implications for conservation. Biological Conservation 107:71–81.

Sedgeley, J.A., and C.F.J. O'Donnell. 1999. Roost selection by the long-tailed bat, *Chalinolobus tuberculatus,* in temperate New Zealand rainforest and its implications for the conservation of bats in managed forests. Biological Conservation 88:261–276.

Sherwin, R.E., W.L. Gannon, and S. Haymond. 2000. The efficacy of acoustic techniques to infer differential use of habitat by bats. Acta Chiropterologica 2:145–153.

Swystun, M.B., J.M. Psyllakis, and R.M. Brigham. 2001. The influence of residual tree patch isolation on habitat use by bats in central British Columbia. Acta Chiropterologica 3:197–201.

Thomas, D.W. 1988. The distribution of bats in different ages of douglas-fir forests. Journal of Wildlife Management 52:619–626.

Tibbels, A.E., and A. Kurta. 2003. Bat activity is low in thinned and unthinned stands of red pine. Canadian Journal of Forest Research 33:2436–2442.

Turbill, C., B.S. Law, and F. Geiser. 2003. Summer torpor in a free-ranging bat from subtropical Australia. Journal of Thermal Biology 28:223–226.

Veilleux, J.P., and S.L. Veilleux. 2004. Intra-annual and inter-annual fidelity to summer roost areas by female eastern pipistrelles, *Pipistrellus subflavus.* American Midland Naturalist 152:196–200.

Veilleux, J.P., J.O. Whitaker Jr., and S.L. Veilleux. 2004. Reproductive stage influences roost use by tree roosting female eastern pipistrelles, *Pipistrellus subflavus.* Ecoscience 11:249–256.

Verboom, B., and K. Spoelstra. 1999. Effects of food abundance and wind on the use of tree lines by an insectivorous bat, *Pipistrellus pipistrellus.* Canadian Journal of Zoology 77:1393–1401.

von Frenckell, B., and R.M.R. Barclay. 1987. Bat activity over calm and turbulent water. Canadian Journal of Zoology 65:219–222.

Vonhof, M.J., and R.M.R. Barclay. 1996. Roost-site selection and roosting ecology of forest-dwelling bats in southern British Columbia. Canadian Journal of Zoology 74:1797–1805.

Vonhof, M.J., H. Whitehead, and M.B. Fenton. 2004. Analysis of Spix's disk-winged bat association patterns and roosting home ranges reveal a novel social structure among bats. Animal Behaviour 68:507–521.

Waldien, D.L., and J.P. Hayes. 2001. Activity areas of female long-eared *Myotis* in coniferous forests in western Oregon. Northwest Science 75: 307–314.

Whitaker, J.O., Jr. 1972. Food habits of bats from Indiana. Canadian Journal of Zoology 50:877–883.

———. 1994. Food availability and opportunistic versus selective feeding in insectivorous bats. Bat Research News 35:75–77.

Willis, C.K.R., and R.M. Brigham. 2004. Roost switching, roost sharing and social cohesion: forest-dwelling big brown bats, *Eptesicus fuscus,* conform to the fission-fusion model. Animal Behaviour 68:495–505.

———. 2005. Physiological and ecological aspects of roost selection by reproductive female hoary bats (*Lasiurus cinereus*). Journal of Mammalogy 86:85–94.

Willis, C.K.R., K.A. Kolar, A.L. Karst, M.C. Kalcounis-Rüppell, and R.M. Brigham. 2003. Long term reuse of trembling aspen cavities as roosts by big brown bats (*Eptesicus fuscus*). Acta Chiropterologica 5:85–90.

Wilson, J.M. 2004. Foraging behaviour of insectivorous bats during an outbreak of western spruce budworm. MS thesis. University of Calgary, Calgary, Alberta.

Winhold, L., E. Hough, and A. Kurta. 2005. Long-term fidelity by tree-roosting bats to a home area. Bat Research News 46:9–10.

Zielinski, W.J., and S.T. Gellman 1999. Bat use of remnant old-growth redwood stands. Conservation Biology 13:160–167.

Zimmerman, G.S., and W.E. Glanz. 2000. Habitat use by bats in eastern Maine. Journal of Wildlife Management 64:1032–1040.

ECOLOGY AND BEHAVIOR OF BATS ROOSTING IN TREE CAVITIES AND UNDER BARK

2

Robert M. R. Barclay and Allen Kurta

Bats spend more time roosting than in any other activity. Thus, it is not surprising that some researchers have argued that roost availability influences the diversity of bat communities (Humphrey 1975), that destruction of roosts is responsible for population declines (Evelyn et al. 2004; Lunney et al. 1988), or that, in this book, there are several chapters devoted to the roosting ecology and behavior of forest-dwelling bats. Whether roosts are used during the day or night, or during hibernation, they must provide protection from predators and ambient environmental conditions, as well as space for social interaction (O'Shea and Bogan 2003). To maximize their fitness, individual bats should select roosts that provide these features, although the relative importance of each feature may vary among species, populations, genders, and reproductive conditions.

Tremendous strides have been made in the past 10 years in documenting and understanding the roosting behavior and ecology of bats, and the implications of forest harvesting and management for bat populations. Our goal in this chapter is to examine current knowledge and search for patterns in roost selection and behavior among species, both from a basic biological perspective and to develop general recommendations regarding forest management and future research. We focus on diurnal roosts in tree cavities and under bark, because other types of roosts are dealt with in subsequent chapters.

GENERAL BIOLOGY OF CAVITY- AND BARK-ROOSTING BATS

Although globally bats vary in size from approximately 2 g to more than 1 kg (Nowak 1994), all species in Canada and the United States are relatively small (<50 g), and this has profound implications for many aspects of their biology, including roosting. Their small size allows bats to occupy small spaces as roosts. Because of their large surface-area-to-volume ratio, bats have high mass-specific metabolic rates, and maintaining a high body temperature is energetically costly (Speakman and Thomas 2003). In addition, flight is an expensive mode of transportation, at least per unit time (Speakman and Thomas 2003). And, lastly, even though females typically have litters of only one or two young, raising them to full size is also costly (Barclay and Harder 2003). In combination, these features mean that bats have large energetic demands and any roosting behaviors that result in energy savings should be favored by natural selection.

The ability to use torpor as a means of conserving energy is a feature of the physiology of many species of bats (Speakman and Thomas 2003). Torpor involves a controlled reduction in body temperature (T_b) and the ability to rewarm using endogenous sources of heat (Wang and Wolowyck 1988). Even shallow torpor, involving declines of <10°C in T_b, can result in significant energy savings (Studier 1981; Webb et al. 1993). Although this benefit applies to all bats, regardless of gender or reproductive condition, the costs of using torpor vary. Torpid bats are incapable of flight and are thus more vulnerable to predation. Additionally, torpor slows gestation (Racey 1973; Racey and Swift 1981), milk production (Wilde et al. 1999), and juvenile growth, all of which should be particularly costly to females and their young, especially in areas with short growing seasons (Hoying and Kunz 1998).

During spring and summer, male and female tree-roosting bats typically segregate in different areas, either in different roosts (Hamilton and Barclay 1994; Kunz and Lumsden 2003), local areas (Barclay 1991), or larger-scale geographical ranges (Cryan 2003). Pregnant females of most tree-roosting bats form maternity colonies involving several to hundreds of individuals (Kunz and Lumsden 2003). Males and nonreproductive females, in contrast, often roost solitarily or in small groups. When one considers the different costs of torpor and different requirements for diurnal roosts, it is not surprising that males, nonreproductive females, and reproductive females often use different types of roosts and behave differently (Broders and Forbes 2004; Hamilton and Barclay 1994; Mattson et al. 1996; Wilkinson and Barclay 1997). Once parturition occurs, females leave their pups in the diurnal roosts while they forage, but may return to nurse the young at night. This may limit the distance lactating females can travel to forage (Henry et al. 2002), and result in different roost-selection criteria than during pregnancy (Chruszcz and Barclay 2002; Lausen and Barclay 2002).

The aggregating behavior of reproductive female bats in maternity roosts likely has evolved for various reasons and involves both benefits and costs. Although a limited supply of appropriate roosts could result in passive aggregations, this seems unlikely for tree-roosting bats in most situations. However, although the number of potential roosts in most natural, unmanaged forests appears to be large, little information is available on their quality. High-quality sites may be in short supply in some areas, forcing bats to aggregate more than they would otherwise.

Individual females in maternity colonies may benefit by living in a group in several ways. Information sharing about roosts and foraging areas has been suggested for several species (Kerth and Reckardt 2003; Wilkinson 1992). Coloniality and associated behaviors also may reduce predation risk (Fenton et al. 1994; Kalcounis and Brigham 1994), although larger groups may be easier for some predators to detect. Finally, there are significant thermoregulatory advantages to roosting communally. The colony may influence the roost microclimate (Burnett and August 1981), and by clustering, individuals reduce exposure to ambient conditions and save energy otherwise used to maintain a high body temperature (Hollis 2004; Kurta 1985; Roverud and Chappell 1991; Trune and Slobodchikoff 1976). Clustering also conserves water by reducing evaporative water loss (Kurta et al. 1989; Webb 1995). These energetic and water-related advantages may be especially important to reproductive females and young.

Colonial roosting also entails costs, including increased transmission of parasites and diseases, and competition for resources (Krebs and Davies 1993). Given the variation in the degree to which tree-roosting bats form aggregations (Kunz and Lumsden 2003), the trade-off between the costs and benefits of coloniality likely varies among species.

The biology of tree-roosting bats leads to the rather obvious prediction that not all potential roosts are equally suitable. Individuals should select tree, stand, and landscape features that minimize the energetic costs of thermoregulation and foraging, protect the bats from predators, and allow the optimum group size to form. Although some researchers initially suggested that the characteristics selected by bats should be similar across species (Vonhof and Barclay 1996), and at a coarse level they are (Hayes 2003), studies on the morphological, ecological, and behavioral differences among species of bats indicate that such differences likely favor selection of different roost characteristics, among and within species. Indeed, this may be one way in which coexisting species partition resources (Foster and Kurta 1999; Humphrey 1975; Sedgeley 2003).

STUDIES ON ROOST SELECTION AND ROOSTING BEHAVIOR

Until recently, most published studies on tree-roosting bats were anecdotal, often describing a single tree roost discovered by accident (Barclay and Cash 1985; Christian 1956; Cope et al. 1974; Humphrey et al. 1977; Kurta et al. 1993a; Parsons et al. 1986). Studies on roosting focused on bats in buildings or caves because large numbers of individuals predictably could be found at specific sites. Advances in radiotelemetry, in particular, reductions in transmitter mass, have allowed a dramatic increase in studies on more species of bats, and, in the past decade, there has been a greater focus on bats roosting in trees.

Although miniature radiotransmitters now allow tracking of most North American bats to their diurnal roosts, there are limitations. Small transmitters are costly, restricting sample sizes and the number of studies. Time and logistical constraints, as well as limited research budgets, typically do not allow all individuals in a colony to be tagged, and thus some questions regarding movements among roosts, colony coherence, and social behavior are not easily addressed. Small sample sizes also often limit the power of conclusions that can be drawn (Lacki and Baker 2003). In addition, the batteries in small transmitters only last for a few weeks and thus season-long tracking of particular individuals is not possible. Multi-year studies on particular individuals, colonies, or roost-trees are rare.

Of the 45 species of bats in Canada and the United States, most are not known to roost in tree cavities or under bark. However, there are at least some data on the roosting ecology and behavior of 16 species that do roost in cavities or under bark (table 2.1). Information is limited on

Table 2.1. Characteristics of the trees used for roosting by various cavity- and bark-roosting bats in Canada and the United States

Species	Location	Sex	Reproductive condition	Number — Bats	Number — Trees	Predominant tree species	Roost type	Decay stage	Mean dbh (cm)	Mean Tree ht (m)	Mean Roost ht (m)	Mean Canopy closure (%)	Mean Bark (%)	Source
Antrozous pallidus	Arizona	F	reproductive	3		Pinus ponderosa		dead						Rabe et al. 1998
Corynorhinus rafinesquii	Louisiana	F	postlactating	9	4	Nyssa sylvatica	cavity		59–103					Lance et al. 2001
	Louisiana				44	Nyssa aquatica	basal hollow	live	120	25	1.3			Gooding and Langford 2004
	Texas	F/M	rep/nonrep		5	N. sylvatica	basal hollow	live						Mirowsky et al. 2004
Eptesicus fuscus	Saskatchewan	F	reproductive	11	27	Populus tremuloides	woodpecker	live	36	26	8.4			Kalcounis and Brigham 1998
	Saskatchewan	F	reproductive			P. tremuloides								Willis et al. 2003
	Maryland	F	reproductive		1	Quercus sp.	cavity				12			Christian 1956
	South Dakota	F	reproductive	3	8	P. ponderosa	cavity	dead	42*	22*	12.5			Cryan et al. 2001
	Oregon	F	reproductive	4	7	P. ponderosa	cavity	live/dead	76	18		25		Betts 1996
	British Columbia	F	reproductive	8	15	P. tremuloides, P. ponderosa*	woodpecker, cavity	early*	large*	tall*			more*	Vonhof 1996
	British Columbia	F	reproductive	33	8	P. ponderosa	cavity	dead	63					Brigham 1991
	Arizona	F	reproductive	6		P. ponderosa		dead						Rabe et al. 1998
Idionycteris phyllotis	Arizona	F	reproductive	21	18	P. ponderosa		dead						Rabe et al. 1998
Lasionycteris noctivagans	Washington	F/M		13	15	P. ponderosa, Pinus monticola*		>3	45*	tall*		less*		Campbell et al. 1996
	South Dakota	F	reproductive	9	10	P. ponderosa	woodpecker	5*	44*	14.2	10.2			Mattson et al. 1996
		F/M	nonrep	7	25	P. ponderosa	bark	4*	37*		3.4			
	Alberta	F	both	6	11	P. tremuloides*	cavity	early	43	22*	11		96	Crampton and Barclay 1998
	Ontario	F	reproductive		1	Tilia americana	woodpecker		38		5.4			Parsons et al. 1986
	Saskatchewan	F	reproductive		1	Populus balsamifera	woodpecker				5			
	Oregon	F	reproductive	10	17	P. ponderosa, Larix sp.	cavity	4,5*	60	27*	16.5*	41.5	57	Betts 1998b
	British Columbia	F	lactating		4		woodpecker							Vonhof and Barclay 1996

	British Columbia	F	reproductive	10	12	*P. tremuloides, Pinus contorta**	woodpecker, cavity	random	large*	tall*			Vonhof 1996
Myotis auriculus	Arizona	F	reproductive	1		*Quercus gambelii*	cavity	dead					Rabe et al. 1998
	Arizona	F	lactating	15	39	*Q. gambelii, P. ponderosa*	cavity	live	46.4*	10.7*	6.2	96*	Bernardos et al. 2004
Myotis austroriparius	Texas	F/M	rep/nonrep	8		*N. sylvatica, N. aquatica*	basal hollow	live					Mirowsky et al. 2004
	Illinois	F	lactating	2	1	*N. aquatica*	basal hollow		105			35	Hofmann et al. 1999
	Louisiana	F/M		50	1	*N. aquatica*	basal hollow		108	26.5			Gooding and Langford 2004
Myotis californicus	British Columbia	F	reproductive	9	19	*Pseudotsuga menziesii, P. ponderosa**	bark, cavity	4,5	56*	27		42*	Brigham et al. 1997
Myotis evotis	Oregon	F	reproductive	21	20	*P. menziesii*	snags	1,2*	93	34		83	Waldien et al. 2000
				41	41	*Tsuga heterophylla**	stumps		59	tall*	1		Waldien et al. 2003
	Oregon	F/M	rep/nonrep	24	24	*P. menziesii*	stumps		<100			>40	
	British Columbia	F/M	nonrep	17	17	*Pinus spp.**	bark of stumps		large*				Vonhof and Barclay 1997
	New Mexico	F	reproductive	1		*Juniperus spp.*	cavity						Chung-MacCoubrey 1996
Myotis lucifugus	Arizona	F	reproductive	15	30	*P. ponderosa, Q. gambelii*		dead					Rabe et al. 1998
	Alberta	F	both	9	16	*P. tremuloides**	cavity	early	41	22*	11	93	Crampton and Barclay 1998
	Manitoba	F	reproductive	1		*P. tremuloides*	cavity	live			2		Barclay and Cash 1985
	British Columbia	F	reproductive	2		*P. tremuloides*	cavity	live	42				Parsons et al. 2002
	Saskatchewan	F	reproductive	1		*P. tremuloides*	woodpecker	live					Kalcounis and Hecker 1996
		M		2	5	*P. tremuloides, Picea glauca*	woodpecker, stump	dead/live	33	22.3	8.4		
	New Brunswick	F		2	2	*Picea rubens*		mid	33			62	Broders and Forbes 2004
		M		12	48	*P. rubens*	bark, cavity	all	33			72	

(continued)

Table 2.1. Continued

Species	Location	Sex	Reproductive condition	Number: Bats	Number: Trees	Predominant tree species	Roost type	Decay stage	dbh (cm)	Mean: Tree ht (m)	Mean: Roost ht (m)	Mean: Canopy closure (%)	Mean: Bark (%)	Source
Myotis septentrionalis	Kentucky	F	reproductive	13	43	*Pinus echinata, Quercus* spp.	cavity, bark	4	29		5	91		Lacki and Schwierjohann 2001
		F/M			14	*Oxydendrum arboreum*	cavities	live	12					
	West Virginia	F	lactating	7	12	*Robinia pseudoacacia**	cavity	4	29	18.7	10.8*	92		Menzel et al. 2002
	Michigan	F	reproductive	11	32	*Acer saccharinum, Fraxinus pennsylvanica*	bark, cavity	live/dead	65	23	10.7	44		Foster and Kurta 1999
	Indiana	F	reproductive		1	*Ulmus americana*	bark	dead						Mumford and Cope 1964
	South Dakota	F	reproductive	9	21	*P. ponderosa*	cavity	dead	39*	9	4.2			Cryan et al. 2001
	New Hampshire	F	reproductive	26	47	*Fagus grandifolia, Acer saccharum*	cavity	early*	41*	15*		83	78*	Sasse and Pekins 1996
	New Brunswick	F	rep/nonrep	16	55	*A. saccharum,* Betula alleghaniensis**	cavity	mid*	44			74	79	Broders and Forbes 2004
		M		16	57	*P. rubens*	bark, cavity	mid*	32			67	74	
Myotis sodalis	Indiana	F	reproductive	16	2	*Carya cordiformis, C. ovata*	bark	dead						Humphrey et al. 1977
	Indiana	M		4	12	*Quercus* spp., *Pinus* spp., *Ulmus americana*	bark	dead	38			49		Whitaker and Brack 2002
		F	reproductive		17	*Populus deltoides, U. americana*	bark		62					Cope et al. 1974
	Indiana	F	reproductive		1	*U. americana*	bark							
	Illinois	F	reproductive	2	1	*Platanus occidentalis*	cavity	dead	56	16	5	0	0	Kurta et al. 1993b
	Michigan	F	reproductive	23	23	*F. pennsylvanica*	bark	dead	40.9	25.1	9.9	0		Kurta et al. 1996
	Michigan	F	reproductive	32	38	*Acer* spp., *Fraxinus* spp., *U. americana*	bark, cavity	dead	42*	18	10	31		Kurta et al. 2002

	Location	Sex	Status	n	n	Tree species	Roost type	Decay	Height	dbh	Roost ht	%		Reference
	North Carolina/Tennessee	F	reproductive	6	8	*Pinus spp., Quercus rubra*	bark	dead	49*	18			46	Britzke et al. 2003
	Kentucky	M		60	280	*Pinus spp., Quercus spp., Carya spp*		dead	30.3					Gumbert et al. 2002
	Missouri	F	reproductive	64	54	*Quercus spp., Carya ovata*	bark	dead	58.4			less*	70	Callahan et al. 1997
Myotis thysanodes	California	F/M	all	9	23	*P. menziesii*	bark	2,3*	121*	40.5*		78.5*	74	Weller and Zabel 2001
	South Dakota	F	lactating	2	9	*P. ponderosa*	cavity	dead	43*	7				Cryan et al. 2001
	New Mexico	F	reproductive	15	3	*P. ponderosa*	cavity							Chung-MacCoubrey 1996
	Arizona	F	reproductive	16		*P. ponderosa*		dead						Rabe et al. 1998
Myotis volans	Oregon	F	reproductive	6	41	*P. menziesii*		1,2	100	40*				Ormsbee and McComb 1998
	South Dakota	F	both	2	7	*P. ponderosa*	cavity	dead	43*	11				Cryan et al. 2001
	British Columbia	F	reproductive	2	2	*P. tremuloides*	cavity	live	45					Parsons et al. 2002
	New Mexico	F	reproductive	13	2	*P. ponderosa, Pinus edulis*	cavity, bark							Chung-MacCoubrey 1996
	Arizona	F	reproductive	16		*P. ponderosa*		dead						Rabe et al. 1998
Myotis yumanensis	California	F/M	both	18	18	*Sequoia sempervirens, Quercus agrifolia**	cavity	alive	115*	19.6*				Evelyn et al. 2004
Nycticeius humeralis	South Carolina	F	reproductive	3	14	*Pinus spp.*	cavity	live	31	21		54		Menzel et al. 2001
		J	nonrep	4	6	*Pinus spp.*	bark	dead	31	18		28		
	Georgia	F/M		3	14	*Pinus spp., Quercus virginiana*	cavities		42	18.7	8.2	67		Menzel et al. 1999
Multiple species	British Columbia	F/M		21	21	*Pinus spp., Larix occidentalis**	bark, woodpecker	4,5*	large*	tall*	18.4	less*		Vonhof and Barclay 1996
	California	F/M		26		*S. sempervirens*	basal hollows	live						Gellman and Zielinski 1996
	Newfoundland	F/M		7	8	*Abies balsamea*	woodpecker, cavity	dead	29	8.7*	4.2		55	Grindal 1999
	New Mexico	F	reproductive	15	15	*P. ponderosa*	cavity							Chung-MacCoubrey 2003
	Arizona	F	reproductive	76	121	*P. ponderosa*	bark, cavity	dead	69*	18		50	more*	Rabe et al. 1998

Means are presented when available.
dbh, diameter at breast height; rep, reproductive; nonrep, nonreproductive.
*Denotes the trait was not selected at random from among available trees or roosts.

many of these species; however, a few species have been the focus of attention including the Indiana myotis (*Myotis sodalis*) and the silver-haired bat (*Lasionycteris noctivagans*). In total, we synthesized information from almost 50 studies on the roosting ecology and behavior of cavity- and bark-roosting bats in Canada and the United States, and others from other parts of the world. Unfortunately many studies are plagued by small samples, and consequently some ignore potential interspecific differences and combine observations for up to eight different species (Grindal 1999; Rabe et al. 1998). Virtually all published reports pool data from adult females in various reproductive states, and many also add data from adult males and juveniles. These methodological inconsistencies may obscure patterns specific to individual species or to various categories of age, sex, or reproductive condition; they make it difficult to test specific hypotheses concerning ecology and behavior. Most studies concentrate on roost selection and data concerning roost fidelity or dispersal often are rudimentary.

Another problem that restricts our ability to synthesize what is known about tree-roosting bats is that different researchers measure different variables regarding the roost, forest stand, or landscape, or do so in different ways. Additionally, the geographic scale used (i.e., tree, stand, or landscape) may well influence the ability to detect roost selection. Despite these limitations, some broad trends in the roosting ecology and behavior of tree-crevice and bark-roosting bats are evident (Kalcounis-Rüppell et al. 2005; Lacki and Baker 2003).

ROOSTING ECOLOGY AND ROOST SELECTION

Forest-dwelling bats use various diurnal roosts, including those in caves, rock crevices, buildings, the foliage of trees, tree cavities and crevices, and spaces under bark. Tree cavities may be created originally by primary excavators (woodpeckers) or through natural decay, and crevices can be formed via lightning strikes or other natural processes. Although cavities and crevices may differ in their characteristics, not enough information is available to treat them separately and many authors do not define or distinguish among them.

Among North American bats, some species appear more flexible than others in the types of roosts they use. For example, female long-eared myotis (*Myotis evotis*) roost in cavities in snags (Waldien et al. 2000) and behind the bark of stumps (Vonhof and Barclay 1997; Waldien et al. 2000), as well as in rock crevices in forested areas (Caceres 1998; Solick 2004). On the other hand, silver-haired bats consistently roost in abandoned woodpecker holes (Mattson et al. 1996; Parsons et al. 1986; Vonhof 1996; Vonhof and Barclay 1996), and Indiana myotis typically roost behind bark (Britzke et al. 2003; Cope et al. 1974; Humphrey et al. 1977; Kurta et al. 1993b, 2002; Whitaker and Brack 2002).

Whether species-specific differences in the general type of roost used are sufficient to reduce competition for roost sites and allow coexistence is unknown. Sedgeley (2003) has suggested, however, that selection of more specific traits of roosts by different species of bats could result in resource partitioning at a small scale. At a more general level (e.g., crevices versus bark roosts), species may appear to overlap in their requirements.

Numerous studies have provided evidence that tree-roosting bats select roost trees with particular attributes from among the trees that are available (Kalcounis-Rüppell et al. 2005; Lacki and Baker 2003). Roost selection, however, has been investigated at several scales. Roost sites and roost trees have been compared with other "available" sites or trees, stands containing the roosts have been compared with other nearby stands, and the larger landscapes that bats roost in have been compared with similar areas more distant from the roost. Not all studies take all three perspectives, although consideration of each scale provides valuable insight.

At a broad scale, reproductive female bats appear to roost at lower elevations than do nonreproductive females or males (Brack et al. 2002; Cryan et al. 2000; Thomas 1988), although this may vary with the ecology of the particular species (Barclay 1991). One hypothesis to explain this distributional pattern is that reproductive females require warmer roosting conditions to minimize thermoregulatory costs, whereas males and nonreproductive females can use cooler, high-elevation sites that allow greater use of torpor as an energy-saving mechanism (Barclay 1991; Cryan et al. 2000; Thomas 1988).

Proximity to resources also may play a role in larger-scale selection of roosting areas by bats. For example, some have argued that proximity to water may be important if such areas provide high-quality foraging opportunities or sites for drinking (Carter et al. 2002; Waldien et al. 2000; Waldien and Hayes 2001). By roosting close to such areas, bats could minimize commuting times and distances, thereby saving time and energy. However, only seven studies have tested whether roosts are closer to water than random points, and only three found that they were (Evelyn et al. 2004; Gellman and Zielinski 1996; Ormsbee and McComb 1998). In one study, roosts were further from water (Mattson et al. 1996). The importance of water likely varies with the ecology of the bat species and the aridity of the habitat. Additionally, commuting costs may be trivial (Kurta et al. 2002; Lumsden et al. 2002; Mattson et al. 1996), especially when compared with the large distances bats travel each night. For example, in 15 studies, the mean distance between where a bat was captured while foraging and where it roosted was 1.5 km (table 2.2). In light of this, it seems likely that the characteristics of the roost itself are more important than relatively small differences in the time and energy required of bats to access other resources.

Selection of roosting sites and trees

Selection of roosting sites and trees by bats has been investigated by comparing the attributes of roosts and roost trees with those of other, randomly selected, "available" trees in the immediate vicinity of the roost or in the same stand (Evelyn et al. 2004; Kurta et al. 2002; Menzel et al. 2002; Vonhof and Barclay 1996). At least two complications occur with such

Table 2.2. Distance between capture location and first roost, and colony size for North American tree-roosting bats

Species	Location	Sex	Reproductive condition	Mean capture to roost (km)	Group size mean (range)	Source
Corynorhinus rafinesquii	Louisiana				(1–80)	Gooding and Langford 2004
Eptesicus fuscus	Saskatchewan	F	reproductive		21 (2–43)	Kalcounis and Brigham 1998
	Saskatchewan	F	reproductive		27 (11–37)	Willis et al. 2003
	Maryland	F	reproductive		50	Christian 1956
	South Dakota	F	reproductive	1.5		Cryan et al. 2001
	Oregon	F	reproductive		max 69	Betts 1996
	British Columbia	F	reproductive	1.8	107	Brigham 1991
Lasionycteris noctivagans	Washington	F/M		1.4		Campbell et al. 1996
	South Dakota	F	reproductive		22 (6–55)	Mattson et al. 1996
		F/M	nonrep		1	
	Alberta	F	both		9	Crampton and Barclay 1998
	Ontario	F	reproductive		23	Parsons et al. 1986
	Saskatchewan	F	reproductive		3	
	Oregon	F	reproductive		max 16	Betts 1998
	British Columbia	F	lactating	0.4	max 21	Vonhof and Barclay 1996
	British Columbia	F	reproductive		12	Vonhof 1996
Myotis auriculus	Arizona	F	lactating	1.4	1–43	Bernardos et al. 2004
Myotis austroriparius	Illinois	F	lactating	5.7	101	Hofmann et al. 1999
	Lousiana	F/M	postrep		50	Gooding and Langford 2004
	Texas	F	reproductive		several 100	Mirowsky et al. 2004
Myotis californicus	British Columbia	F	reproductive	1.5	14 (1–52)	Brigham et al. 1997
Myotis evotis	Oregon	F	reproductive	0.66		Waldien et al. 2000
	Oregon	F	reproductive	0.52		Waldien and Hayes 2001
	British Columbia	F/M	nonrep		1	Vonhof and Barclay 1997
	New Mexico	F	reproductive		5	Chung-MacCoubrey 1996
Myotis lucifugus	Alberta	F	both		15	Crampton and Barclay 1998
	Manitoba	F	reproductive		24	Barclay and Cash 1985
	Saskatchewan	F	reproductive		23	Kalcounis and Hecker 1996
		M			1	
	New Brunswick	M			(1–5)	Broders and Forbes 2004
Myotis septentrionalis	Kentucky	F	reproductive		max 65	Lacki and Schwierjohann 2001
		F/M			1	
	West Virginia	F	lactating		31 (11–65)	Menzel et al. 2002
	Michigan	F	reproductive		17	Foster and Kurta 1999
	Indiana	F	reproductive		30	Mumford and Cope 1964
	South Dakota	F	reproductive	2.2	4	Cryan et al. 2001
	New Hampshire	F	reproductive	0.6	max 36	Sasse and Pekins 1996
	New Brunswick	F	both		10 (1–26)	Broders and Forbes 2004
		M			1	
Myotis sodalis	Indiana	F	reproductive		max 28	Humphrey et al. 1977
	Indiana	F	reproductive		max 384	Whitaker and Brack 2002
	Illinois	F	reproductive	2.2	max 95	Kurta et al. 1993a
	Michigan	F	reproductive		max 45	Kurta et al. 1993b
	Michigan	F	reproductive		max 31	Kurta et al. 2002
	North Carolina/ Tennessee	F			max 81	Britzke et al. 2003
	Kentucky	M			max 10	Gumbert et al. 2002

Table 2.2. Continued

Species	Location	Sex	Reproductive condition	Mean capture to roost (km)	Group size mean (range)	Source
Myotis thysanodes	California	F/M	all	0.4 (0.03–0.98)	31 (1–88)	Weller and Zabel 2001
	South Dakota	F	lactating	1	9	Cryan et al. 2001
	New Mexico	F	reproductive		max 40	Chung-MacCoubrey 1996
Myotis volans	Oregon	F	reproductive		<300	Ormsbee and McComb 1998
	South Dakota	F	both	1.9	4	Cryan et al. 2001
	New Mexico	F	reproductive		max 200	Chung-MacCoubrey 1996
Myotis yumanensis	California	F/M	both	2		Evelyn et al. 2004
Nycticeius humeralis	South Carolina	F	reproductive		11	Menzel et al. 2001
		J	nonrep		1	

Sample sizes are provided in table 2.1.
nonrep, nonreproductive; postrep, postreproductive

analyses. First, different authors define "available trees" differently in terms of such variables as minimum height, presence of cavities, and decay stage (Bernardos et al. 2004; Betts 1998b; Crampton and Barclay 1998; Kurta et al. 2002; Ormsbee and McComb 1998; Vonhof and Barclay 1996). Restricting the random sample to trees that actually have cavities or loose bark and could thus potentially house bats seems logical, because if all trees were used as the comparison group, the basic conclusion might be simply that bats select trees in which they can roost. However, variation in the definition of what constitutes an available tree may result in differences in conclusions among studies, making comparisons difficult.

The second problem with the "available tree" approach is that researchers often assume that the sampled pool of available trees is not used by bats. Typically this cannot be confirmed, especially because of the frequent roost changing that occurs in most species. In one study (Kalcounis and Brigham 1998), all "random" trees showed evidence of use by bats. Thus, if "available" trees are used by bats at some point, the ability to detect differences between used and unused trees is either reduced or the measured response is incorrect.

Bats preferred particular species of trees in 12 of 15 studies that examined use of roosts with respect to tree species. This presumably reflects selection of tree or roost characteristics rather than species of tree per se (Cryan et al. 2001). Indeed, the list of tree species used as roosts is extensive and includes many deciduous and coniferous species. Indiana myotis alone have been radiotracked to trees of 44 species (Kurta, in press). Nonetheless, some species or genera of tree appear to have characteristics that make them more suitable as roosts. For example, many studies report bats roosting in pines (*Pinus;* table 2.1), perhaps reflecting the fact that the bark of many pines loosens in large sheets as the tree decays, thereby providing roosting spaces (Brigham et al. 1997; Rabe et al. 1998; Vonhof and Barclay 1996). Other trees, such as trembling aspen (*Populus tremuloides*), are used because the decay process results in internal

Table 2.3. Characteristics of roost trees examined in studies of forest-dwelling bats in Canada and the United States, and whether the characteristic was used randomly compared with what was available on the landscape

Characteristic	Studies in which characteristic was tested (n)	Studies in which characteristic was not used randomly (n)	Studies in which characteristic was not used randomly (%)
Tree species	15	12	80
Tree dbh	23	18	78
Tree height	21	15	71
Tree decay stage	11	9	82
Canopy closure	9	5	56
Amount of bark	10	4	40
Distance to water	7	4	57
Entrance orientation	6	3	50

dbh, diameter at breast height

cavities in the heartwood with entrances at branch scars or woodpecker holes (Crampton and Barclay 1998; Kalcounis and Brigham 1998).

Of the characteristics measured by researchers, tree size (height and diameter) is often a significant variable associated with the roost preferences of bats (tables 2.1 and 2.3). In 18 of 23 studies, roost trees had a larger diameter at breast height (dbh) than randomly selected available trees, and in 15 of 21 studies, roost trees were taller. For example, in northern Alberta, little brown myotis (*Myotis lucifugus*) and silver-haired bats roosted preferentially in large trees despite the fact they were rare, especially in young and medium-aged stands (Crampton and Barclay 1998).

Various reasons have been proposed to explain selection of large trees by bats. Large trees have more cavities and thus more roosting opportunities (Evelyn et al. 2004; Lindenmayer et al. 1993). They tend to be in more open areas or extend above the canopy, thereby making detection and access easier, and perhaps increasing the amount of solar radiation they receive (Brigham et al. 1997; Vonhof and Barclay 1996; Waldien et al 2000). Larger trees have thicker bark that may provide greater insulation for roosts (Rabe et al. 1998). Finally, larger trees may provide larger cavities that allow occupation by greater numbers of individuals (fig. 2.1).

In most studies (9 of 11) that assessed preference for trees of a particular decay stage, bats did not use trees at random, but typically preferred trees in early-to-mid stages of decay (table 2.1). Although most roost trees were dead, in some cases they were alive but with defects that allowed bats access to cavities. For example, large basal hollows in live redwoods (*Sequoia sempervirens*) were used by various species of bats in coastal California (Gellman and Zielinski 1996), cavities in trembling aspen were used in boreal forests in Saskatchewan and Alberta (Crampton and Barclay 1998; Kalcounis and Brigham 1998; Kalcounis and Hecker 1996), and small, curled pieces of bark on the trunks of living shagbark hickories (*Carya ovata*) were consistently used by Indiana myotis in the eastern United States (Callahan et al. 1997; Gardner et al. 1991; Whitaker and Brack 2002).

Bats frequently prefer roost trees that are in relatively open sites. For example, in five of nine studies that estimated canopy closure around roost trees, female bats roosted in trees with less closure than found at randomly chosen, available trees (table 2.1). Similarly, several studies found that roost trees were further from other tall trees than were available trees (Betts 1998b; Brigham et al. 1997; Evelyn et al. 2004; Vonhof 1996). Several hypotheses have been proposed to explain selection for these attributes. Trees in open areas may receive greater solar radiation, resulting in warmer roosts that reduce thermoregulatory costs for reproductive females and their young (Vonhof and Barclay 1996). More open trees also may be easier for bats to access. Roosts of northern myotis (*Myotis septentrionalis*), however, occur in areas of dense canopy cover (Foster and Kurta 1999) or near taller trees (Menzel et al. 2002). The northern myotis is particularly maneuverable, and cluttered roost trees may pose little hindrance to this species (Foster and Kurta 1999), but this does not explain the possible benefit of close, taller trees (Menzel et al. 2002).

Characteristics of specific roosts are often difficult to measure because roosts tend to be high on dead, unstable snags that are dangerous to climb. Roosts vary in height above the ground from those in basal hollows (Gellman and Zielinski 1996; Mirowsky et al. 2004) to those above 10–15 m (table 2.1). In two studies, roost cavities were significantly higher off the ground than were randomly chosen, available cavities (silver-haired bats, Betts 1998b; northern myotis, Menzel et al. 2002). Bats in northern Alberta preferred deep cavities (Crampton and Barclay 1998), and in a few studies (3 of 6; table 2.3), the orientation of cavity entrances was not random, with entrances most often facing south (Kalcounis and Brigham 1998; Mattson et al. 1996; Vonhof and Barclay 1997).

Specific characteristics of roost cavities may reflect preference by bats, or may result from preferences of woodpeckers that originally created the cavities subsequently occupied by bats. Cavities that are high off the ground and have small entrances may be favored because they reduce predation risk (Betts 1998b). However, high, south-facing cavities also may increase exposure to solar radiation. On the other hand, various species of woodpeckers create hollows with particular orientations, heights, and entrances (Crockett and Hadow 1975; Inouye 1976; Inouye et al. 1981), and this may limit the options available to cavity-roosting bats (Crampton and Barclay 1998; Kalcounis and Brigham 1998; Kalcounis and Hecker 1996; Mattson et al. 1996).

Stand-level characteristics associated with roost selection have been measured less often than tree characteristics. Some researchers have compared tree and snag density around roosts with those in other parts of stands, but bats do not appear to have a consistent preference. In some cases, the area around roosts was more open, with lower tree density (Campbell et al. 1996; Cryan et al. 2001; Erickson and West 2003), although this could be a consequence of selection of large trees rather than selection for openness per se (Brigham et al. 1997; Vonhof and Barclay

Figure 2.1.
Examples of roost trees used by *Myotis* species in forests of western Canada and Michigan: (a) a tall snag in a canopy gap (roosting site indicated by marker); (b) a tall snag in a closed canopy; (c) a large plate of exfoliating bark on a snag typical of roosts used as maternity sites; and (d) a cavity in a live tree. *Photos by R. M. R. Barclay and A. Kurta*

1996). In other studies, tree density was greater around roosts than in other stands (Mattson et al. 1996; Rabe et al. 1998). This variation may reflect the relative density of different types of forest, because studies in which roosts were in more dense stands involved generally more open forests. Snags, or trees considered to be available to bats as roosts sites, are often more common around roost trees than in other stands (Campbell et al. 1996; Cryan et al. 2001; Lacki and Baker 2003; Rabe et al. 1998; Waldien et al. 2000), and roost plots contain larger-diameter trees than do other areas (Miller et al. 2002; Sasse and Pekins 1996).

One interpretation of the stand characteristics favored by roosting bats is that roosts tend to be in older stands, although stand age usually is not measured. In northern Alberta, roosts of female little brown myotis and silver-haired bats were all in old stands, even though the bats were caught originally in stands of various ages (Crampton and Barclay 1998). Likewise in South Dakota, the big brown bat (*Eptesicus fuscus*) roosted in mature forest stands, although no such preference was detected for several other species (Cryan et al. 2001). Although it is a less-direct measure of roosting preference, some studies found greater bat activity, as measured by captures or with ultrasonic detectors, in older stands than in younger ones (Crampton and Barclay 1998; Miller et al. 2002; Perkins and Cross 1988; Thomas 1988). Thomas (1988) interpreted the activity patterns as evidence that bats roosted in old stands but then left them to forage elsewhere.

If bats choose to roost in large, early-decay-stage trees in relatively open canopies as many studies have demonstrated (Kalcounis-Rüppell et al. 2005), then roosting in older stands simply may reflect the greater abundance of suitable roost trees in those stands. Indeed, Crampton and Barclay (1998) showed that few trees of the appropriate size and decay stage were available in young or medium age stands of boreal mixed-wood forests in Alberta.

At least one species of bat in North America, the long-eared myotis, appears regularly to use tree stumps as roosts when other roosting options are limited (Vonhof and Barclay 1997; Waldien et al. 2000, 2003). Although only nonreproductive female and male bats used stumps in clearcuts in southern British Columbia (Vonhof and Barclay 1997), reproductive females also used them in young forests in Oregon when large snags were in low abundance (Waldien et al. 2000). Although this indicates the flexibility of this species, the stump characteristics that were selected suggest that stumps remain viable as roost sites for only a short time. Bats roosted under bark and selected tall, large-diameter stumps on steep slopes. Stumps were accessible because of their height and because they were in recent clearcuts with short vegetation. Waldien et al. (2000) estimated that such stumps would be available from the time the bark started to peel until the stump was overgrown by surrounding vegetation, a period typically lasting only about five years.

Why do bats select particular roosts?

Selection by roosting bats of particular cavities, trees, or forest stands has been explained in various ways, although most evidence is observational rather than experimental (Hayes 2003). Some characteristics of roosts may reduce predation risk or commuting costs, but the weight of evidence suggests that roost microclimate and its impact on thermoregulation are the primary factors involved in roost selection by bats. Indeed, many authors have invoked thermoregulation and its costs as reasons for selection of particular roost traits (Evelyn et al. 2004; Humphrey et al. 1977; Kurta et al. 2002; Menzel et al. 2001; Ormsbee and McComb 1998; Vonhof and Barclay 1996; Waldien et al. 2000; Weller and Zabel 2001), although few provide any data to support the suggestion and no study in Canada or the United States has experimentally tested the hypothesis.

Support for the hypothesis that roost microclimate, in particular, temperature, influences roost selection comes in several forms. First, a few studies measured temperature in tree cavities used by bats. Second, the behavior of tree-roosting bats occasionally suggests that temperature is important. Finally, conditions in other types of roosts (e.g., buildings, bat houses, rock crevices, and foliage) have been measured and in a few cases manipulative experiments have been conducted (Kunz and Lumsden 2003).

In the few studies conducted in North America and New Zealand that have measured temperatures in tree roosts, roost temperatures are typically more stable than external ambient temperatures are, so that roosts are warmer at night and in the morning but cooler in the heat of the day (Kalcounis and Brigham 1998; Kalcounis and Hecker 1996; Sedgeley 2001, 2003; Vonhof and Barclay 1997). Unfortunately, roosts rarely have been compared with randomly selected tree cavities, although Kalcounis and Brigham (1998) found that roosts in aspen trees were thermally different than cavities in conifers. Warm roost temperatures at night may be especially important after parturition when young are left behind in the roosts (Chruszcz and Barclay 2002), whereas roosts that remain cooler than ambient in the daytime may prevent overheating (Vonhof and Barclay 1997). Sedgeley (2001) measured both temperature and humidity in roost cavities of long-tailed bats (*Chalinolobus tuberculatus*) in New Zealand and suggested that the relatively stable and high humidity in maternity roosts was important in reducing water loss.

Several studies observed behaviors of bats that are consistent with the hypothesis that roosts are selected on the basis of their microclimate. For example, roosting groups of northern myotis are larger during pregnancy than lactation (Foster and Kurta 1999; Lacki and Schwierjohann 2001). Likewise, groups of long-eared myotis are larger in cool, mountain areas than in warmer prairie areas in Alberta (Solick 2004). Roosting in a large group may be a mechanism for reproductive females to reduce thermoregulatory costs by clustering, especially in spring when ambient temperatures are low. In only two studies has selection of roosts been correlated with current environmental conditions. On cool days, reproductive female Indiana myotis used roosts in trees that were in the open, but they used shaded trees in forest interiors on hot days or when it had rained (Callahan et al. 1997); similar changes in roost selection by Indiana myotis on warm or rainy days were also noted (Humphrey et al. 1977).

Bats roosting in structures other than tree cavities or bark, such as buildings, bridges, or rock crevices, appear to select roosts with particular thermal characteristics (Ormsbee et al., chap. 5 in this volume). Both long-eared myotis and big brown bats using rock crevices selected roosts during pregnancy that warmed rapidly from solar radiation (Chruszcz and Barclay 2002; Lausen and Barclay 2003). During lactation, females moved to roosts that were buffered against fluctuations in ambient temperature and that stayed warm at night when nonvolant pups were alone. Likewise, bats in buildings select sites that maintain warm temperatures during the day (Hamilton and Barclay 1994; Hoying and Kunz 1998; Lourenço and Palmeirim 2004; Williams and Brittingham 1997) and have relatively large thermal gradients that allow individuals flexibility in their thermoregulatory strategy (Williams and Brittingham 1997). Even foliage-roosting bats, such as eastern red bats (*Lasiurus borealis*) and hoary bats (*L. cinereus*), appear to roost in sites with particular thermal characteristics that minimize energy costs (Constantine 1966; Hutchinson and Lacki 2001; Willis and Brigham 2005).

Two experimental tests of the hypothesis that bats select diurnal roosts on the basis of temperature have been conducted. During pregnancy, female Bechstein's bats (*Myotis bechsteini*) selected bat boxes that were cooler than other boxes, but during lactation they chose warmer boxes (Kerth et al. 2001b). Reproductive female soprano pipistrelles (*Pipistrellus pygmaeus*) also selected bat boxes that provided the warmest temperatures, although they switched boxes when the temperature in the box rose above 40°C (Lourenço and Palmeirim 2004). Although experiments in which roost microclimate is manipulated while other variables are controlled are needed for tree cavities and bark roosts, the logistics of such experiments will be difficult and experiments may be more practical using other types of roosts.

Does roost quality and selection matter?

The influences of the number or quality of roosts on reproductive success and population abundance rarely have been addressed, although the implicit assumption is that number and quality of roosts influence fitness (Lunney et al. 1988). In two areas in New Zealand, roosts of long-tailed bats differed in characteristics associated with quality (Sedgeley and O'Donnell 2004). The authors suggested that roost differences were at least partly responsible for differences in the reproductive success of the bats in the two areas. Loss of building roosts was linked to lower reproductive success of big brown bats in Ontario (Brigham and Fenton 1986). Although these studies suggest that the availability of quality roosts influences the reproductive success (i.e., fitness) of tree-roosting bats, further evidence, including experimental evidence that quality roosts are indeed limited, is required.

Variation in roost selection by gender and reproductive condition

Virtually all studies on roost selection by bark- and cavity-roosting bats focus on reproductive females or combine data for males and females (table 2.1). The rationale has been that roost selection to minimize thermoregulatory or commuting costs is more critical for females than for males. The range of roosts suitable for females is thus viewed as more restricted (Crampton and Barclay 1998; Vonhof and Barclay 1996). Additionally, reproductive females are perceived as more important to the viability of bat populations (Cryan et al. 2001; Hayes 2003). Both arguments may be correct, to a degree, but if the roosting requirements of males or nonreproductive females differ from those of reproductive females, this information has important implications for forest management because it might mean that a broader range of wildlife trees or forest ages are required to maintain complete populations of bats. Similarly, if females require different roost conditions during different stages of reproduction, this too must be understood.

Regardless of sex or reproductive condition, diurnal roosts must provide bats with protection from predators and the elements. Roosts used by males and nonreproductive females should also provide conditions conducive to the use of torpor, whereas those used by reproductive females should minimize the costs of maintaining high embryonic and juvenile growth rates. Furthermore, diurnal roosts used by lactating females must provide appropriate conditions for their nonvolant young at night after the mothers leave to forage. Thus, different roost conditions may be preferred by female bats at different stages of reproduction (Chruszcz and Barclay 2002; Lausen and Barclay 2003; Veilleux et al. 2004).

Limited data suggest that, in at least some species, males and nonreproductive females do have different roost-selection criteria than reproductive females do, and that pregnant and lactating females also differ. For example, in South Dakota, maternity colonies of silver-haired bats used tree cavities exclusively, while solitary individuals were more flexible and often roosted under bark, suggesting that solitary males were more opportunistic in their roost choice (Mattson et al. 1996). Likewise, Cryan et al. (2001) found differences in the characteristics of roosts used by reproductive and nonreproductive females of several species in South Dakota, and Broders and Forbes (2004) found differences in the roosts used by male and reproductive female northern myotis in New Brunswick. Across their range, adult female Indiana myotis and their young occupy larger-diameter trees than do males (Kurta, in press). Lacki and Schwierjohann (2001) found that colonies of northern myotis were in taller trees that were surrounded by more snags than trees used by individuals. On the other hand, Evelyn et al. (2004) found no difference between roosts of male and female Yuma myotis (*Myotis yumanensis*), although roosts of reproductive females were not examined. While Waldien et al. (2000) found no difference in roost traits of long-eared myotis of different reproductive condition, sample sizes were relatively small in their study. More information is required on the roosts selected by males and nonreproductive females before conclusions or management recommendations can be made.

TEMPORAL USE AND SPATIAL DISTRIBUTION OF ROOSTS
NUMBER OF ROOSTS

Two of the most ubiquitous aspects of the lives of forest bats in North America are that individuals frequently change roosts and that they require multiple trees during a season. In 21 studies involving radiotracking of 10 species, bats changed roosts on average every 2.5 (SE = 0.2) days (table 2.4), sometimes even during daylight (Kurta et al. 2002). Although transmitters remain attached and functional for only a short time (7 ± 0.8 days; $n = 8$ studies from 6 species), the number of roost trees used by an individual and the duration of the tracking period are correlated with

Table 2.4. Summary of radiotracking studies on North American bats roosting in cavities or under bark that provide data on residency time and spatial distribution of roosts

Species	No. and type of roost*	Type of individual tracked†	Mean length of continuous residency (days)	Mean distance to next roost (m)	Mean transmitter life (days)	Source
C. rafinesquii	44 cavity	?	"moved . . . every few days"			Gooding and Langford 2004
C. townsendii	5 bridge, 4 cavity	9F	"switched frequently"	range, 70–2,500		Lance et al. 2001
E. fuscus	8 cavity		2.5			Brigham 1991
	8 tree	1P, 2L	3.25	1,100	8.7	Cryan et al. 2001
	54 cavity	61F	1.7		5.9	Willis and Brigham 2004
L. noctivagans	17 cavity	10L or P	2.9	280	12	Betts 1998a, b
	11 cavity, 1 bark	2P, 1L, 2PL, 1J	2.67	280		Crampton and Barclay 1998
M. auriculus	11 cavity	18 L	2.0	308		Bernardos et al. 2004
M. californicus	19 bark?	9P or L		401		Brigham et al. 1997
M. evotis	rock	F	"switched roosts regularly"			Chruszcz and Barclay 2002
	41 stump, 30 tree, 2 log	4P, 9L, 6PL, 2N	1.2			Waldien et al. 2000
M. lucifugus	16 cavity	1P, 4L, 3N, 1F	3.67	1,050		Crampton and Barclay 1998
M. septentrionalis	21 (most crevice or cavity)	6P, 1L, 2N	3.25	600	3.9	Cryan et al. 2001
	15 bark, 8 crevice, 6 cavity	5P, 2L, 1PL, 2J	2	414	5.6	Foster and Kurta 1999
	11 cavity, 1 bark	7L	5.3			Menzel et al. 2002
	7 bark	2PL, 1M, 1FJ	1.6			Kurta 2000
	47 tree	26?	"tended to move often"	"some roosts were clustered"		Sasse and Pekins 1996
M. sodalis	54 bark	48F, 16J		roosts "not widely dispersed"		Callahan et al. 1997
	tree (mostly bark)	4F, 56M	2.21			Gumbert et al. 2002
	23 bark	6P, 7L, 5PL/N, 5J	2.9	74.2		Kurta et al. 1996
	32 bark, 6 crevice	P, L, PL, N, J	2.4	686	7.4	Kurta et al. 2002
M. thysanodes	27 rock, 9 cavity	2P, 11L, 2M	1.79	500		Cryan et al. 2001
	15 bark, 2 crevice		1.7	254	6.3	Weller et al. 2001
M. volans	9 rock, 7 tree	4L, 1PL, 4F, 1M	3.44	1,100	6.2	Cryan et al. 2001
	40 tree, 1 rock	16F	2.16	412.5		Ormsbee 1996
N. humeralis	14 cavity	3P, 4L, 4N	1.8	roosts "in close proximity"		Duchamp et al. 2004
	13 cavity, 7 bark	3L/PL, 4J	2.3			Menzel et al. 2001

*Bridge, under bridge; cavity, tree cavity; crevice, crevice in tree; rock, rock crevice; tree, site not specified
†P, pregnant; L, lactating; PL, postlactating; N, nonreproductive adult female; F, adult female of unknown reproductive condition; M, adult male; J, juvenile; and ?, unspecified individuals

the number of different trees occupied by a single bat, usually ranging from one to six (Foster and Kurta 1999; Kronwitter 1988; Kurta et al. 1996, 2002; O'Donnell and Sedgeley 1999; Ormsbee 1996). The relationship between number of trees used and days tracked, however, may not be linear over longer periods, that is, the number of different trees used may not increase continually, or at least the rate of accretion may slow. For example, Gumbert et al. (2002) radiotracked individual male Indiana myotis, some for more than 40 days, and the relationship between number of trees used and days tracked appeared to generate an asymptote at less than 20 trees per bat.

The number of trees used by individuals is often reported, whereas the number of trees required by an entire maternity colony remains largely unknown, despite the need for such information for effective management. Most radiotracking studies use a "shotgun" approach, whereby individual bats from different colonies are tracked (Betts 1998a, b); hence, the needs of a colony are seldom the focus. The Indiana myotis is one of the few tree-roosting species in North America for which individual colonies have been monitored throughout the reproductive season and over multiple years (Callahan et al. 1997; Gardner et al. 1991; Humphrey et al. 1977; Kurta et al. 1996, 2002). Radiotracking members of nine maternity colonies in Illinois, Indiana, Missouri, and Michigan for an entire season indicated that at least 8–25 trees were used by each colony in any one year (Callahan et al. 1997; Carter 2003; Kurta et al. 1996, 2002; Sparks 2003). Not every bat in each colony could be radiotracked continuously and simultaneously, however, so it is unlikely that every tree used for roosting was found. In addition, many trees were visited by only one bat (Kurta et al. 1996, 2002), and if individual females in a colony use as many trees as males over the course of a month or more (Gumbert et al. 2002), then the total number of trees used by a colony likely is much higher than 8–25 (see also Willis and Brigham 2004).

The number of different trees used by a colony during one season probably varies by an array of factors, including availability of alternate roosts, time spent continuously at one tree, and frequency with which bats return to specific trees after roosting elsewhere. Reuse of trees by the same individual or members of the same colony within the same season has been documented for silver-haired bats (Mattson et al. 1996; Vonhof and Barclay 1996) and big brown bats (Willis and Brigham 2004) that occupied tree cavities, long-legged myotis (*Myotis volans*) that roosted in rock crevices or tree cavities (Cryan et al. 2001), Indiana myotis that lived under bark (Gardner et al. 1996), and northern myotis that used either tree cavities or bark (Foster and Kurta 1999; Menzel et al. 2002). Most published studies do not note whether roosts are reoccupied after a period of absence, however, presumably because specific roosts typically are not monitored after the radiotagged bat leaves or its transmitter fails.

ARE ALL TREES USED TO THE SAME EXTENT?

Bats of many species use certain roosts more than others, and some trees consistently have higher populations of bats than others, suggesting that these well-used trees are of higher quality (Chung-MacCoubrey 2003). For example, of 44 trees used by members of a maternity colony of Indiana myotis in Michigan over four years, 15 trees (39%) were occupied by only one bat on one day, whereas only eight trees (21%) had emergence counts greater than 10 bats. These eight trees were used frequently and accounted for 95% of the bat-days (i.e., one bat seen leaving a tree on 1 night = 1 bat-day) during the four years (Kurta et al. 2002).

Some well-used trees may be occupied constantly during a maternity season, but this seems rare. For example, northern myotis in New Hampshire used 17% of their roosts "continuously" (Sasse and Pekins 1996), whereas there was no continual use of any roost by this species in Michigan (Foster and Kurta 1999). Similar intraspecific variation has been reported for Indiana myotis, with some trees occupied all summer (Humphrey et al. 1977) or for prolonged periods (Kurta et al. 2002), while other trees are used infrequently (Kurta et al. 1993b, 1996). Thus, the tendency to concentrate activity at one, two, or more roosts appears to vary within a species of bat and presumably is related to quality of a specific tree and to the quality or availability of alternative roosts.

INDIVIDUAL OR COLONIAL MOVEMENTS?

Results of a few studies suggest that an entire colony of bats simultaneously will leave a tree and move en masse to another (Betts 1998a; Cryan et al. 2001; Rabe et al. 1998). If such behavior is consistent, it implies strong and continual social cohesion among all members of a colony. There is little evidence for such extreme bonding in most species, however. In each study that suggested cohesive movement of an entire colony, bats were not marked individually, and the duration of each observation was short relative to an entire season. Although all bats in each study vacated a tree over one night on at least one occasion, individuals of many species constantly changed roosts and chance alone could result in all animals moving on the same day, especially if few bats resided in the tree. Moreover, the only evidence for group cohesion typically was that an individual with a transmitter was found in a different roost the next day with a similar number of bats. It is possible that members of a small group occasionally remain together as they move to a new tree, but this behavior does not seem to be the norm for North American species. A possible exception is the silver-haired bat; radiotracking all females from two small colonies (16 bats) of this species in British Columbia indicated that group cohesion apparently was maintained during roost changes, at least during lactation (B. Betts, in litt.).

Even though roost changes typically do not involve all members of a colony, some individuals do preferentially associate with each other and often change roosts together. This relationship is not exclusive, however, as specific individuals are not always found together. Although preferential associations have been demonstrated for only one species in Canada, the big brown bat (Willis and Brigham 2004), such behavior also has been documented for two species of temperate, cavity-roosting bats elsewhere—Bechstein's myotis in Germany (Kerth and König 1999) and long-tailed bats in New Zealand (O'Donnell 2000). Genetic analyses of Bechstein's myotis indicated that relatedness was not the basis for the consistent associations (Kerth and König 1999). In both big brown bats and Bechstein's myotis, however, nonreproductive individuals preferentially associated with reproductive bats, suggesting the possibility of cooperative breeding (Kerth and König 1999; Willis and Brigham 2004). Long-term studies of known individuals for other species in North America may help elucidate the reasons for such bonding.

Although individuals may roost with preferred associates, there are many examples of reproductive females or juveniles roosting alone for one or more nights (Brigham et al. 1997; Butchkoski and Hassinger 2002; Chung-MacCoubrey 2003; Mattson et al. 1996; Vonhof and Barclay 1997). In northern myotis, for instance, 25% of 65 emergence counts at various roosts yielded only a single animal (Foster and Kurta 1999), and in Indiana myotis, 12% of 170 observations were of only one bat (Kurta et al. 1996). Even lactating females occasionally roost alone (e.g., Indiana myotis, Gardner et al. 1991, Kurta et al. 1996; silver-haired bats, Mattson et al. 1996; and long-tailed bats in New Zealand, O'Donnell and Sedgeley 1999).

DISTANCE MOVED OVERNIGHT

When a bat changes roost trees, the distance traveled, in general, is short and energetically insignificant. The average distance moved in 14 studies involving nine species was 497 ± 97 m, ranging from means of 74 to 1,100 m (table 2.4). If the typical flight speed of a commuting bat is 15 km/h (Patterson and Hardin 1969), the time required for the average move is only 2 min. Although flight is energetically costly, temperate species usually spend 3–8 h per night in flight (Barclay 1989; Murray and Kurta 2004) and, thus, the energetic cost of a roost change is trivial relative to total expenditure on flight per night.

Distances moved between roosts vary greatly within species. For example, multiple studies on four species—fringed myotis (*Myotis thysanodes*, Cryan et al. 2001; Weller and Zabel 2001), long-legged myotis (Cryan et al. 2001; Ormsbee 1996), northern myotis (Foster and Kurta 1999; Menzel et al. 2002), and Indiana myotis (Kurta et al. 1996, 2002)—found mean travel distances that differed by factors of two to nine within species (table 2.4). Moreover, the largest (5,800 m) and smallest (1 m) overnight changes observed for any bat were exhibited by members of the same species (Indiana myotis, Kurta et al. 1996, 2002).

Intraspecific variation in distance moved may be related to differences in habitat within the home range of various colonies (Betts 1998a; Waldien et al. 2000). A colony of Indiana myotis that moved an average of only 74 m roosted entirely within a 5-ha forested wetland with more than 100 potential roost trees (dead with peeling bark), whereas a colony that moved an average of 686 m inhabited a fragmented agricultural area that lacked such a large, single concentration of roosts (Kurta et al. 1996, 2002). The disparity in average distance traveled by these two colonies may reflect this apparent difference in availability of roosts. In addition,

Murray and Kurta (2002, 2004) hypothesized that new trees used as day roosts were discovered when a bat first uses a tree within their foraging area as a night roost. Thus, the average distance a bat moves to new day roosts could be affected by the distance to a colony's foraging sites, as well as by local abundance of potential roosts.

FIDELITY TO ROOSTING AREAS

Bats often display fidelity to a group of roosts that are clumped in the environment (i.e., a roosting area). This tendency has been suggested for a number of tree-roosting species from Africa, New Zealand, and Australia (Fenton 1983; Lunney et al. 1988; O'Donnell and Sedgeley 1999; Taylor and Savva 1988), as well as for some North American species (Callahan et al. 1997; Clark 2003; Cryan et al. 2001; Foster and Kurta 1999; Gumbert et al. 2002; Kurta et al. 1996; Sasse and Pekins 1996), including those using natural structures other than trees (Chruszcz and Barclay 2002; Lausen and Barclay 2003). Although statistical validation of clumping is lacking, in general, roost trees used by a maternity colony of Indiana myotis were significantly closer to each other than random points within their roosting habitat were, even though individual roost trees used by members of that colony were up to 8 km apart (Kurta et al. 2002). Individuals and colonies often have more than one cluster of roost trees (Gumbert et al. 2002; Kurta et al. 2002).

Clumping of roosts used by bark- or cavity-roosting species results in part because clumped trees tend to be similar in age, and they simultaneously become exposed to windstorms, disease vectors, or rotting agents. Consequently, nearby trees of similar size often simultaneously develop cavities or die. Nevertheless, species that roost in foliage also show clumping of roosts (Mager and Nelson 2001; Menzel et al. 1998; Veilleux et al. 2003b), suggesting that there may be benefits to roosting in a familiar area or perhaps the microclimate near a cluster is more favorable than elsewhere within the bat's home range.

ROOST-SWITCHING AND FISSION-FUSION SOCIETIES

There are three broad types of roost switching that likely are exhibited by most species: episodic, emergency, and recurrent. Episodic changes are the most predictable and perhaps the most understandable; they occur annually at the same stage of the reproductive cycle and may be a response to changing energetic needs, population size, or other consistent factors. Episodic switching likely would involve many, and often all, members of a colony. Emergency switching, in contrast, involves movement of a colony, or of individuals, to a new roost in response to an immediate crisis or unusual event; it likely is the rarest form of roost switching. Recurrent switching involves frequent, as-yet unpredictable, movement of individuals at any time of season.

EPISODIC ROOST SWITCHING

Two studies on northern myotis indicated that use of certain trees was highest in spring, when females were pregnant, and that the colony apparently splintered into smaller groups before parturition (Foster and Kurta 1999; Sasse and Pekins 1996). Roosts used early in the season may act as staging sites (Foster and Kurta 1999), similar to those described for migratory birds (Sparling and Krapu 1994). Large populations at staging roosts could provide enhanced thermoregulatory benefits through clustering at a cold time of year or increased access to information concerning productive foraging sites when food is scarce (Foster and Kurta 1999; Willis 2003).

Some colonies disperse or fragment soon after young become volant. For example, populations of Indiana myotis at specific roosts decreased after all young took flight in late July (Brack 1983; Humphrey et al. 1977; Kurta et al. 1993a, b). This type of episodic change in roost is not unique to forest-dwelling bats; it also occurs in species of bats roosting in buildings (Humphrey and Cope 1976). Breakup of a maternity colony may reduce competition or represent the beginning of migration toward hibernacula. Alternatively, movement at this time of year may involve mothers acquainting their young with known roosts before migration begins.

Individual roosts typically vary in structural and thermal characteristics, and some episodic switching may result from bats selecting different types of trees during pregnancy, lactation, and postlactation, as energetic requirements of the females and young change. Although some tree-roosting species, such as Indiana myotis and big brown bats, use different sites during pregnancy and lactation (Kurta et al. 2002; Willis 2003), there currently is no evidence that thermal characteristics of the trees are the reason for this type of episodic switching, perhaps because of the difficulty and expense of monitoring the microclimate at a large number of poorly accessible sites (e.g., under the bark high on a dead tree). Nevertheless, the use of roosts with different thermal characteristics during pregnancy and lactation is documented for several species that roost in rock crevices (Chruszcz and Barclay 2002; Lausen and Barclay 2003; Lewis 1996) and that use bat houses in Germany (Kerth et al. 2001b), and we suspect that bats roosting under bark or in cavities change roosts for similar reasons.

EMERGENCY ROOST SWITCHING

In contrast to the predictable changes associated with episodic switching, bats also may switch trees in response to unpredictable factors. For example, a colony of southeastern myotis (*Myotis austroriparius*) roosting in the basal cavity of a large tupelo gum tree (*Nyssa aquatica*) in a slough was forced to move when water levels rose and began blocking the entrance (Hofmann et al. 1999). Destruction of bark roosts due to storms or deterioration also may force abandonment in midseason (Gardner et al. 1991; Kurta 1994), and roost trees may be felled by humans or strong winds (Belwood 2002; Willis et al. 2003). In addition, disturbance by terrestrial or avian predators (Sparks et al. 2003; Veilleux et al. 2003a) or competitors (Kurta and Foster 1995; Mason et al. 1972) may cause all bats to vacate a roost temporarily or permanently.

Recurrent switching is the most common type of roost changing for many species. It involves repeated movement of individuals among trees and often includes movement into and out of a roost that remains occupied by other bats. Evidence for such movement comes from two main sources—emergence counts and radiotracking. The number of bats emerging from a particular tree often fluctuates from night to night. For example, some trees used by Rafinesque's big-eared bats (*Corynorhinus rafinesquii*) might shelter "one bat one day, 50 bats the next, and zero bats a few days later" (Gooding and Langford 2004). At roosts of California myotis (*Myotis californicus*), counts fluctuated from 12 to 5 and 5 to 27 bats on consecutive nights (Brigham et al. 1997), and the number of Indiana myotis at a roost differed by up to 30 animals over a few days (Humphrey et al. 1977; Kurta et al. 1993a, b). Fluctuations in population size at colonies that are consistently monitored at sunset suggest that individuals are moving in or out of the tree. This type of roost switching also likely is the basis for reports that bats "did not appear to form stable colonies" (long-legged myotis, Weller and Zabel 2001), and "there is no such thing as a stable colony" (California myotis, Brigham et al. 1997).

Although minor deviations in population size at evening emergence might be construed as observer error (Betts 1998a) or reflect movement of migratory transients early or late in the year (Humphrey et al. 1977), the suggestion of a steady flux of individuals at focal trees of most species is supported by the behavior of radiotracked individuals. Preferred trees of Indiana myotis were occupied by a group of bats for long periods, but concurrent radiotracking of some colony members indicated that individuals moved in and out, even though the overall population at the focal tree appeared stable (Callahan 1997; Kurta et al. 1993a, 1996, 2002). Radiotracking of other species revealed that group composition regularly changes (Crampton and Barclay 1998; Cryan et al. 2001; Lewis 1996; Willis and Brigham 2004), that recurrent roost switching involves all classes of individuals, including adult males, reproductive females, nonreproductive females, and juveniles, and that recurrent switching can occur at any time of year (Gumbert et al. 2002; Kurta et al. 1996, 2002).

Are there differences in the amount of recurrent roost switching among reproductive conditions, seasons of the year, or other parameters? Although Campbell et al. (1996) found that adult silver-haired bats did not change roosts, but juveniles did, location of the bats was only checked twice per week and roost changes by adults may have been missed. Several other studies found no statistical difference in frequency of switching among ages, sexes, or reproductive conditions (Brigham 1991; Lewis 1996; Menzel et al. 2001; Waldien et al. 2000). Most of these studies involved three-to-six age/sex groups, however, often with sample sizes of only two to four animals per category, and the resulting statistical power was quite low.

Studies with adequate samples have yielded inconsistent differences in the rate of roost switching among categories of age, sex, or reproductive condition. For example, there was no difference in roost-changing frequency among "pre-pregnant," pregnant, lactating, and nonreproductive big brown bats (Willis and Brigham 2004), but pregnant Indiana myotis changed roosts more often than lactating individuals (Kurta et al. 1996, 2002). In New Zealand, there was no difference in roost-changing frequency among different categories of adult female long-tailed bats, but adult females moved less often than juveniles or adult males (O'Donnell and Sedgeley 1999).

The cause of recurrent movement is not immediately apparent, and there may be no single explanation. Roost switching may be a means of disrupting arthropod life cycles and decreasing ectoparasite loads by leaving behind parasitic stages that typically remain in the roost when bats leave to forage (Lewis 1996). However, this cannot explain why there is often a continual influx and efflux of individuals at roosts that remain occupied by some bats, or why abandoned roosts are reoccupied after only a few days. Pallid bats (*Antrozous pallidus*) that moved frequently had more ectoparasites than those that moved less frequently (Lewis 1996), but if an antiparasite strategy was operative, bats that move frequently should have fewer, not more, ectoparasites than more-sedentary individuals. Although it is conceivable that bats abandon roosts because of severe parasitic infestations, this type of roost change would be classified as either emergency switching, if it were a rare event, or episodic switching, if it coincided every year with particular times of a parasite's life cycle. For example, population size of a macronyssid mite (*Steatonyssus occidentalis*) always is highest when nonvolant young are present in the roosts of big brown bats (Miller et al. 1973). We suggest, however, that ectoparasitic infestation is an unlikely explanation for recurrent switching.

Recurrent movement between roosts may be an antipredator strategy (Lewis 1995). Although predation attempts might result in emergency switching (Sparks et al. 2003; Veilleux et al. 2003a), avoidance of predators as an explanation of recurrent switching is not consistent with continual movement of bats into and out of an occupied focal roost, or the return of bats to a particular tree after only a day or two. In addition, the clumped nature of many roosts makes it unlikely that short-distance roost changes have a significant effect on avian predators or persistent, tree-climbing, mammalian carnivores, especially when one considers that a clump of dead trees lacking leaves provides an almost unobstructed view for predators (Kurta et al. 1996).

Frequent changes of roost in response to changes in local food supply (i.e., bats moving closer to productive foraging sites, Lewis 1995) may make sense for frugivorous species that feed on a widely dispersed, yet highly clumped, resource that is available for a short time (Brooke et al. 2000), but there is no evidence for such behavior in the insectivorous species occupying trees in North America. Distances involved in changes of roost are consistently so small that they are energetically insignificant (Brigham and Fenton 1986; Kurta et al. 1996; Lewis 1996; Mattson et al. 1996).

If thermal constraints influence roost choice by females in various reproductive states and result in episodic switching, can daily changes in ambient temperature or rainfall explain the typical, short-term, recurrent roost switching of individuals? No obvious relationship between temperature and roost changing was noted for fringed myotis in California (Weller and Zabel 2001), or between emergence counts at roosts of Indiana myotis in Michigan and either maximum or minimum ambient temperatures (Kurta et al. 1996). However, radiotracked Indiana myotis in Missouri were more likely to use alternate roosts, many of which were living shagbark hickories, in the interior of a forest on days with warm mornings or days following rain (Callahan et al. 1997). Humphrey et al. (1977) also noted that Indiana myotis in Indiana roosted more often in a living shagbark hickory on warm days and during periods of rain, although they provided no statistical analysis. Based on a combined sample from multiple species, Vonhof and Barclay (1996) indicated that bats were less likely to change roosts during periods of rain, and Lewis (1996) reported the same for pallid bats using rock crevices. Although weather may influence patterns of recurrent roost switching in some species, it is not the only explanation, because many bats change roosts more frequently than even the weather, and individuals of many species remain at focal roosts while other colony members change location.

The ephemeral nature of bark and cavity roosts, relative to buildings and caves, may promote recurrent roost switching (Kurta et al. 1996, 2002; Lewis 1995; Weller and Zabel 2001). Bats that are aware of alternate trees are less likely to suffer from the sudden destruction of a favored roost through predatory attempts, storms, or other natural events (Gardner et al. 1991; Kurta 1994; Willis 2003). Similarly, tree bats vacate their summer grounds and migrate to suitable hibernation sites, and knowledge of multiple roosts that were visited in August or September may be important when bats return in spring, stressed by a prolonged hibernation period and a potentially long migration (Kurta and Murray 2002; Kurta and Rice 2002). When they return, females are pregnant, the weather often is cool and wet, and flying insects are scarce. Knowledge of multiple roosts might mitigate the inevitable loss of some trees over winter and ensure that pregnant females immediately find suitable shelter.

If ephemerality affects the degree of fidelity, cavity-dwelling bats should change less often than bark-dwelling bats (Lewis 1995). This pattern is not readily apparent in the reports that we surveyed (table 2.4), but the studies are confounded by combining various reproductive conditions, sexes, and ages, by being performed in habitats with differing availability of roosts, and by some species using multiple types of roost. Although the tendency for members of the same species to occupy different kinds of roosts (e.g., bark, tree cavity, rock crevice, or building) confounds interspecific comparisons, it potentially allows insightful intraspecific comparisons. For instance, radiotagged silver-haired bats remained for longer periods in cavity roosts than in bark roosts (Vonhof and Barclay 1996), and individual pregnant Indiana myotis had longer residency times when using crevices in trees than when roosting under bark

(Kurta et al. 2002). Although these studies support the ephemerality hypothesis as an explanation of recurrent switching, some bats roost in cavities that are very long lived (O'Donnell and Sedgeley 1999; Willis et al. 2003) or in rock crevices (Lausen and Barclay 2002), yet these bats also frequently switch roosts. Hence, even ephemerality is not the total answer. Roost changing induced by ephemerality of roosts may be tempered over evolutionary time by the immediate availability of potential roosts, with more frequent changes occurring when a large number of quality roosts are available (Chung-MacCoubrey 2003; Sherwin et al. 2003). Studies of tree bats, however, rarely quantify the density of available roosts in a particular area, making it difficult to test this hypothesis. Nevertheless, Indiana myotis roosting in a forested wetland in Michigan with a high density of roosts moved once every 2.9 days, whereas those living in a region with a lower density of roosts changed trees with similar frequency, suggesting that availability was not an important factor (Kurta et al. 1996, 2002). The colony with the lower density of available roosts, however, had much longer roost-change distances, suggesting that bats may respond to reduced availability, up to a point, by expanding the size of their roosting area rather than reducing the frequency of switching.

FISSION-FUSION SOCIETIES

Recurrent roost switching and fluctuating composition of the group at any particular tree suggest the existence of a fission-fusion society (Kurta et al. 2002). In this type of society, members frequently coalesce to form a group (fusion), but composition of that group is in perpetual flux, with individuals frequently departing to be solitary or to form smaller groups (fission) for a variable time before returning to the main unit. Individuals often preferentially associate with some members of the larger group and may even avoid associating with other members.

This type of flexible social organization is common among cetaceans (Conner 2000) and primates (McGrew et al. 1996; Terborgh and Janson 1986), but it also occurs in other mammals such as spotted hyenas (*Crocuta crocuta*, Holekamp et al. 1997) and kinkajous (*Potos flavus*, Kays and Gittleman 2001). In whales, all individuals in the society are members of a pod, and in hyenas, this society is termed a clan; in bats, members of the fission-fusion society collectively form what biologists historically have called the "colony." Although many members of a colony may reside in one tree at any one time, other members roost elsewhere as solitary individuals or in small subgroups of fluctuating composition. Such a fission-fusion society has been suggested for a few species of bats (Kerth and König 1999; Kurta et al. 2002; O'Donnell 2000; Willis and Brigham 2004), and further research may show that this type of social organization is common.

Group formation in most fission-fusion societies typically is viewed as an antipredator mechanism, whereas fission occurs in response to fluctuating availability of patchy food (Boesch et al. 2002; Conner 2000; Conner et al. 2000; Terborgh and Janson 1986). In temperate bats, colony formation by reproductive females in summer often is assumed to be related to energetic concerns, with grouped animals able to cluster and reduce thermoregulatory costs (Kurta 1985), although other factors, such as information transfer, may play a role in some species (Wilkinson 1992; but see Kerth et al. 2001a). The reason for repeated fission of bat colonies (i.e., recurrent roost switching) is not clear and, as discussed above, it is probably not related to any single factor.

DIFFERENCES IN SWITCHING—FINAL COMMENTS

We speculate that the interplay between ephemerality and availability of roosts is a major determinant of the frequency of recurrent roost switching by most species during the reproductive season. Actual frequency of movement, however, likely is affected by intrinsic factors (reproductive condition, sex, age, and size of young), local conditions (roost quality, and weather), the occurrence of episodic changes (warmer roosts for lactating females or fragmentation during postlactation), and emergency movements (response to flooding, roost damage, and predators). Investigators attempting to explain recurrent switching must understand that the overall frequency of switching, as typically reported in the literature, represents the sum of episodic, emergency, and recurrent roost changes.

Almost every study that has documented roost changes by bats has done little more than reiterate the potential reasons for roost switching that were assembled by Lewis (1995). Of course, little else can be done when samples are small and are confounded with uncontrolled variables. Henshaw (1970, 188) once expressed dismay at the continual cataloging of metabolic rates and body temperatures of bats and noted that the literature contained a "paucity of integrative experimentation but an abundance of integrative speculation." The situation is similar with the literature on the behavior of tree-roosting bats. After almost 20 years of cataloging switching frequencies and fluctuating emergence counts for various species, it is time to move on to more fruitful, experimental approaches that test specific hypotheses (Kerth et al. 2001a, b; Willis 2003).

INTERANNUAL FIDELITY TO ROOSTS

Interannual fidelity (philopatry) of bats can be associated with use of particular roost trees, general roosting areas, or broad foraging areas and commuting corridors and involve fidelity of either specific individuals or reuse by members of the same population (Gumbert et al. 2002). Such fidelity, if it exists, is important from a management perspective, because it would allow resource managers to target their actions more specifically, especially when dealing with rare species. Nevertheless, the demonstration of interannual fidelity is uncommon, in part, because many studies do not last more than one year.

Interannual reuse of specific roost trees is known for six species of bats in North America (table 2.5). A few observations represent recaptures of banded individuals, but most reflect reuse by other individuals of the same species that were radiotracked to the tree or captured as they emerged for foraging. There has been, up to now, no consistent pattern as to what type of tree is more likely to be occupied in multiple years.

Table 2.5. Use of roost trees among years by North American bark- and cavity-roosting bats

Species of bat	Species of tree	Alive or dead*	Cavity/ crevice or bark†	Band recapture or other observation‡	Maximum number of consecutive years	Source	Comment
E. fuscus	American beech, Fagus grandifolia	A	C	O	≥20	Kurta 1980	
	Trembling aspen, Populus tremuloides	A	C	B and O	10	Willis et al. 2003	
M. austroriparius	Tupelo gum, Nyssa aquatica	A	C	O	2	Hofmann et al. 1999	
M. californicus	Ponderosa pine, Pinus ponderosa	D	both	O	5	Barclay and Brigham 2001	multiple trees
M. septentrionalis	American beech, Fagus grandifolia	D	C	O	2	Sasse and Pekins 1996; D. Sasse, in litt.	2 trees
	Silver maple, Acer saccharinum	A	C	B	2	Foster and Kurta 1999	
		D	C	O	2	Foster and Kurta 1999	
		D	B	O	2	Foster and Kurta 1999	
M. sodalis	Bitternut hickory, Carya cordiformis	D	B	B	2	Humphrey et al. 1977	
	Black ash, Fraxinus nigra	D	B	O	2	Kurta et al. 2002; A. Kurta, unpubl. data	2 trees
	Cottonwood, Populus deltoides	D	C	O	4	Kurta et al. 2002; A. Kurta, unpubl. data	
		D	B	O	3	Brack 1983	
	Green ash, Fraxinus pennsylvanica	D	B	O	2–4	Kurta et al. 1996; Kurta and Foster 1995	5 trees
	Red oak, Quercus rubra	D	B	O	2	Gardner et al. 1991	
	Red maple, Acer rubrum	D	B	O	2	Kurta et al. 2002; A. Kurta, unpubl. data	
	Post oak, Quercus stellata	D	B	O	3	Gardner et al. 1991	
	Shagbark hickory, Carya ovata	A	B	B and O	2	Gardner et al. 1991; Humphrey et al. 1977	2 trees
		D	B	O	2–6	Gardner 1991; Whitaker and Brack 2002; J. O. Whitaker, Jr. in litt.	2 trees
	Short-leaf pine, Pinus echinata	?	B	O	3	Gumbert et al. 2002	
	Silver maple, Acer saccharinum	D	B	B	2	Kurta et al. 2002; A. Kurta, unpubl. data	
	Slippery elm, Ulmus rubra	D	C	O	2	Kurta et al. 2002; A. Kurta, unpubl. data	

*A, alive; D, dead
†C, cavity/crevice; B, bark
‡B, band recapture; O, other observation

Coniferous and deciduous trees are reused, roosting sites are either in cavities or under bark, and both reoccupied cavities and bark roosts occur in living and dead trees.

Some trees are used for prolonged periods. A colony of Indiana myotis, for instance, roosted in a dead shagbark hickory for at least six consecutive years (J. O. Whitaker Jr., in litt.), whereas bark roosts of California myotis were still in use five years after discovery, although the number of bats occupying the trees had decreased, presumably reflecting a decline in quality of the remaining bark (Barclay and Brigham 2001). Cavity roosts of big brown bats in trembling aspen, in contrast, were used for three consecutive years and showed no decrease in number of bats during that time; some cavities were occupied 9 and 10 years after initial discovery, with no noticeable decline in structural quality (Willis et al. 2003). Additionally, big brown bats probably used a living, hollow beech

(*Fagus grandifolia*) for at least 20 years (Kurta 1980). Although availability of suitable roosts has been suggested as a factor in determining whether a particular site is reused between years (Chung-MacCoubrey 2003; Lewis 1995; Sherwin et al. 2003), Indiana myotis and northern myotis showed between-year fidelity to specific trees in a small forested wetland despite the presence of over 100 apparently suitable roosts (table 2.5) (Foster and Kurta 1999; Kurta et al. 1996). The number of consecutive years that a roost is used strongly reflects length of the individual study, resulting in underestimates of how long specific trees continue to be occupied.

Among North American tree-roosting bats, the best evidence of philopatry is for the Indiana myotis. Over four years, Kurta and Murray (2002) banded 29 adult females at a colony in Michigan. Forty-one percent were recaptured within the home range of the colony in subsequent summers. Five individuals were caught at exactly the same location where they were initially netted, either at specific roost trees or along a wooded fence line used as a commuting corridor, and members of the colony consistently used this corridor over a 9-year period (Winhold et al. 2005). During winter, individuals were found hibernating up to 532 km from their summer roost (Kurta and Murray 2002).

Although mounting evidence indicates that tree bats can be as loyal to their home areas (table 2.5) as are bats that roost in caves or buildings (Brenner 1968; Humphrey and Cope 1976), cavity- and bark-roosting species, out of necessity, must change the focus of their roosting activity as years go by. Preferred roost trees inevitably decline in quality—living trees die, and dead trees lose bark and fall over. A few species of bats, such as Rafinesque's big-eared bat and southeastern myotis (Clark 2003; Gooding and Langford 2004; Mirowsky et al. 2004), occupy cavities in living trees, and potentially these sites are available for many years. Most forest-dwelling bats, however, rely primarily on dead trees for roosting, and because appropriate dead trees often are clumped in the environment, a colony of bats likely will change the location of its focal roost tree and focal cluster of trees many times during the life of these long-lived mammals (Wilkinson and South 2002). For instance, the focal roosting area for a colony of Indiana myotis gradually moved 2 km over a 3-year period (Kurta et al. 2002). Foraging areas and commuting paths probably persist longer than do many roosting sites, at least in undisturbed habitats, so even though bats move their roosting sites across the landscape, their home range should be more stable, perhaps facilitating management.

RECOMMENDATIONS
FUTURE RESEARCH

Although knowledge regarding the roosting ecology and behavior of bats has increased rapidly in the past 10 years, significant gaps still exist in our understanding. Most crevice- and bark-roosting species in Canada and the United States have been studied, at least to a certain degree, although there is relatively little known regarding the southeastern myotis, eastern small-footed myotis (*Myotis leibii*), Keen's myotis (*M. keenii*), or western pipistrelle (*Pipistrellus hesperus*). Geographically, western North America has been the focus of attention (Hayes 2003). Forests in eastern Canada and the United States are not only dominated by different species of bats and trees, they also have been influenced by humans and under management for much longer. More attention should be paid to these forests. Although this may merely confirm the general patterns apparent from studies that already have been conducted, differences among species and geographic variation within species do occur. For regional or local management to be most effective, local knowledge is important. For example, almost all research on the thermal characteristics of bat roosts has been conducted in relatively cool, temperate areas (Chruszcz and Barclay 2002; Hamilton and Barclay 1994; Kerth et al. 2001b; Lausen and Barclay 2003; Sedgeley 2001; Vonhof and Barclay 1997). We need to know whether bats in warmer environments have different selection criteria and how this influences the types of trees that are required (Law and Anderson 2000).

In addition to filling species and geographic gaps, future research needs to address several more general questions. How do roost preferences and behavior of males and nonreproductive females compare to those of reproductive females? How do the preferences and behavior of pregnant and lactating females differ? Only with knowledge regarding all gender, age, and reproductive classes will management be able to protect entire bat populations. Given that colonies usually exist as diffuse entities and that we have yet to follow all individuals of a colony, or even any one individual for an entire season, a more intensive study is needed to determine the roosting requirements and behavior of entire colonies. Is there a minimum viable colony size? How do new colonies form or individuals locate new roost sites? To address management issues at the appropriate scale, we need to understand better the scale at which bat populations operate. How large is the roosting home range of a colony and how many trees does the colony require in a season? How does habitat fragmentation influence the roosting ecology and behavior of colonies? The degree of forest fragmentation differs between eastern and western North America, and comparing the different effects may be instructive.

While the above research will improve our knowledge, understanding the ecology and behavior of forest-dwelling bats requires an experimental approach (Hayes 2003). Almost no manipulative experiments have been conducted, but to understand such important behaviors as roost selection and switching, they will be essential. How important is roost temperature versus humidity in determining the quality of a roost? What roost and site features produce optimal roost microclimates? What are the impacts on individuals and populations of reduced roost availability or quality? For logistical reasons, some of these questions will be difficult to answer using natural tree-crevice and bark roosts, but we may learn much from experiments using other types of roosts that are easier to manipulate, such as bat houses (Kerth et al. 2001b; Lourenço and Palmeirim 2004).

Given the current knowledge of roosting habits of bats, we can propose some general forest management recommendations (see also Hayes 2003; Hayes and Loeb, chap. 8 in this volume). Forest management that includes bats must take a landscape perspective (Barclay and Brigham 1996; Lumsden et al. 2002). Unlike the home ranges of small birds and terrestrial mammals, those of bats are large (Callahan et al. 1997). In addition, the roost-switching behavior prevalent among most populations means that bat colonies require forest stands, at a minimum, not just single roost trees. Landscapes must provide not only sufficient numbers of quality roost trees, but also recruits to replace roost trees that are routinely lost (Barclay and Brigham 2001; Britzke et al. 2003; Callahan et al. 1997; Gumbert et al. 2002; Kurta et al. 2002). Snag densities are generally much higher in unmanaged stands than in managed ones (Mattson et al. 1996); thus, managed forests likely provide fewer trees of the types used by bats (Campbell et al. 1996; Russo et al. 2004; Sedgeley 2003). To provide roosting habitat for bats, management must retain large-diameter, tall, early-decay-stage trees, especially those with natural cavities or sloughing bark. For species that roost behind bark, roost sites may be particularly ephemeral and bats may move more often, thereby requiring even greater numbers of suitable trees (Kurta et al. 2002).

Although habitat fragmentation potentially has negative effects on bat populations and the functioning of colonies, we cannot draw any definitive conclusions about fragmentation effects at this time (Duchamp et al., chap. 9 in this volume). Indeed, some activities, especially in second-growth stands, may have positive effects by creating a more natural forest structure. Until data on the impact of forest fragmentation are available, a cautious approach would be to maximize the size of contiguous forest patches, thereby increasing the likelihood that the home range of a colony could be encompassed by a single patch. If fragmentation must occur, then wooded corridors should be maintained between adjacent patches of forest, because many species of bat consistently follow tree-lined paths rather than cross large open areas (Murray and Kurta 2004; Verboom and Huitema 1997; Winhold et al., 2005). Once species- and region-specific roosting home ranges have been established, more specific recommendations regarding patch size and snag density will be possible.

ACKNOWLEDGMENTS

We thank B. Sasse, D. Waldien, J. Veilleux, J. O. Whitaker Jr., R. M. Brigham, and M. Kalcounis-Rüppell for supplying unpublished data and preprints, and R. M. Brigham, B. J. Betts, S. Loeb, and J. P. Hayes for constructive comments on earlier drafts of the manuscript. RMRB's research has been funded by the Natural Sciences and Engineering Research Council of Canada, the University of Calgary, the British Columbia Ministries of Forests and Water, Land and Air Protection, Forestry Canada, the British Columbia Habitat Conservation Fund, and Bat Conservation

International. Previously unreported data from AK were obtained with funding from the Nongame Program of the Michigan Department of Natural Resources, U.S. Forest Service (Huron-Manistee National Forests), and Michigan Chapter of the Nature Conservancy.

LITERATURE CITED

Barclay, R.M.R. 1989. The effect of reproductive condition on the foraging behavior of female hoary bats, *Lasiurus cinereus*. Behavioral Ecology and Sociobiology 24:31–37.

———. 1991. Population structure of temperate zone insectivorous bats in relation to foraging behaviour and energy demand. Journal of Animal Ecology 60:165–178.

Barclay, R.M.R., and R.M. Brigham, eds. 1996. Bats and forests symposium. Research Branch, British Columbia Ministry of Forests, Victoria, BC.

———. 2001. Year-to-year re-use of tree-roosts by California bats (*Myotis californicus*) in southern British Columbia. American Midland Naturalist 146:80–85.

Barclay, R.M.R., and K.J. Cash. 1985. A non-commensal maternity roost of the little brown bat (*Myotis lucifugus*). Journal of Mammalogy 66:782–783.

Barclay, R.M.R., and L.D. Harder. 2003. Life histories of bats: life in the slow lane, pp. 209–253, *in* Bat ecology (T.H. Kunz and M.B. Fenton, eds.). University of Chicago Press, Chicago, IL.

Belwood, J.J. 2002. Endangered bats in suburbia: observations and concerns for the future, pp. 193–198, *in* The Indiana bat: biology and management of an endangered species (A. Kurta and J. Kennedy, eds.). Bat Conservation International, Austin, TX.

Bernardos, D.A., C.L. Chambers, and M.J. Rabe. 2004. Selection of gambel oak roosts by southwestern myotis in ponderosa pine-dominated forests, northern Arizona. Journal of Wildlife Management 68:595–601.

Betts, B.J. 1996. Roosting behaviour of silver-haired bats (*Lasionycteris noctivagans*) and big brown bats (*Eptesicus fuscus*) in northeast Oregon, pp. 55–61, *in* Bats and forests symposium (R.M.R. Barclay and R.M. Brigham, eds.). Research Branch, British Columbia Ministry of Forests, Victoria, B.C.

———. 1998a. Variation in roost fidelity among reproductive female silver-haired bats in northeastern Oregon. Northwestern Naturalist 79:59–63.

———. 1998b. Roosts used by maternity colonies of silver-haired bats in northeastern Oregon. Journal of Mammalogy 79:643–650.

Boesch, C., G. Hohmann, and L.F. Marchant. 2002. Behavioural diversity in chimpanzees and bonobos. Cambridge University Press, Cambridge, U.K.

Brack, V., Jr. 1983. The non-hibernating ecology of bats in Indiana with emphasis on the endangered Indiana bat, *Myotis sodalis*. Ph.D. dissertation, Purdue University, West Lafayette, IN.

Brack, V., Jr., C.W. Stihler, R.J. Reynolds, C.M. Butchkoski, and C.S. Hobson. 2002. Effect of climate and elevation on distribution and abundance in the mideastern United States, pp. 21–28, *in* The Indiana bat: biology and management of an endangered species (A. Kurta and J. Kennedy, eds.). Bat Conservation International, Austin, TX.

Brenner, F.J. 1968. Three-year study of two breeding colonies of the big brown bat, *Eptesicus fuscus*. Journal of Mammalogy 49:775–778.

Brigham, R.M. 1991. Flexibility in foraging and roosting behaviour by the big brown bat (*Eptesicus fuscus*). Canadian Journal of Zoology 69:117–121.

Brigham, R.M., and M.B. Fenton. 1986. The influence of roost closure on the roosting and foraging behaviour of *Eptesicus fuscus* (Chiroptera, Vespertilionidae). Canadian Journal of Zoology 64:1128–1133.

Brigham, R.M., M. Vonhof, R.M.R. Barclay, and J.C. Gwilliam. 1997. Roosting behavior and roost-site preferences of forest-dwelling California bats (*Myotis californicus*). Journal of Mammalogy 78:1231–1239.

Britzke, E.R., M.J. Harvey, and S.C. Loeb. 2003. Indiana bat (*Myotis sodalis*) maternity roosts in the southern United States. Southeastern Naturalist 2:235–242.

Broders, H.G., and G.J. Forbes. 2004. Interspecific and intersexual variation in roost-site selection of northern long-eared and little brown bats in the Greater Fundy National Park ecosystem. Journal of Wildlife Management 68:602–610.

Brooke, A.P., C. Solek, and A. Tualaulelei. 2000. Roosting behavior of colonial and solitary flying foxes in American Samoa. Biotropica 32:338–350.

Burnett, C.D., and P.V. August. 1981. Time and energy budgets for dayroosting in a maternity colony of *Myotis lucifugus*. Journal of Mammalogy 62:758–766.

Butchkoski, C.M., and J.D. Hassinger. 2002. Ecology of a maternity colony roosting in a building, pp. 130–142, *in* The Indiana bat: biology and management of an endangered species (A. Kurta and J. Kennedy, eds.). Bat Conservation International, Austin, TX.

Caceres, M.C. 1998. The summer ecology of *Myotis* species bats in the interior wet-belt of British Columbia. M.S. thesis, University of Calgary, Calgary, Alberta.

Callahan, E.V., R.D. Drobney, and R.L. Clawson. 1997. Selection of summer roosting sites by Indiana bats (*Myotis sodalis*) in Missouri. Journal of Mammalogy 78:818–825.

Campbell, L.A., J.G. Hallett, and M.A. O'Connell. 1996. Conservation of bats in managed forests: use of roosts by *Lasionycteris noctivagans*. Journal of Mammalogy 77:976–984.

Carter, T.C. 2003. Summer habitat use of roost trees by the endangered Indiana bat (*Myotis sodalis*) in the Shawnee National Forest of southern Illinois. Ph.D. dissertation, Southern Illinois University, Carbondale, IL.

Carter, T.C., S.K. Carroll, J.E. Hofmann, J.E. Gardner, and G.A. Feldhamer. 2002. Landscape analysis of roosting habitat in Illinois, pp. 160–164, *in* The Indiana bat: biology and management of an endangered species (A. Kurta and J. Kennedy, eds.). Bat Conservation International, Austin, TX.

Christian, J.J. 1956. The natural history of a summer aggregation of the big brown bat, *Eptesicus fuscus fuscus*. The American Midland Naturalist 55:66–94.

Chruszcz, B.J., and R.M.R. Barclay. 2002. Thermoregulatory ecology of a solitary bat, *Myotis evotis*, roosting in rock crevices. Functional Ecology 16:18–26.

Chung-MacCoubrey, A.L. 1996. Bat species composition and roost use in pinyon-juniper woodlands of New Mexico, pp. 118–123, *in* Bats and forests symposium (R.M.R. Barclay and R.M. Brigham, eds.). Research Branch, British Columbia Ministry of Forests, Victoria, B.C.

———. 2003. Monitoring long-term reuse of trees by bats in pinyon-juniper woodlands of New Mexico. Wildlife Society Bulletin 31:73–79.

Clark, M.K. 2003. Status and monitoring of rare bats in bottomland hardwood forests, pp. 79–90, *in* Monitoring trends in bat populations of the United States and territories: problems and prospects (T.J. O'Shea and M.A. Bogan, eds.). U.S. Geological Survey Information and Technology Report ITR-2003–0003.

Connor, R.C. 2000. Group living in whales and dolphins, pp. 199–218, *in* Cetacean societies: field studies of dolphins and whales (J.L. Mann, R.C. Connor, P.L. Tyack, and H. Whitehead, eds.). University of Chicago Press, Chicago, IL.

Connor, R.S., R.S. Wells, J. Mann, and A.J. Read. 2000. The bottlenose dolphin: social relationships in a fission-fusion society, pp. 91–126, *in* Cetacean societies: field studies of dolphins and whales (J.L. Mann, R.C. Connor, P.L. Tyack, and H. Whitehead, eds.). University of Chicago Press, Chicago, IL.

Constantine, D.G. 1966. Ecological observations of lasiurine bats in Iowa. Journal of Mammalogy 47:34–41.

Cope, J.B., A.R. Richter, and R.S. Mills. 1974. A summer concentration of the Indiana bat, *Myotis sodalis,* in Wayne County, Indiana. Proceedings of the Indiana Academy of Science. 83:482–484.

Crampton, L.H., and R.M.R. Barclay. 1998. Selection of roosting and foraging habitat by bats in different-aged mixedwood stands. Conservation Biology 12:1347–1358.

Crockett, A.B., and H.H. Hadow. 1975. Nest site selection by Williamson and red-naped sapsuckers. Condor 77:365–368.

Cryan, P.M. 2003. Seasonal distribution of migratory tree bats (*Lasiurus* and *Lasionycteris*) in North America. Journal of Mammalogy 84:579–593.

Cryan, P.M., M.A. Bogan, and J.S. Altenbach. 2000. Effect of elevation on distribution of female bats in the Black Hills, South Dakota. Journal of Mammalogy 81:719–725.

Cryan, P.M., M.A. Bogan, and G.M. Yanega. 2001. Roosting habits of four bat species in the Black Hills of South Dakota. Acta Chiropterologica 3:43–52.

Duchamp, J.E., D.W. Sparks, and J.O. Whitaker Jr. 2004. Foraging-habitat selection by bats at an urban-rural interface: comparisons between a successful and a less successful species. Canadian Journal of Zoology 82:1157–1164.

Erickson, J.L., and S.D. West. 2003. Associations of bats with local structure and landscape features of forested stands in western Oregon and Washington. Biological Conservation 109:95–102.

Evelyn, M.J., D.A. Stiles, and R.A. Young. 2004. Conservation of bats in suburban landscapes: roost selection by *Myotis yumanensis* in a residential area in California. Biological Conservation 115:463–473.

Fenton, M.B. 1983. Roosts used by the African bat, *Scotophilus leucogaster* (Chiroptera: Vespertilionidae). Biotropica 15:129–132.

Fenton, M.B., I.L. Rautenbach, S.E. Smith, C.M. Swanepoel, J. Grosell, and J. van Jaarsveld. 1994. Raptors and bats: threats and opportunities. Animal Behaviour 48:9–18.

Foster, R., and A. Kurta. 1999. Roosting ecology of the northern bat (*Myotis septentrionalis)* and comparisons with the endangered Indiana bat (*Myotis sodalis*). Journal of Mammalogy 80:659–672.

Gardner, J.E., J.D. Garner, and J.E. Hofmann. 1991. Summer roost selection and roosting behavior of *Myotis sodalis* (Indiana bat) in Illinois. Unpublished report, Illinois Natural History Survey, Champaign, IL.

Gardner, J.E., J.E. Hofmann, and J.D. Garner. 1996. Summer distribution of the federally endangered Indiana bat (*Myotis sodalis*) in Illinois. Transactions of the Illinois Academy of Science 89:187–196.

Gellman, S.T., and W.J. Zielinski. 1996. Use by bats of old-growth redwood hollows on the northern coast of California. Journal of Mammalogy 77:255–265.

Gooding, G., and J.R. Langford. 2004. Characteristics of tree roosts of Rafinesque's big-eared bat and southeastern bat in northeastern Louisiana. Southwestern Naturalist 49:61–67.

Grindal, S.D. 1999. Habitat use by bats, *Myotis* spp., in western Newfoundland. Canadian Field-Naturalist 113:258–263.

Gumbert, M.W., J.M. O'Keefe, and J.R. MacGregor. 2002. Roost fidelity in Kentucky, pp. 143–152, *in* The Indiana bat: biology and management of an endangered species (A. Kurta and J. Kennedy, eds.). Bat Conservation International, Austin, TX.

Hamilton, I.M., and R.M.R. Barclay. 1994. Patterns of daily torpor and day-roost selection by male and female big brown bats (*Eptesicus fuscus*). Canadian Journal of Zoology 72:744–749.

Hayes, J.P. 2003. Habitat ecology and conservation of bats in western coniferous forests, pp. 81–119, *in* Mammal community dynamics in coniferous forests of western North America: management and conservation. (C.J. Zabel and R.G. Anthony, eds.). Cambridge University Press, Cambridge, MA.

Henry, M., D.W. Thomas, R. Vaudry, and M. Carrier. 2002. Foraging distances and home range of pregnant and lactating little brown bats (*Myotis lucifugus*). Journal of Mammalogy 83:767–774.

Henshaw, R.E. 1970. Thermoregulation in bats, pp. 188–232, *in* About bats: a chiropteran symposium. (B.H. Slaughter and D.W. Walton, eds.). Southern Methodist University Press, Dallas, TX.

Hofmann, J.E., J.E. Gardner, J.K. Krejca, and J.D. Garner. 1999. Summer records and a maternity roost of the southeastern myotis (*Myotis austroriparius*) in Illinois. Transactions of the Illinois Academy of Science 92:95–107.

Holekamp, K.E., S.M. Cooper, C.I. Katona, N.A. Berry, L.G. Frank, and L. Smale. 1997. Patterns of association among female spotted hyenas (*Crocuta crocuta*). Journal of Mammalogy 78:55–64.

Hollis, L.M. 2004. Thermoregulation by big brown bats (*Eptesicus fuscus*): ontogeny, proximate mechanisms, and dietary influences. Ph.D. dissertation, University of Calgary, Calgary, Alberta.

Hoying, K.M., and T.H. Kunz. 1998. Variation in size at birth and post-natal growth in the insectivorous bat *Pipistrellus subflavus* (Chiroptera: Vespertilionidae). Journal of Zoology (London) 245:15–27.

Humphrey, S.R. 1975. Nursery roosts and community diversity of Nearctic bats. Journal of Mammalogy 56:321–346.

Humphrey, S.R., and J.B. Cope. 1976. Population ecology of the little brown bat, *Myotis lucifugus,* in Indiana and north-central Kentucky. American Society of Mammalogists, Special Publication 4:1–81.

Humphrey, S.R., A.R. Richter, and J.B. Cope. 1977. Summer habitat and ecology of the endangered Indiana bat, *Myotis sodalis.* Journal of Mammalogy 58:334–346.

Hutchinson, J.T., and M.J. Lacki. 2001. Selection of day roosts by red bats in mixed mesophytic forests. Journal of Wildlife Management 64:87–94.

Inouye, D.W. 1976. Non-random orientation of entrance holes to woodpecker nests in aspen trees. Condor 78:101–102.

Inouye, R.S., N.J. Huntly, and D.W. Inouye. 1981. Non-random orientation of gila woodpecker nest entrances in saguaro cacti. Condor 83:88–89.

Kalcounis, M.C., and R.M. Brigham. 1994. Impact of predation risk on emergence by little brown bats, *Myotis lucifugus* (Chiroptera: Vespertilionidae), from a maternity colony. Ethology 98:201–209.

———. 1998. Secondary use of aspen cavities by tree-roosting big brown bats. Journal of Wildlife Management 62:603–611.

Kalcounis, M.C., and K.R. Hecker. 1996. Intraspecific variation in roost-site selection by little brown bats (*Myotis lucifugus*), pp. 81–90, *in* Bats and forests symposium (R.M.R. Barclay and R.M. Brigham, eds.). Research Branch, British Columbia Ministry of Forests, Victoria, BC.

Kalcounis-Rüppell, M.C., J.M. Psyllakis, and R.M. Brigham. 2005. Tree roost selection by bats: an empirical synthesis using meta-analysis. Wildlife Society Bulletin 33:1123–1132.

Kays, R.W., and J.L. Gittleman. 2001. The social organization of the kinkajou *Potos flavus* (Procyonidae). Journal of Zoology (London) 253:491–504.

Kerth, G., and B. König. 1999. Fission, fusion, and nonrandom associations in female Bechstein's bats (*Myotis bechsteinii*). Behaviour 136:1187–1202.

Kerth, G., and K. Reckardt. 2003. Information transfer about roosts in female Bechstein's bats: an experimental field study. Proceedings of the Royal Society, London B 270:511–517.

Kerth, G., M. Wagner, and B. König. 2001a. Roosting together, foraging apart: information transfer about food is unlikely to explain sociality in female Bechstein's bats (*Myotis bechsteinii*). Behavioral Ecology and Sociobiology 50:283–291.

Kerth, G., K. Weissmann, and B. König. 2001b. Day roost selection in female Bechstein's bats (*Myotis bechsteini*): a field experiment to determine the influence of roost temperature. Oecologia 126:1–9.

Krebs, J.R., and Davies, N.B. 1993. An introduction to behavioural ecology. Blackwell Scientific, Oxford, U.K.

Kronwitter, F. 1988. Population structure, habitat use and activity patterns of the noctule bat, *Nyctalus noctula,* Schreb., 1774 (Chiroptera: Vespertilionidae) revealed by radio-tracking. Myotis 26:23–85.

Kunz, T.H., and L.F. Lumsden. 2003. Ecology of cavity and foliage roosting bats, pp. 3–89, *in* Bat ecology (T.H. Kunz and M.B. Fenton, eds.). University of Chicago Press, Chicago, IL.

Kurta, A. 1980. The bats of southern Lower Michigan. MS thesis, Michigan State University, East Lansing, MI.

———. 1985. External insulation available to a non-nesting mammal, the little brown bat (*Myotis lucifugus*). Comparative Biochemistry and Physiology 82A:413–420.

———. 1994. Bark roost of a male big brown bat, *Eptesicus fuscus.* Bat Research News 35:63.

———. 2000. The bat community in northwestern Lower Michigan, with emphasis on the Indiana bat and eastern pipistrelle. Unpublished report, Huron-Manistee National Forests, Cadillac, MI.

———. In press. Roosting ecology and behavior of Indiana bats (*Myotis sodalis*) in summer, *in* Proceedings of the Indiana Bat and coal mining: a technical interactive forum (K.C. Vories and A. Harrington, eds.). Office of Surface Mining, U.S. Department of the Interior, Alton, IL.

Kurta, A., G.P. Bell, K.A. Nagy, and T.H. Kunz. 1989. Water balance of free-ranging little brown bats (*Myotis lucifugus*) during pregnancy and lactation. Canadian Journal of Zoology 67:2468–2472.

Kurta, A., and R. Foster. 1995. The brown creeper (Aves: Certhiidae): a competitor of bark-roosting bats? Bat Research News. 36:6–7.

Kurta, A., J. Kath, E.L. Smith, R. Foster, M.W. Orick, and R. Ross. 1993a. A maternity roost of the endangered Indiana bat (*Myotis sodalis*) in an unshaded, hollow, sycamore tree (*Platanus occidentalis*). American Midland Naturalist 130:405–407.

Kurta, A., D. King, J.A. Teramino, J.M. Stribley, and K.J. Williams. 1993b. Summer roosts of the endangered Indiana bat (*Myotis sodalis*) on the northern edge of its range. American Midland Naturalist 129:132–138.

Kurta, A., and S.W. Murray. 2002. Philopatry and migration of banded Indiana bats (*Myotis sodalis*) and effects of radio transmitters. Journal of Mammalogy 83:585–589.

Kurta, A., S.W. Murray, and D.H. Miller. 2002. Roost selection and movements across the summer landscape, pp. 118–129, *in* The Indiana bat: biology and management of an endangered species (A. Kurta and J. Kennedy, eds.). Bat Conservation International, Austin, TX.

Kurta, A., and H. Rice. 2002. Ecology and management of the Indiana bat in Michigan. Michigan Academician 33:361–376.

Kurta, A., K.J. Williams, and R. Mies. 1996. Ecological, behavioural, and thermal observations of a peripheral population of Indiana bats (*Myotis sodalis*), pp. 102–117, *in* Bats and forests symposium (R. M. R. Barclay and R. M. Brigham, eds.). Research Branch, British Columbia Ministry of Forests, Victoria, B.C.

Lacki, M.J., and M.D. Baker. 2003. A prospective power analysis and review of habitat characteristics used in studies of tree-roosting bats. Acta Chiropterologica 5:199–208.

Lacki, M.J., and J.H. Schwierjohann. 2001. Day-roost characteristics of northern bats in mixed mesophytic forest. Journal of Wildlife Management 65:482–488.

Lance, R.F., B. Hardcastle, A. Talley, and P.L. Leberg. 2001. Day-roost selection by Rafinesque's big-eared bats (*Corynorhinus rafinesquii*) in Louisiana forests. Journal of Mammalogy 82:166–172.

Lausen, C.L., and R.M.R. Barclay. 2002. Roosting behaviour and roost selection of female big brown bats (*Eptesicus fuscus*) roosting in rock crevices in southeastern Alberta. Canadian Journal of Zoology 80:1069–1076.

———. 2003. Thermoregulation and roost selection by reproductive female big brown bats (*Eptesicus fuscus*) roosting in rock crevices. Journal of Zoology (London) 260:235–244.

Law, B.S., and J. Anderson. 2000. Roost preferences and foraging ranges of the eastern forest bat *Vespedelus pumilus* under two disturbance histories in northern New South Wales, Australia. Australian Ecology 25:352–367.

Lewis, S.E. 1995. Roost fidelity of bats: a review. Journal of Mammalogy 76:481–496.

———. 1996. Low roost-site fidelity in pallid bats: associated factors and effect on group stability. Behavioral Ecology and Sociobiology 39:335–344.

Lindenmayer, D.B., R.B. Cunningham, C.F. Donnelly, M.T. Tanton, and H.A. Nix. 1993. The abundance and development of cavities in eucalyptus trees: a case study in the montane forests of Victoria, southeastern Australia. Forest Ecology and Management 60:77–104.

Lourenço, S.I., and J.M. Palmeirim. 2004. Influence of temperature in roost selection by *Pipistrellus pygmaeus* (Chiroptera): relevance for the design of bat boxes. Biological Conservation 119:237–243.

Lumsden, L.F., A.F. Bennett, and J.E. Silins. 2002. Location of roosts of the lesser long-eared bat *Nyctophilus geoffroyi* and Gould's wattled bat *Chalinolobus gouldii* in a fragmented landscape in south-eastern Australia. Biological Conservation 106:237–249.

Lunney, D., J. Barker, D. Priddel, and M. O'Connell. 1988. Roost selection by Gould's long-eared bat, *Nyctophilus gouldi* Tomes (Chiroptera: Vespertilionidae), in logged forest on the south coast of New South Wales. Australian Wildlife Research 15:375–384.

Mager, K.J., and T.A. Nelson. 2001. Roost-site selection by eastern red bats (*Lasiurus borealis*). American Midland Naturalist 145:120–126.

Mason, C.F., R.E. Stebbings, and G.P. Winn. 1972. Noctules and starlings competing for roosting holes. Journal of Zoology (London) 166:467.

Mattson, T.A., S.W. Buskirk, and N.L. Stanton. 1996. Roost sites of the silver-haired bat (*Lasionycteris noctivagans*) in the Black Hills, South Dakota. Great Basin Naturalist 56:247–253.

McGrew, W.C., L.F. Marchant, and T. Nishida, eds. 1996. Great ape societies. Cambridge University Press, Cambridge, U.K.

Menzel, M.A., T.C. Carter, B.R. Chapman, and J. Laerm. 1998. Quantitative comparison of tree roosts used by red bats (*Lasiurus borealis*) and Seminole bats (*L. seminolus*). Canadian Journal of Zoology 76:630–634.

Menzel, M.A., T.C. Carter, W.M. Ford, and B.R. Chapman. 2001. Tree-roost characteristics of subadult and female adult evening bats (*Nycticeius humeralis*) in the Upper Coastal Plain of South Carolina. American Midland Naturalist 145:112–119.

Menzel, M.A., D.M. Krishon, T.C. Carter, and J. Laerm. 1999. Notes on tree roost characteristics of the northern yellow bat (*Lasiurus intermedius*), the seminole bat (*Lasiurus seminolus*), the evening bat (*Nycticeius humeralis*), and the eastern pipistrelle (*Pipistrellus subflavus*). Florida Scientist 62:185–193.

Menzel, M.A., S.F. Owen, W.M. Ford, J.W. Edwards, P.B. Wood, B.R. Chapman, and K.V. Miller. 2002. Roost tree selection by northern long-eared bat (*Myotis septentrionalis*) maternity colonies in an industrial forest of the central Appalachian Mountains. Forest Ecology and Management 155:107–114.

Miller, J.R., G.E. Jones, and K.C. Kin. 1973. Populations and distribution of *Steatonyssus occidentalis* (Ewing) (Acarina: Macronyssidae) infesting the big brown bat (Chiroptera: Vespertilionidae). Journal of Medical Entomology 10:606–613.

Miller, N.E., R.D. Drobney, R.L. Clawson, and E.V. Callahan. 2002. Summer habitat in northern Missouri, pp. 165–171, *in* The Indiana bat: biology and management of an endangered species (A. Kurta and J. Kennedy, eds.). Bat Conservation International, Austin, TX.

Mirowsky, K.M., P.A. Horner, R.W. Maxey, and S.A. Smith. 2004. Distributional records and roosts of southeastern myotis and Rafinesque's big-eared bat in eastern Texas. Southwestern Naturalist 49:294–298.

Mumford, R.E., and J.B. Cope. 1964. Distribution and status of the Chiroptera of Indiana. American Midland Naturalist. 72:473–489.

Murray, S.W., and A. Kurta. 2002. Spatial and temporal variation in diet, pp. 182–192, *in* The Indiana bat: biology and management of an endangered species (A. Kurta and J. Kennedy, eds.). Bat Conservation International, Austin, TX.

———. 2004. Nocturnal activity of the endangered Indiana bat (*Myotis sodalis*). Journal of Zoology (London) 262:197–206.

Nowak, R.M. 1994. Walker's bats of the World. Johns Hopkins University Press, Baltimore, MD.

O'Donnell, C. 2000. Cryptic local populations in a temperate rainforest bat *Chalinolobus tuberculatus* in New Zealand. Animal Conservation 3:287–297.

O'Donnell, C., and J. Sedgeley. 1999. Use of roosts by the long-tailed bat, *Chalinobus tuberculatus,* in temperate rainforest in New Zealand. Journal of Mammalogy 80:913–923.

Ormsbee, P.C. 1996. Characteristics, use, and distribution of day roosts selected by female *Myotis volans* (long-legged myotis) in forested habitat of the central Oregon Cascades, pp. 124–131, *in* Bats and forests symposium (R.M.R. Barclay and R.M. Brigham, eds.). Research Branch, British Columbia Ministry of Forests, Victoria, BC.

Ormsbee, P.C., and W.C. McComb. 1998. Selection of day roosts by female long-legged myotis in the central Oregon Cascade Range. Journal of Wildlife Management 62:596–603.

O'Shea, T.J., and M.A. Bogan. 2003. Introduction, pp. 69–77, *in* Monitoring trends in bat populations of the United States and Territories: problems and prospects (T.J. O'Shea and M.A. Bogan, eds.). U.S. Geological Survey Information and Technology Report ITR-2003-0003.

Parsons, H.J., D.A. Smith, and R.F. Whittam. 1986. Maternity colonies of silver-haired bats, *Lasionycteris noctivagans,* in Ontario and Saskatchewan. Journal of Mammalogy 67:598–600.

Parsons, S., K.J. Lewis, and J.M. Psyllakis. 2002. Relationships between roosting habitat of bats and decay of aspen in the sub-boreal forests of British Columbia. Forest Ecology and Management 177:559–570.

Patterson, A.P., and J.W. Hardin. 1969. Flight speeds of five species of vespertilionid bats. Journal of Mammalogy 50:152–153.

Perkins, J.M., and S.P. Cross. 1988. Differential use of some coniferous forest habitats by hoary and silver-haired bats in Oregon. Murrelet 69:21–24.

Rabe, M.J., T.E. Morrell, H. Green, J.C. deVos Jr., and C.R. Miller. 1998. Characteristics of ponderosa pine snag roosts used by reproductive bats in northern Arizona. Journal of Wildlife Management 62:612–621.

Racey, P.A. 1973. Environmental factors affecting the length of gestation in heterothermic bats. Journal of Reproduction and Fertility (Supplement) 19:175–189.

Racey, P.A., and S.M. Swift. 1981. Variation in gestation length in a colony of pipistrelle bats (*Pipistrellus pipistrellus*) from year to year. Journal of Reproduction and Fertility 61:123–129.

Roverud, R.C., and M.A. Chappell. 1991. Energetic and thermoregulatory aspects of clustering behavior in the Neotropical bat, *Noctilio albiventris.* Physiological Zoology 64:1527–1541.

Russo, D., L. Cistrone, G. Jones, and S. Mazzoleni. 2004. Roost selection by barbastelle bats (*Barbastella barbastellus,* Chiroptera: Vespertilionidae) in beech woodlands of central Italy: consequences for conservation. Biological Conservation 117:73–81.

Sasse, D.B., and P.J. Pekins. 1996. Summer roosting ecology of northern long-eared bats (*Myotis septentrionalis*) in the White Mountain National Forest, pp. 91–101, *in* Bats and forests symposium (R.M.R. Barclay and R.M. Brigham, eds.). Research Branch, British Columbia Ministry of Forests, Victoria, BC.

Sedgeley, J.A. 2001. Quality of cavity microclimate as a factor influencing selection of maternity roosts by a tree-dwelling bat, *Chalinolobus tuberculatus,* in New Zealand. Journal of Applied Ecology 38:425–438.

———. 2003. Roost site selection and roosting behaviour in lesser short-tailed bats (*Mystacina tuberculata*) in comparison with long-tailed bats (*Chalinolobus tuberculatus*) in *Nothofagus* forest, Fiordland. New Zealand Journal of Zoology 30:227–241.

Sedgeley, J.A., and C.F.J. O'Donnell. 2004. Roost use by long-tailed bats in South Canterbury: examining predictions of roost-site selection in a highly fragmented landscape. New Zealand Journal of Ecology 28:1–18.

Sherwin, R.E., W.L. Gannon, and J.S. Altenbach. 2003. Managing complex systems simply: understanding inherent variation in the use of roosts by Townsend's big-eared bat. Wildlife Society Bulletin 31:62–72.

Solick, D.I. 2004. Differences in the morphology and behaviour of western long-eared bats (*Myotis evotis*) within and between environments. MS thesis, University of Calgary, Calgary, Alberta.

Sparks, D.W. 2003. How does urbanization impact bats? Ph.D. dissertation, Indiana State University, Terre Haute, IN.

Sparks, D.W., M.T. Simmons C.L. Gummer, and J.E. Duchamp. 2003. Disturbance of roosting bats by woodpeckers and raccoons. Northeastern Naturalist 10:105–108.

Sparling, D.W., and G.L. Krapu. 1994. Communal roosting and foraging behavior of staging sandhill cranes. The Wilson Bulletin 106:62–77.

Speakman, J.R., and D.W. Thomas. 2003. Physiological ecology and energetics of bats, pp. 430–490, *in* Bat ecology (T.H. Kunz and M.B. Fenton, eds.). University of Chicago Press, Chicago, IL.

Studier, E. 1981. Energetic advantages of slight drops in body temperature in little brown bats *Myotis lucifugus.* Comparative Biochemistry and Physiology A 70:537–540.

Taylor, R.J., and N.M. Savva. 1988. Use of roost sites by four species of bats in state forest in south-eastern Tasmania. Australian Wildlife Research 15:637–645.

Terborgh, J., and C.H. Janson. 1986. The socioecology of primate groups. Annual Review of Ecology and Systematics 17:111–135.

Thomas, D.W. 1988. The distribution of bats in different ages of Douglas-fir forests. Journal of Wildlife Management 52:619–626.

Trune, D.R., and C.N. Slobodchikoff. 1976. Social effects of roosting on metabolism of the pallid bat (*Antrozous pallidus*). Journal of Mammalogy 57:656–663.

Veilleux, J.P., J.O. Whitaker Jr., and S.L. Veilleux. 2003b. Tree-roosting ecology of reproductive female eastern pipistrelles, *Pipistrellus subflavus,* in Indiana. Journal of Mammalogy 84:1068–1075.

———. 2004. Reproductive stage influences roost use by tree roosting female eastern pipistrelles, *Pipistrellus subflavus.* Ecoscience 11:249–256.

Veilleux, S.L., J.P. Veilleux, J. Duchamp, and J.O. Whitaker Jr. 2003a. Possible predation attempt at a roost tree of evening bats (*Nycticeius humeralis*). Bat Research News 44:186–187.

Verboom, B., and H. Huitema. 1997. The importance of linear landscape elements for the pipistrelle *Pipistrellus pipistrellus* and the serotine bat *Eptesicus serotinus*. Landscape Ecology 12:117–125.

Vonhof, M.J. 1996. Roost-site preferences of big brown bats (*Eptesicus fuscus*) and silver-haired bats (*Lasionycteris noctivagans*) in the Pend d'Oreille Valley in southern British Columbia, pp. 62–80, *in* Bats and forests symposium (R.M.R. Barclay and R.M. Brigham, eds.). Research Branch, B. C. Ministry of Forests, Victoria, BC.

Vonhof, M.J., and R.M.R. Barclay. 1996. Roost-site selection and roosting ecology of forest-dwelling bats in southern British Columbia. Canadian Journal of Zoology 74:1797–1805.

———. 1997. Use of tree stumps as roosts by the western long-eared bat. Journal of Wildlife Management 61:674–684.

Waldien, D.L., and J.P. Hayes. 2001. Activity areas of female long-eared myotis in coniferous forests in western Oregon. Northwest Science 75: 307–314.

Waldien, D.L. J.P. Hayes, and E.B. Arnett. 2000. Day-roosts of female long-eared myotis in western Oregon. Journal of Wildlife Management 64:785–796.

Waldien, D.L., J.P. Hayes, and B.E. Wright. 2003. Use of conifer stumps in clearcuts by bats and other vertebrates. Northwest Science 77:64–71.

Wang, L.C.H., and M.W. Wollowyk. 1988. Torpor in mammals and birds. Canadian Journal of Zoology 66:133–137.

Webb, P.I. 1995. The comparative ecophysiology of water balance in microchiropteran bats. Symposium of the Zoological Society of London 67:203–218.

Webb, P.I., J.R. Speakman, and P.A. Racey. 1993. The implication of small reductions in body temperature for radiant and convective heat loss in resting endothermic brown long-eared bat, *Plecotus auritus*. Journal of Thermal Biology 18:131–135.

Weller, T.J., and C.J. Zabel. 2001. Characteristics of fringed myotis day roosts in northern California. Journal of Wildlife Management 65:489–497.

Whitaker, J.O., Jr., and V. Brack Jr. 2002. Distribution and summer ecology in the state of Indiana, pp. 48–54, *in* The Indiana bat: biology and management of an endangered species (A. Kurta and J. Kennedy, eds.). Bat Conservation International, Austin, TX.

Wilde, C.J., C.H. Knight, and P.A. Racey. 1999. Influence of torpor on milk protein composition and secretion in lactating bats. Journal of Experimental Zoology 284:35–41.

Wilkinson, G.S. 1992. Information transfer at evening bat colonies. Animal Behaviour 44:501–518.

Wilkinson, G.S., and J.M. South. 2002. Life history, ecology and longevity in bats. Aging Cell 1:124–131.

Wilkinson, L.C., and R.M.R. Barclay. 1997. Differences in the foraging behaviour of male and female big brown bats (*Eptesicus fuscus*) during the reproductive period. Ecoscience 4:279–285.

Williams, L.M., and M.C. Brittingham. 1997. Selection of maternity roosts by big brown bats. Journal of Wildlife Management 61:359–368.

Willis, C.K.R. 2003. Physiological ecology of roost selection in female, forest-living big brown bats (*Eptesicus fuscus*) and hoary bats (*Lasiurus cinereus*). Ph.D. dissertation, University of Regina, Regina, Saskatchewan.

Willis, C.K.R., and R.M. Brigham. 2004. Roost switching, roost sharing and social cohesion: forest-dwelling big brown bats, *Eptesicus fuscus,* conform to the fission-fusion model. Animal Behaviour 68:495–505

———. 2005. Physiological and ecological aspects of roost selection by reproductive female hoary bats (*Lasiurus cinereus*). Journal of Mammalogy 86:85–94.

Willis, C.K.R., K.A. Kolar, A.L. Karst, M.C. Kalcounis-Rüppell, and R.M. Brigham. 2003. Medium-and long-term reuse of trembling aspen cavities as roosts by big brown bats (*Eptesicus fuscus*). Acta Chiropterologica 5:85–90.

Winhold, L., E. Hough, and A. Kurta. 2005. Long-term fidelity of tree-roosting bats to a home area. Bat Research News 46:9–10.

BEHAVIOR AND DAY-ROOSTING ECOLOGY OF NORTH AMERICAN FOLIAGE-ROOSTING BATS

3

Timothy C. Carter and Jennifer M. Menzel

Bats depend on roosts for hibernation, mating, rearing of young, and protection from predators and adverse weather conditions. Consequently, roosts play a critical role in the survival of all species of bats (Kunz and Lumsden 2003), and destruction of roosts is a key factor in the decline of many bat populations throughout the world, including North America (Pierson 1998). Many species of bats roost in forests during summer and winter, and although composition and aerial extent of North American forests have changed dramatically over the past centuries (MacCleery 1992; Williams 1989), biologists have little knowledge of how such changes have affected the lives of forest-dwelling bats. Nevertheless, understanding how bats use forests is vital for developing effective management strategies for both forests and the community of bats that lives within them (O'Shea and Bogan 2003).

Many studies have examined use of forests by bats that roost in tree cavities or under bark (Barclay and Kurta, chap. 2 in this volume), but relatively few reports have focused on foliage-roosting species. Almost 20 percent ($n = 8$) of the North American species of bats typically roost among leaves, including the hoary bat (*Lasiurus cinereus;* fig. 3.1), eastern red bat (*L. borealis;* fig. 3.2), western red bat (*L. blossevillii*), Seminole bat (*L. seminolus*), northern yellow bat (*L. intermedius*), southern yellow bat (*L. ega*), western yellow bat (*L. xanthinus*), and eastern pipistrelle (*Pipistrellus subflavus*). Bats that roost among leaves differ from other forest-dwelling bats in various aspects of morphology, ecology, behavior, and life history; therefore, foliage-roosting bats have different requirements and rely on different resources of the forest than do bats that use cavities or bark. Every forest in North America, except those in the extreme north, probably shelters one or more species of foliage-roosting bat, and any complete plan for management of a forest should consider the needs of these species, as well as those bats that seek shelter in cavities or behind bark.

We begin this review with some background about the life history, general biology, and patterns of summer roosting behavior for the eight species of North American foliage-roosting bats. We then examine features of the forest that may have an impact on the selection of roosts at three spatial scales: the tree, the forest stand, and the landscape. We use

Figure 3.1. (left) A hoary bat (*Lasiurus cinereus*) roosting in the canopy of a tree. *Photo by T. Carter*

Figure 3.2. (right) A red bat (*Lasiurus borealis*) roosting among leaves. *Photo by T. Carter*

data from the literature to test the hypothesis that foliage-roosting bats show species-specific preferences for roosting sites and to determine if there are any generalizations evident in habitat use among these species. We conclude with some management recommendations and suggestions for future research.

BIOLOGY OF FOLIAGE-ROOSTING BATS
GEOGRAPHIC RANGE

A rudimentary understanding of the geographic distributions of foliage-roosting bats is important for a number of reasons. First, most of these bats are rarely studied; therefore, their distributions are not widely known (Cryan and Veilleux, chap. 6 in this volume). Second, some foliage-roosting bats consistently roost in ecological settings that are unique to certain parts of the continent. Finally, the geographic distributions of a few species vary considerably among seasons.

Within the United States and Canada, lasiurines have some of the largest and smallest geographic ranges of any species of bat (Wilson and Ruff 1999). The hoary bat, for example, lives throughout most of North America (Shump and Shump 1982a), and the eastern red bat is found in most forested areas east of the Rocky Mountains (Shump and Shump 1982b). In contrast, the western red bat occurs mainly along the west coast, from southern British Columbia, south through California, and east to western Texas (Constantine 1959; Cryan 2003; Genoways and

Baker 1988). The three yellow bats primarily live in Mexico and Central America, and all reach the northern limit of their distributions in the southern United States. Within the United States, the western yellow bat ranges from southern California to western Texas, but the southern yellow bat is an uncommon resident along the Gulf Coast of Texas (Barbour and Davis 1969; Dixon 1997; Kurta and Lehr 1995). The northern yellow bat occurs in the Southeast, mainly from Texas to Florida and the Carolinas, but it is more restricted to coastal regions than the Seminole bat, which resides in the southeastern United States (Webster et al. 1980; Wilkins 1987).

Lasiurines that live in Canada and the northern United States migrate south for winter, up to 2,000 km from where they spend the summer months (Cryan and Veilleux, chap. 6 in this volume; Cryan et al. 2004; Davis and Lidicker 1956; Findley and Jones 1964). Some individuals may travel into Mexico, but members of each species of *Lasiurus* apparently remain in the southern United States throughout the winter (Davis and Lidicker 1956; Findley and Jones 1964; Kurta and Lehr 1995; Webster et al. 1980). Although these bats may spend considerable time in torpor during winter, they often arouse and feed (Boyles et al. 2003; Whitaker et al. 1997), and consequently, land managers in southern forests should consider the year-round needs of these bats in terms of foraging and roosting habitat.

The eastern pipistrelle is found throughout the United States, east of the Great Plains, and northward into southeastern Canada. In contrast to lasiurines, eastern pipistrelles rarely migrate more than 50 km, and throughout most of their range these bats typically hibernate in a mine or a cave where they remain throughout winter (Fujita and Kunz 1984). In the Deep South, eastern pipistrelles spend the winter in more exposed sites, such as bridges or buildings, and in these locations pipistrelles often arouse from torpor and feed, as do lasiurines (Layne 1992; Sandel et al. 2001). The range of the eastern pipistrelle extends into Central America, regions where hibernation in winter is less likely, and we suspect that some eastern pipistrelles may roost in trees during winter in the extreme southern United States, such as southern Texas or south-central Florida.

BODY SIZE AND REPRODUCTIVE PATTERNS

Lasiurines are relatively large bodied when compared with most species of bats in North America, ranging in adult mass from 9–15 g for the eastern red bat to 25–35 g for the hoary bat, the largest bat throughout much of the North American continent (Wilson and Ruff 1999). This variation in size is greater than observed for most other genera of bats in the United States and Canada, perhaps indicating greater intrageneric than intergeneric competition. Lasiurines roost in relatively exposed locations compared with other tree-roosting bats and have the highest values for insulation of all species of bats in North America (Shump and Shump 1980). Additionally, they have a high aspect ratio and high wing loading (Elliot and Wong 1969), and as expected from their morphology, lasiurines typically forage in open habitats, often along forest edges (Mager and Nelson 2001).

Lasiurines usually give birth to two young per litter each season, but these bats are unique among North American species in having four mammary glands. Consequently, litters of three or four young are not unusual, especially among red bats, and litters containing five offspring have been recorded (Wilson and Ruff 1999). Young lasiurines develop more slowly than the offspring of crevice-roosting species, because foliage roosts do not offer as much thermal protection as bark or tree hollows, leading to a greater use of torpor by foliage-roosting bats (Koehler and Barclay 2000). Nevertheless, many lasiurines migrate long distances to warmer regions in late summer and begin hibernation later than other temperate vespertilionids, and this delay presumably provides young lasiurines with a longer time to develop (Feldhamer et al. 2003; Koehler and Barclay 2000). Other than family units consisting of an adult female and her dependent young, lasiurines are usually solitary. Although groups of northern and southern yellow bats have been reported, further study is needed to decide whether these groups represent interacting members of a colony or are simply passive aggregations caused by a scarcity of some resource (Cryan and Veilleux, chap. 6 in this volume).

In contrast to lasiurines, the eastern pipistrelle is one of the smallest bats in North America, weighing only 6–8 g (Fujita and Kunz 1984). Like other foliage-roosting species, eastern pipistrelles typically have two young per litter per year (Jennings 1958), with the two neonates forming relatively one of the largest litter masses of all species of bats, weighing up to 54% of the mother's postpartum mass (Hoying and Kunz 1998). Unlike lasiurines, eastern pipistrelles that roost in trees consistently form maternity colonies, although these colonies are quite small and contain an average of only four adults (Veilleux and Veilleux 2004).

ROOST-SWITCHING AND SPATIAL DISTRIBUTION OF ROOSTS
ROOST-SITE FIDELITY

Radiotracking studies have demonstrated that foliage-roosting bats frequently switch roost trees. Seminole bats, for example, spent an average of 1.5 days in each roost tree in South Carolina (Menzel et al. 2000). Similarly, eastern red bats used individual trees for 1.2–3.3 days in Kentucky, Mississippi, and South Carolina (Elmore et al. 2004; Hutchinson and Lacki 2000; Leput 2004; Menzel et al. 2000). Variation in the frequency and extent of roost switching is quite large, even within a species. For instance, Mager and Nelson (2001) reported that eastern red bats rarely returned to the same roost tree on consecutive days, whereas Leput (2004) tracked a female eastern red bat to the same tree for 22 consecutive days. Although eastern red bats change trees often, they occasionally return to a previously used tree rather than locating a new roosting site each day (Mager and Nelson 2001).

In hoary bats, reproductive condition apparently affects the frequency of roost switching. Although pregnant and postlactating bats often change trees, lactating females do not; one lactating female occupied the same branch of the same tree for 14 consecutive days and another did so for 27 days (Willis 2003; Willis and Brigham 2005). Little is known about roost fidelity of other lasiurines, but it is likely that they too switch roosts periodically.

Like hoary and eastern red bats, eastern pipistrelles use more than one roost. In South Carolina, eastern pipistrelles switched trees every 3.3 days (Leput 2004). In Indiana, duration of continuous occupancy of a day roost by adult females varied from 1 to 17 days, with an average of 3.9 days (Veilleux et al. 2003). These pipistrelles occupied either clusters of dead leaves or clusters of living leaves, but there was no difference in length of residency between types of roost. Additionally, the frequency of roost switching was not correlated with a variety of roost parameters, such as canopy cover, amount of clutter, or species of tree, although roost switching occurred less often during warmer weather. Reuse of roost trees was common, with 44% of adult females returning to a previously occupied tree during the length of time (average = 9 days) that transmitters functioned.

In her comprehensive review of roost-site fidelity, Lewis (1995) suggested that bats that use abundant or ephemeral roosting sites should change roosts more frequently than bats that occupy more stable, less common sites. Consequently, one would predict that bats that roost among leaves should change trees more often than bats that live in tree hollows or under bark. An analysis of 21 studies, involving 10 species of North American bats that inhabit cavities or roost behind bark, indicated that roost switching typically occurred every 2.5 days, with 20 of the studies reporting averages of 1.2–3.4 days (Barclay and Kurta, chap. 2 in this volume). These rates are similar to the averages of 1.2–3.9 days described for foliage-roosting lasiurines and eastern pipistrelles (Elmore et al. 2004; Hutchinson and Lacki 2000; Leput 2004; Menzel et al. 2000; Veilleux et al. 2003). These data indicate no clear difference in the frequency of roost switching by foliage-roosting bats when compared with species of bats that use bark or hollows, suggesting that availability and ephemerality may not be the primary explanations for roost switching by tree-roosting bats (Barclay and Kurta, chap. 2 in this volume).

Information concerning interannual fidelity to a specific tree or roosting area is rare, perhaps because few studies last more than one year. Nonetheless, Willis (2003) reported that one of 20 roost trees occupied by female hoary bats in Saskatchewan was reused in successive years. Veilleux and Veilleux (2004) recaptured two eastern pipistrelles in consecutive summers, and although these bats were not radiotracked to a tree that they used the previous year, they did roost within 23 and 73 m of trees that they occupied the summer before. Additionally, there was anecdotal evidence of natal philopatry for both hoary bats and eastern pipistrelles. A female hoary bat, banded as a juvenile, was found roosting in a tree only 30 m from the tree used by her mother the year before (Willis 2003), and a female eastern pipistrelle roosted in a tree 141 m from the site where she had been banded as a juvenile during the previous summer (Veilleux and Veilleux 2004).

Trees used by cavity- and bark-roosting bats typically are clumped in the environment, rather than being spread throughout forested habitats (Barclay and Kurta, chap. 2 in this volume), and the same appears true of foliage-roosting bats, although data are available only for three species. For example, roost trees of radiotagged Seminole bats in Georgia were located within an average area of 0.2 ha (Menzel et al. 1998). Eastern red bats in Georgia, Kentucky, and Mississippi had average roosting areas that ranged from less than 0.004 to 2.6 ha (Elmore et al. 2004; Hutchinson and Lacki 2000; Menzel et al. 1998). In contrast, Mager and Nelson (2001) calculated an average roosting area of 90 ha for eastern red bats in Illinois; an average that is more than an order of magnitude higher than any other value reported in the literature. Such an unusually large roosting area likely reflects the urban nature of that study site.

Roosts of the eastern pipistrelle also occur close together. In Indiana, roosting areas of six females, which used at least three trees each, averaged 0.23 ha, with a mean distance between successive roosts of only 60 m (Veilleux et al. 2003). In South Carolina, Leput (2004) reported an area of 0.02 ha enclosing three trees used by a single eastern pipistrelle. An adult male in Georgia used a roosting area of 0.7 ha, which amounted to only 0.02% of the home range of the bat (Krishon et al. 1997).

The small size of roosting areas reported for foliage-roosting bats may be an artifact of increased wing loading and lower maneuverability resulting from the extra mass of radiotransmitters (Aldridge and Brigham 1988); however, such an effect does not seem likely because bats carrying radiotransmitters frequently forage far from their roost (Lacki et al., chap. 4 in this volume). The small roosting areas do suggest that foliage-roosting bats are not as flexible in their roosting requirements as previously thought. Use of discrete areas of forest for roosting, in conjunction with inter-annual fidelity and natal philopatry to these areas, may indicate that there are specific features of the forest that these bats seek out or perhaps familiarity with a specific area is somehow beneficial to these animals (Veilleux and Veilleux 2004).

CHARACTERISTICS OF ROOSTS
TYPES OF ROOSTS AND SPECIES OF TREE

Little is known about the roost-site characteristics of the western red bat. There is an isolated report of a western red bat roosting in a fig tree (*Ficus* sp.) that occurred in a grove of cottonwoods (*Populus deltoides*) in Arizona (Hargrave 1944). In addition, Constantine (1959) found numerous western red bats in California roosting among the leaves of apricot trees (*Prunus armeniaca*) in orchards and a few in fig trees and orange trees (*Citrus sinensis*). Snow (1996) reported that western red bats roosted in sycamore (*Platanus occidentalis*), cottonwood, and black walnut (*Juglans nigra*), as well as in leafy shrubs (table 3.1).

The eastern red bat, in contrast, is the most studied foliage-roosting species in North America. These small lasiurines are typically found in the outer edge of deciduous trees, on small branches and leaf petioles

Table 3.1. Species of trees, shrubs, and epiphytes used by foliage-roosting bats

Species of plant	Species of bat					
	Lasiurus blossevillii	*Lasiurus borealis*	*Lasiurus cinereus*	*Lasiurus ega, L. intermedius, L. xanthinus*	*Lasiurus seminolus*	*Pipistrellus subflavus*
American basswood *Tilia americana*		5				
American beech *Fagus grandifolia*		5				
Ash, green *Fraxinus pennsylvanica*		12	1			
Ash, white *Fraxinus americana*						16
Banana *Musa* sp.				6, 13		
Black cherry *Prunus serotina*		3	3			
Black gum *Nyssa sylvatica*		10, 12				16
Black walnut *Juglans nigra*	14	3, 9				16
Boxelder *Acer negundo*		3	3			16
Catalpa *Catalpa* sp.		3				
Cottonwood *Populus deltoides*	14					16
Elm *Ulmus* spp.		3, 9, 10, 12	3			2, 16
Giant dagger yucca *Yucca carnersosana*				17		
Hackberry *Celtis occidentalis*		3				16
Hackberry *Celtis reticulata*				6		
Hawthorn *Crataegus* sp.		3				
Hickory *Carya* spp.		4, 5, 9, 10, 12				
Honeylocust *Gleditsia triacanthos*		9				
Maple, red *Acer rubrum*		5, 10, 12				
Maple, silver *Acer saccharinum*		3				
Maple, sugar *Acer saccharum*		5				16
Mexican fan palm *Washingtonia gracilis*				6		
Oak, bur *Quercus macrocarpa*						16
Oak, chestnut *Quercus prinus*		5				
Oak, laurel *Quercus laurifolia*		10, 12				2
Oak, live *Quercus virginiana*		10			10, 11	
Oak, northern red *Quercus rubra*		5				7, 16
Oak, overcup oak *Quercus lyrata*		12				

(continued)

Table 3.1. Continued

Species of plant	Lasiurus blossevillii	Lasiurus borealis	Lasiurus cinereus	Lasiurus ega, L. intermedius, L. xanthinus	Lasiurus seminolus	Pipistrellus subflavus
Oak, post *Quercus stellata*		10, 12	15			
Oak, southern red *Quercus falcata*		10, 12				
Oak, swamp chestnut *Quercus michauxii*		10, 12				2
Oak, water *Quercus nigra*		10, 12				
Oak, white *Quercus alba*		5, 10, 12	15			7, 16
Oak, willow *Quercus phellos*		12				
Ohio buckeye *Aesculus glabra*						16
Osage orange *Maclura pomifera*			3			
Pine, loblolly *Pinus taeda*		4, 10			10, 11	
Pine, longleaf *Pinus palustris*		10, 12			10	
Pine, pond *Pinus serotina*					10, 11	
Pine, shortleaf *Pinus echinata*			15			
Pine, slash *Pinus elliottii*					10, 11	
Red mulberry *Morus rubra*		3				
Spanish moss *Tillandsia usneoides*				11	10	11
Sumac *Rhus* sp.		3				
Sweet birch *Betula lenta*		5				
Sweetgum *Liquidambar styraciflua*		4, 9, 10, 12				2
Sycamore *Platanus occidentalis*	14	5, 10				
Texas sabal palm *Sabal texana*				13		
Washington palm *Washingtonia filifers*				6, 13		
Water tupelo *Nyssa aquatica*		10, 12				
White spruce *Picea glauca*			18			
Wild plum *Prunus* sp.		3	3			
Willow *Salix* sp.		3				
Yellow poplar *Liriodendron tulipifera*		5, 10, 12				16

*Numerals indicate sources: 1, Barclay 1984; 2, Carter et al. 1999; 3, Constantine 1966; 4, Elmore et al. 2004; 5, Hutchinson and Lacki 2000; 6, Kurta and Lehr 1995; 7, Kurta et al. 1999; 8, Leput 2004; 9, Mager and Nelson 2001; 10, Menzel et al. 1998; 11, Menzel et al. 1999; 12, Menzel et al. 2000; 13, Mirowsky 1997; 14, Snow 1996; 15, R. Thill, pers. comm.; 16, Veilleux et al. 2003; 17, Weyandt et al. 2001; and 18, Willis and Brigham 2005.

(table 3.1; Barbour and Davis 1969; Hutchinson and Lacki 2000; Mager and Nelson 2001). In Kentucky, these bats usually roosted in portions of the upper canopy where they often were surrounded by clumps of dead leaves (Hutchinson and Lacki 2000), but another study emphasized the dark-green foliage used by eastern red bats for roosting (McClure 1942).

Eastern red bats use a variety of deciduous trees for roosting, and this diversity probably reflects, in part, the geographic distribution of the bat. For example, 64% of 64 trees used by eastern red bats in Georgia were sweetgum (*Liquidambar styraciflua*), black oak (*Quercus nigra*), white oak (*Q. alba*), and black gum (*Nyssa sylvatica;* Elmore et al. 2004; Leput 2004; Mager and Nelson 2001; Menzel et al. 1998, 2000). In contrast, 68% of 44 roosts in Kentucky were in hickories (*Carya* spp.), yellow poplar (*Liriodendron tulipifera*), white oak, and American beech (*Fagus grandifolia;* Hutchinson and Lacki 2000). Although Hutchinson and Lacki (2000) indicated that the mix of trees used by eastern red bats in their study was not different from a random sample of available hardwood species, Menzel et al. (1998) showed that eastern red bats in the South used some species of tree less (red maple, *Acer rubrum*) or more (white oak) often than expected, based on their availability in the surrounding forest, suggesting some selectivity on the part of these bats (see also Mager and Nelson 2001). Eastern red bats occasionally used conifers in Iowa (McClure 1942), but they used conifers less often than expected based on availability in Kentucky and Georgia (Hutchinson and Lacki 2000; Menzel et al. 1998). Nevertheless, in Mississippi, 30% of 141 roost trees were in loblolly pine (*Pinus taeda*), reflecting availability of conifers in the adjacent woods (Elmore et al. 2004).

Not all roosting sites of eastern red bats are in trees (Hutchinson and Meisenburg 2004; McClure 1942). Mager and Nelson (2001) radio-tracked 12 eastern red bats to 75 roosting locations; although 79% of the sites were in foliage, 10% were on the bole of deciduous trees, 7% were in grass or leaf litter, and 4% were under shingles of a house. In addition, eastern red bats often roosted in the foliage of vines that were supported by deciduous trees (Constantine 1966; Elmore et al. 2004).

The hoary bat commonly hangs from branches of either deciduous or coniferous trees (table 3.1; Barclay 1984; Constantine 1966; Kalcounis 1994; Willis and Brigham 2005). In Arkansas, R. Thill (pers. comm.) hypothesized that shortleaf pine (*Pinus echinata*) was an important roosting resource for hoary bats because of widespread use of this tree species. In Saskatchewan, Willis and Brigham (2005) used radiotracking to locate 32 maternity roosts of hoary bats, 31 of which were in white spruce (*Picea glauca*). However, visual searches by Constantine (1966) in Iowa indicated that hoary bats often roosted in American elm (*Ulmus americana*) and box elder (*Acer negundo*). There also are records of hoary bats roosting in a squirrel nest (Neill 1952), underneath driftwood (Conner 1971), on the side of a building (Bowers et al. 1968), and in a woodpecker hole (Cowan and Guiguet 1965).

For many years, the preferred roosting site of Seminole bats was thought to be within clumps of Spanish moss (*Tillandsia usneoides;* fig.

a

b

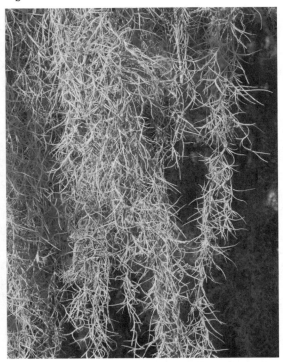

Figure 3.3. Potential roosting sites for lasiurine bats typical of coastal areas of southeastern United States including (a) a deciduous tree covered in Spanish moss (*Tillandsia usneoides*) and (b) a close-up of Spanish moss demonstrating its ability to provide cover for roosting bats. *Photos by A. Sherman and G. Marks, respectively*

3.3), an epiphyte that grows on many trees in the southeastern United States (Barbour and Davis 1969; Constantine 1958; Jennings 1958). However, most records were from winter, and few Seminole bats were found in Spanish moss from July through September, despite search efforts (Constantine 1958; Jennings 1958). Recently, Menzel et al. (1998, 2000) radiotracked Seminole bats during summer in Georgia and South Carolina and found them roosting along the branches of trees, primarily pines (*Pinus* spp.), where the bats often resembled pine cones. These authors suggested that Seminole bats occupied trees in summer and Spanish moss in winter, perhaps as a "response to differing physiological demands," such as pregnancy versus hibernation (Menzel et al. 1998, 633).

Roosts that are reported for western and southern yellow bats reflect the tropical nature of their distributions, with both species roosting in dead fronds of palms (*Washingtonia filifers* and *W. gracilis*) and trees with palmlike leaves (fig. 3.4), such as giant dagger yucca (*Yucca carnersosana*) and banana (*Musa* spp.; table 3.1; Baker et al. 1971; Higginbotham et al. 2000; Kurta and Lehr 1995; Mirowsky 1997; Spencer et al. 1988; Webster et al. 1980; Weyandt et al. 2001). These bats readily take to ornamental palms in residential settings (Mirowsky 1997), and populations of western and southern yellow bats may be expanding as palms are planted for ornamental purposes (Spencer et al. 1988). Yellow bats, however, only use trees in which the petticoat, a ring of dead fronds that encircles the bole below the living foliage, has not been removed, usually for aesthetic reasons (Mirowsky 1997). Although yellow bats in the West may benefit

Figure 3.4.
An example of a palm tree used by lasiurine bats for roosting in southeastern United States.
Photo by R. Sidner

from the high densities of palm and palm-like trees that are planted in urban areas, study of the roosting ecology of these species in urban and rural environments is required to quantify the relative importance of natural versus anthropogenic habitats. Western and southern yellow bats occasionally occupy other sites that resemble large dead leaves, such as dried corn stalks and thatched roofing (Alayo 1958; Baker and Dickerman 1956), and one was found roosting in a hackberry (*Celtis reticulata*) in New Mexico (Mumford and Zimmerman 1963).

Although northern yellow bats also may roost in palm trees (Mirowsky 1997), these bats are commonly found in Spanish moss in the eastern portions of their range. Therefore, northern yellow bats will likely use any species of tree, such as live oak (*Quercus virginiana*), that supports large drapes of the epiphyte (Constantine 1966; Jennings 1958; Webster et al. 1980). This bat also has been reported in large pines in Florida (Sherman 1945).

Eastern pipistrelles occupy a larger variety of locations than do lasiurines, including barns and other buildings, bridges, Spanish moss, living foliage, and clumps of dead leaves suspended in an otherwise healthy tree (Cope et al. 1961; Feldhamer et al. 2003; Findley 1954; Jones and Pagels 1968; Jones and Suttkus 1973; Lane 1946; Leput 2004; Menzel et al. 1999; Veilleux et al. 2003; Whitaker 1998; Winchell and Kunz 1996). Many reported roosts of eastern pipistrelles are from sites other than foliage, especially during winter (Feldhamer et al. 2003; Fujita and Kunz 1984). However, foliage roosts are likely used more often than indicated in the literature because such sites are more cryptic than other roosting locations, making them more difficult for biologists to discover (Carter et al. 1999; Findley 1954; Veilleux et al. 2003). Foliage roosts and roosts in Spanish moss occur in a variety of deciduous trees, but unlike many lasiurines, eastern pipistrelles apparently do not use conifers (table 3.1) (Carter et al. 1999; Kurta et al. 1999; Leput 2004; Veilleux et al. 2003).

In the only comprehensive studies of tree-roosting behavior of eastern pipistrelles, Veilleux et al. (2003, 2004) found that reproductive females roosted most often in groups of dead leaves and preferred certain species of trees in Indiana. For example, they occupied oaks (*Quercus* spp.) more often than expected and maples (*Acer* spp.) less often than expected, based on availability of potential roost trees. Oaks typically possessed a cluster of leaves on the terminal portion of each branch, and if a branch would break, a tight cluster of dead leaves resulted and formed an umbrella-shaped shelter. Roosting among small clusters of dead leaves appears common for the eastern pipistrelle and is similar to the roosting habits of the tube-nosed bat (*Murina florium*) in Australia (Shulz and Hannah 1998).

CHARACTERISTICS OF A ROOSTING SITE WITHIN A TREE

Foliage roosts of all temperate species of bats in North America have some broad similarities with the leaf tents that are made by bats in tropical regions (Kunz et al. 1994). Both types of roosts typically provide dense cover above and around the sides of the bat, and this cover presumably modifies the microclimate, shelters the animal from rain, and conceals the bat from visually oriented predators. In addition, there generally is an opening in the foliage directly below where the bat hangs that allows for easy flight to and from the roosting site (Constantine 1959, 1966; Hutchinson and Lacki 2000; Mager and Nelson 2001; Saugey et al. 1998; Veilleux et al. 2003; Willis and Brigham 2005).

Few studies, however, have examined quantitatively the position of roosting sites within the foliage, and there appears to be no consistent trend among foliage-roosting species in North America. For example, the height of roosting sites used by individual western red bats, eastern red bats, hoary bats, Seminole bats, and eastern pipistrelles varies greatly, ranging from one to more than 20 m above ground (Barclay 1984; Constantine 1958, 1959, 1966; Hutchinson and Lacki 2000; Leput 2004; Mager and Nelson, 2001; Menzel et al. 1998; Saugey et al. 1998; Willis and Brigham 2005). Although early reports, which relied on visual searches to locate roosting bats, suggested that foliage-roosting species generally roosted close to the ground, more recent studies using radiotracking have now demonstrated that these bats also roost very high in the canopy (e.g., Hutchinson and Lacki 2000).

Data on roost height are available only for eastern red bats and eastern pipistrelles. Veilleux et al. (2004) demonstrated that reproductively active, female eastern pipistrelles roosted farther from the top of the canopy than nonreproductive females, and the authors suggested that sites closer to the ground decreased exposure to wind and provided a more buffered thermal environment. On the other hand, Constantine (1966) indicated that families of eastern red bats roosted higher than solitary bats. Constantine (1966) speculated that a greater roosting height for family units provided more concealment from terrestrial predators, minimized disturbance from activity on the ground, and offered more space for young bats to drop during their initial flights.

A few studies have examined the side of the tree (aspect) on which bats roost. Mean compass direction varied among species, and there was no consistent trend, other than the fact that no species of foliage-roosting bat preferred the north side of a tree. A number of reports suggested that eastern red bats, hoary bats, and northern yellow bats typically selected sites with a southeast-to-southwest exposure (Constantine, 1958, 1959, 1966; Menzel at al. 1999; Willis and Brigham 2005). It was assumed that such an orientation provided a warmer roosting environment for these bats because of increased exposure to solar heating, but data on ambient temperature of roosting sites of hoary bats in Saskatchewan did not support this hypothesis (Willis and Brigham 2005). Additionally, foliage roosts of female eastern pipistrelles in Indiana were randomly oriented with respect to compass direction (Veilleux et al. 2003), as were roosts of eastern red bats and Seminole bats in Georgia (Menzel et al. 1998). Constantine (1959, 15) believed that western red bats preferred sites with a southern exposure during cool spring weather and that the bats "chose any side of a tree during summer." Consequently, future studies of roost aspect should quantitatively compare the aspect that is selected during different seasons or at least use ambient temperature as a covariate in analyses.

Various authors have speculated that the position of a roost within a tree (e.g., aspect, height, or distance from trunk) affects the microclimate to which bats are exposed. Although early reports were qualitative (Constantine 1966), two recent studies have quantified roost microclimate. Hutchinson and Lacki (2001, 206) compared temperatures between sites occupied by eastern red bats and exposed settings in the canopy with exposed sites defined as open spaces not occupied or shielded by canopy foliage. Exposed sites typically experienced higher and more variable daytime temperatures than roosting sites; temperatures recorded in exposed settings frequently exceeded 40°C and occasionally reached 50°C, presumably because of the effects of direct solar radiation on the temperature sensors. Although equivocal, these data indicate that roosting among leaves by eastern red bats may provide a more moderate environment than roosting in more open areas of the canopy.

In the second quantitative study, Willis and Brigham (2005) examined the microclimate of roosting sites of female hoary bats and made comparisons with similar sites on the side of the tree opposite the roost. Although ambient temperatures did not differ between roosting and nonroosting sites, wind speed was slightly lower at roosting sites. Using a convective cooling model and data on metabolic rate from the literature, Willis and Brigham (2005) concluded that protection from wind saved females and young a small but significant amount of energy. Similar multifaceted studies that incorporate variables in addition to ambient temperature, such as wind velocity, the effects of sun flecking (dabs of reflected sunlight; McComb and Noble 1981; Veilleux et al. 2004), and radiative exchange with surrounding leaves and the ground, are needed to help understand possible energetic benefits of foliage roosts and ultimately patterns of roost selection by foliage-roosting bats.

CHARACTERISTICS OF FOREST STANDS

Foliage-roosting bats have been found in a variety of forested stands, and as with the species of tree used for roosting, some of this variety undoubtedly reflects the specific area of the continent in which the study was completed. Eastern red bats, for example, have been radiotracked to oak-hickory-shortleaf pine stands (Saugey et al. 1998), mature deciduous stands (Mager and Nelson 2001; Menzel et al. 1998), and oak-hickory and cove hardwood stands (Hutchinson and Lacki 2000). Elmore et al. (2004) discovered 83% of 117 red bat roosts in thinned stands of pine (<37 years old), with the remainder in streamside management zones. Seminole bats roosted in mature managed pine and mixed pine-hardwood forests (Menzel et al. 2000). Menzel et al. (1998) suggested that, in summer, Seminole bats prefer pine-dominated communities, whereas the morphologically similar eastern red bats favor hardwood-dominated communities. In autumn, winter, and spring, Jennings (1958) commonly found Seminole bats in and near types of forest that supported clumps of Spanish moss.

In South Carolina, northern yellow bats roosted in communities dominated by mature live oaks (Menzel et al. 1999). In Florida, Jennings (1958) documented northern yellow bats in two different plant communities in different parts of the state. Stands of mature turkey oak (*Quercus laevis*) and stands of longleaf pine, which were thick with Spanish moss, provided roosting habitat in central Florida, whereas the primary type of cover, containing the requisite clumps of Spanish moss, was mature live-oak forests in the lower and flatter areas of Florida.

Eastern pipistrelles roost in a variety of habitats. Carter et al. (1999) tracked one individual that roosted and foraged primarily in a bottomland hardwood forest. However, most eastern pipistrelles seem to prefer upland and riparian habitats for roosting than bottomland forest (Davis and Mumford 1962; Kurta et al. 1999; Menzel et al. 1999; Veilleux et al. 2003). Leput (2004) documented eastern pipistrelles roosting in cove hardwood, pine-hardwood, and pine-dominated stands in South Carolina.

Do foliage-roosting bats exhibit selection within the forest stand? We have already discussed the apparent preference by eastern red bats and eastern pipistrelles for some species of trees over others for roosting (Hutchinson and Lacki 2001; Menzel et al. 1998). However, quantitative comparisons involving variables other than species of tree are rare, and most involve only the eastern red bat. Roost trees of the eastern red bat in South Carolina occurred in areas with a greater height of overstory trees, greater basal area, and denser canopy in the overstory than other areas of the stand (Menzel et al. 2000). Additionally, there was greater plant species diversity in the overstory and understory surrounding roosts than in randomly selected plots within the same stand (Menzel et al. 2000). In Kentucky, roost trees were found in locations that contained a lower density of trees, trees of larger diameter, and a greater basal area than plots surrounding randomly chosen trees (Hutchinson and Lacki 2000). These data indicate that eastern red bats apparently select areas of mature forest for roosting. Elmore et al. (2004) also concluded that stand-level variables had a greater impact on roost selection by eastern red bats than did landscape- or tree-level variables in Mississippi; no landscape variable and only one tree-level variable (diameter) explained variation in roosting patterns, and then only for juvenile females.

Roosts of the eastern red bat and those of the Seminole bat often are in dominant trees that are larger in diameter and taller than surrounding trees (Elmore et al. 2004; Hutchinson and Lacki 2000; Leput 2004; Mager and Nelson 2001; Menzel et al. 1998, 1999, 2000). Bark- and cavity-roosting bats also choose larger and taller trees than those available, and various authors have speculated that such trees allow more roosting opportunities, provide greater insulation (thicker bark), or have larger cavities that could more easily accommodate a colony of bats (Barclay and Kurta, chap. 2 in this volume). Clearly, none of these explanations are relevant to solitary bats roosting in foliage. Another proposed advantage, however, that could also apply to foliage-roosting bats is that taller trees may be easier to locate in a forested environment and provide easier and safer access for a flying bat (Vonhof and Barclay 1996).

In contrast to the eastern red bat, Willis and Brigham (2005) found that trees occupied by hoary bats were similar in diameter and height to trees in the surrounding forest. They hypothesized that reduced height shielded bats from prevailing winds and decreased convective cooling, thus lessening energetic expenditure in the cool Canadian environment. Roost trees also had lower canopy cover and a lower density of trees on the southeast side of the roost tree compared with randomly chosen trees. This latter variable presumably indicated greater solar exposure and a warmer roost during morning hours, although there was no significant difference in ambient temperature between occupied and unoccupied sites at any time of day.

CHARACTERISTICS OF LANDSCAPES

Little information is available on the roosting preferences of foliage-roosting bats at the level of the landscape (Duchamp et al., chap. 9 in this volume). Although eastern red bats often roost in edge habitats along streams or open fields in landscapes dominated by agriculture (Constantine 1958; Kunz 1973; Leput 2004; Mumford 1973), roost trees in extensive forests are not closer to edges than randomly selected trees (Hutchinson and Lacki 2000). Although many authors suggest that bats locate their roosts near sources of water, roosts used by the eastern red bat did not differ from randomly chosen trees in distance to water (Hutchinson and Lacki 2000). Hutchinson and Lacki (2000) reported that most roost trees were on upper slopes close to ridge tops in Kentucky, and Leput (2004) discovered that female eastern red bats selected trees on north- and northwest-facing slopes in South Carolina. Hutchinson and Lacki (2000) postulated that eastern red bats roosting in the canopies of trees higher in elevation might have a survival advantage over bats roosting closer to the ground where exposure to predators is likely higher.

Data are sparse for other lasiurines and insufficient to detect any patterns. Like eastern red bats, hoary bats are commonly found roosting in habitats where the edge of the forest meets fields or riparian areas, and along fencerows (Constantine 1966, Koehler and Barclay 2000). Northern yellow bats in the Southeast are often associated with water, as most roosts appear to be located close to permanent bodies of water, although statistical comparisons with random sites are lacking (Menzel et al. 1999; Webster et al. 1980). Additionally, because most roosts used by northern yellow bats are in Spanish moss, mature forests that provide this roosting resource presumably are required.

Veilleux et al. (2003) hypothesized that eastern pipistrelles used upland forest more than riparian or bottomland habitats because upland forests were comprised of a greater number of trees with preferred characteristics, such as oaks. Although eastern pipistrelles have been observed foraging along edges (Davis and Mumford 1962), roosts were often found within the forest (Carter et al. 1999; Kurta et al., 1999; Menzel et al. 1999). In Indiana, roosts of eastern pipistrelles averaged 52 m from a forest edge (Veilleux et al. 2003), and Leput (2004) reported that roosts of this species were closer to sources of water than unused trees.

PATTERNS AND CONCLUSIONS

Little is known about foliage-roosting bats in North America compared with species of bats that have different roosting requirements, and this may be due in part to the perceived commonness of foliage-roosting bats and their roosting habitats. In general, most financial resources dedicated to research on bats are allocated to the study of species listed as threatened or endangered (e.g., the Indiana myotis, *Myotis sodalis*); however, no foliage-roosting species of bat has such status except the Hawaiian hoary bat (*Lasiurus cinereus semotus*) which is not found on the North American continent (USFWS 1998). Roost sites are often listed as a major limiting resource for bats (Fenton 1997; Lewis 1995) and, because it is easy to assume that foliage roosts are readily available, land managers may presume that populations of these animals are not in need of conservation. Existing data, however, suggest that populations of many foliage-roosting species have declined over the past several decades (Carter et al. 2003). Furthermore, the solitary nature of foliage-roosting bats may contribute to the paucity of knowledge about them. Colonial bats that roost in stable structures, such as buildings or caves, are relatively conspicuous and easy to study because of their numbers and high fidelity to their roosts (Lewis 1995). In contrast, foliage-roosting bats typically roost alone or in small family groups and display low roost fidelity (Lewis 1995). Such behaviors make collecting data on solitary bats living among leaves more time consuming and potentially more difficult.

Additional research is critical to delineating habitat requirements of foliage-roosting species so that sound recommendations for their management can be proposed. Most of what we know about the roosting behavior and ecology of all tree-roosting bats in North America has been obtained in the past 15 years, mostly through studies using small radio-transmitters (Barclay and Kurta, chap. 2 in this volume; Lacki et al., chap. 4 in this volume). However, to date, no study has used this technology on the western red bat or any of the yellow bats. Further, even though the hoary bat is one of the most widespread species in the New World, there is only a single comprehensive study of roosting preferences of this species that used radiotracking techniques (Willis and Brigham 2005).

Although each foliage-roosting species appears to have distinct roosting preferences, there are some commonalities among these bats in their behavior, ecology, and roost selection. Leaves typically shelter the bat from above and from the sides, but there is invariably a clear zone beneath the roost into which the bat can drop when taking flight. Roosts most often are in large trees in mature areas of forest. With the exception of eastern pipistrelles, most foliage-roosting bats roost alone and do not form colonies. Foliage-roosting bats frequently change roosts, usually every 1–4 days. Because roosting sites in leaves presumably are not as protected or as environmentally stable as cavities, bats may be switching roosts in response to changing environmental conditions (Veilleux et al. 2003; Willis and Brigham 2005). Even though foliage-roosting bats frequently change trees, the spatial distribution of their roosts within forested stands is small compared with the size of their home range, suggesting that these bats are selecting specific features of trees or forests for roosting. The concept of selection is supported by reports of interannual fidelity to a roosting area and natal philopatry in some species. These trends support the hypothesis that foliage-roosting bats are more selective, and not necessarily as flexible in their roosting requirements, than previously thought.

Although similarities exist among various studies of the roosting behavior and ecology of foliage-roosting bats, there often are distinct dissimilarities (e.g., roosting heights or selection of tall trees), and some of this variation may well indicate true differences among species or regions. However, most studies suffer from small samples and combine data from juveniles, adult males, and adult females in various reproductive conditions. Hence, the conclusions of any one report may depend on the composition of its study population and the time of year in which fieldwork was conducted. It is not reasonable to assume that the energetic and roosting requirements of bats varying in sex, age, or reproductive condition will be identical (Elmore et al. 2004), or that they are static throughout the year, and future studies must explore these potential differences. The ecology and behavior of foliage-roosting bats remains a fertile area for future research that promises to provide insight into the patterns and processes of roost selection by all species of tree-roosting bats.

LITERATURE CITED

Alayo, D.P. 1958. Lista de los mamiferos de Cuba, vivientes y extinguidos. Universidad de Oriente, Museo Ramsden, Santiago de Ciba.

Aldridge, H.D.J.N., and R.M. Brigham. 1988. Load carrying and maneuverability in a insectivorous bat: a test of the 5% "rule" of radio-telemetry. Journal of Mammalogy 69:379–382.

Baker, R.H., and R.W. Dickerman. 1956. Daytime roost of the yellow bat in Veracruz. Journal of Mammalogy 37:443.

Baker, R.J., T. Mollhagen, and G. Lopez. 1971. Notes on *Lasiurus ega*. Journal of Mammalogy 52:849–852.

Barbour, R.W., and W.H. Davis. 1969. Bats of America. University Press of Kentucky, Lexington, KY.

Barclay, R.M.R. 1984. Observations on the migration, ecology and behavior of bats at Delta Marsh, Manitoba. Canadian Field-Naturalist 98:331–336.

Bowers, J.R., G.A. Heidt, and R.H. Baker. 1968. A late autumn record for the hoary bat in Michigan. Jack-Pine Warbler 46:33.

Boyles, J.G., J.C. Timpone, and L.W. Robbions. 2003. Late-winter observations of red bats, *Lasiurus borealis,* and evening bats, *Nycticeius humeralis,* in Missouri. Bat Research News 44:59–61.

Carter, T.C., M.A. Menzel, B.R. Chapman, and K.V. Miller. 1999. Summer foraging and roosting behavior of an eastern pipistrelle *Pipistrellus subflavus*. Bat Research News 40:5–6.

Carter, T.C., M.A. Menzel, and D.A. Saugey. 2003. Population trends of solitary foliage-roosting bats, pp. 41–47, *in* Monitoring trends in bat populations of the United States and territories: problems and prospects (T.J. O'Shea, and M.A. Bogan, eds.). U.S. Geological Survey Information and Technology Report ITR-2003-0003.

Conner, P.F. 1971. The mammals of Long Island, New York. Bulletin of the New York State Museum, Science Series 416:1–78.

Constantine, D.G. 1958. Ecological observations on lasiurine bats in Georgia. Journal of Mammalogy 39:64–70.

———. 1959. Ecological observations on lasiurine bats in the North Bay area of California. Journal of Mammalogy 40:13–15.

———. 1966. Ecological observations on lasiurine bats in Iowa. Journal of Mammalogy 47:34–41.

Cope, J.B., W. Baker, and J. Confer. 1961. Breeding colonies of four species of bats of Indiana. Proceedings of the Indiana Academy of Science 70:262–266.

Cowan, I. M., and C.J. Guiguet. 1965. The mammals of British Columbia. Handbook of the British Columbia Provincial Museum 11:1–141.

Cryan, P.M. 2003. Seasonal distribution of migratory tree bats (*Lasiurus* and *Lasionycteris*) in North America. Journal of Mammalogy 84:579–593.

Cryan, P.M., M.A. Bogan, R.O. Rye, G.P. Landis, and C.L. Kester. 2004. Stable hydrogen isotope analysis of bat hair as evidence for seasonal molt and long-distance migration. Journal of Mammalogy 85:995–1001.

Davis, W.H., and W.Z. Lidicker Jr. 1956. Winter range of the red bat, *Lasiurus borealis*. Journal of Mammalogy 37:280–281.

Davis, W.H., and R.E. Mumford. 1962. Ecological notes on the bat *Pipistrellus subflavus*. American Midland Naturalist 68:394–398.

Dixon, M. 1997. The thrill of discovery. Bats 15:5.

Elliot, O., and M. Wong. 1969. Aspect ratio, loading, wing span, and membrane areas of bats. Journal of Mammalogy 50:362–367.

Elmore, L.W., D.A. Miller, and F.J. Vilella. 2004. Selection of diurnal roosts by red bats (*Lasiurus borealis*) in an intensively managed pine forest in Mississippi. Forest Ecology and Management 199:11–20.

Feldhamer, G.A., T.C. Carter, A.T. Morzillo, and E.H. Nicholson. 2003. Use of bridges as day roosts by bats in southern Illinois. Transactions of the Illinois State Academy of Science 96:107–112.

Fenton, M.B. 1997. Science and the conservation of bats. Journal of Mammalogy 78:1–14.

Findley, J.S. 1954. Tree roosting of the eastern pipistrelle. Journal of Mammalogy 35:433.

Findley, J.S., and C. Jones. 1964. Seasonal distribution of the hoary bat. Journal of Mammalogy 45:461–470.

Fujita, M.S., and T.H. Kunz. 1984. *Pipistrellus subflavus*. Mammalian Species 228: 1–6.

Genoways, H.H., and R.J. Baker. 1988. *Lasiurus blossevillii* (Chiroptera: Vespertilionidae) in Texas. Texas Journal of Science 40:111–113.

Hargrave, L.L. 1944. A record of *Lasiurus borealis teliotis* from Arizona. Journal of Mammalogy 25:414.

Higginbotham, J.L., M.T. Dixon, and L.K. Ammerman. 2000. Yucca provides roost for *Lasiurus xanthinus* (Chiroptera: Vespertilionidae) in Texas. Southwestern Naturalist 45:338–340.

Hoying, K.M., and T.H. Kunz. 1998. Variation in size at birth and post-natal growth in the insectivorous bat *Pipistrellus subflavus* (Chiroptera: Vespertilionidae). Journal of Zoology (London) 245:15–27.

Hutchinson, J.T., and M.J. Lacki. 2000. Selection of day roosts by red bats in mixed mesophytic forests. Journal of Wildlife Management 64:87–94.

———. 2001. Possible microclimate benefits of roost site selection in the red bat, *Lasiurus borealis*, in mixed mesophytic forests of Kentucky. Canadian Field-Naturalist 115:205–209.

Hutchinson, J.T., and M. Meisenburg. 2004. Two winter roost sites of lasiurines in north-central Florida. Bat Research News 45:90–91.

Jennings, W.L. 1958. The ecological distribution of bats in Florida. Ph.D. dissertation, University of Florida, Gainesville, FL.

Jones, C., and J. Pagels. 1968. Notes on a population of *Pipistrellus subflavus* in southern Louisiana. Journal of Mammalogy 49:134–139.

Jones, C., and R.D. Suttkus. 1973. Colony structure and organization of *Pipistrellus subflavus* in southern Louisiana. Journal of Mammalogy 54:962–968.

Kalcounis, M.C. 1994. Selection of tree roost sites by big brown (*Eptesicus fuscus*), little brown (*Myotis lucifugus*), and hoary (*Lasiurus cinereus*) bats in Cypress Hills, Saskatchewan. Bat Research News 35:103.

Koehler, C.E., and R.M.R. Barclay. 2000. Post-natal growth and breeding biology of the hoary bat (*Lasiurus cinereus*). Journal of Mammalogy 81:234–244.

Krishon, D.M., M.A. Menzel, T.C. Carter, and J. Laerm. 1997. Notes on the range of four species of vespertilionid bats (Chiroptera) on Sapelo Island Georgia. Georgia Journal of Science 55:215–223.

Kunz, T.H. 1973. Resource utilization: temporal and spatial components of bat activity in central Iowa. Journal of Mammalogy 54:14–32.

Kunz, T.H., M.S. Fujita, A.P. Brooke, and G.F. McCracken. 1994. Convergence in tent architecture and tent-making behavior among Neotropical and Paleotropical bats. Journal of Mammalian Evolution 2:57–78.

Kunz, T.H., and L.F. Lumsden. 2003. Ecology of cavity and foliage roosting bats, pp. 3–89, *in* Bat ecology (T.H. Kunz and M.B. Fenton, eds.). University of Chicago Press, Chicago, IL.

Kurta, A., and G.C. Lehr. 1995. *Lasiurus ega*. Mammalian Species 515:1–7.

Kurta, A., C. Schumacher, M. Kurta, and S. DeMers. 1999. Roosting sites of an eastern pipistrelle during late-summer swarming. Bat Research News 40:8–9.

Lane, H.K. 1946. Notes on *Pipistrellus subflavus subflavus* during the season of parturition. Proceedings of the Pennsylvania Academy of Science 20:57–61.

Layne, J.N. 1992. Status of the eastern pipistrelle *Pipistrellus subflavus* at its southern range limit in eastern United States. Bat Research News 33:43–46.

Leput, D.W. 2004. Eastern red bat (*Lasiurus borealis*) and eastern pipistrelle (*Pipistrellus subflavus*) maternal roost selection: implications for forest management. M.S. thesis, Clemson University, Clemson, SC.

Lewis, S.E. 1995. Roost fidelity of bats: a review. Journal of Mammalogy 76:481–496.

MacCleery, D.W. 1992. American forests: a history of resiliency and recovery. Forest Historic Society, Durham, NC.

Mager, K.J., and T.A. Nelson. 2001. Roost-site selection by eastern red bats (*Lasiurus borealis*). American Midland Naturalist 145:120–126.

McClure, H.E. 1942. Summer activities of bats (genus *Lasiurus*) in Iowa. Journal of Mammalogy 23:430–434.

McComb, W.C., and R.E. Noble. 1981. Microclimates of nest boxes and natural cavities in bottomland hardwoods. Journal of Wildlife Management 45:284–289.

Menzel, M.A, T.C. Carter, B.R. Chapman, and J. Laerm. 1998. Quantitative comparison of tree roosts use by red bats (*Lasiurus borealis*) and Seminole bats (*Lasiurus seminolus*). Canadian Journal of Zoology 76:630–634.

Menzel, M.A, T.C. Carter, W.M. Ford, B.R. Chapman, and J. Ozier. 2000. Summer roost tree selection by eastern red (*Lasiurus borealis*), Seminole (*L. seminolus*), and evening (*Nycticeius humeralis*) bats in the Upper Coastal Plain of South Carolina. Proceedings of the Annual Conference of the Southeastern Association of Fish and Wildlife Agencies 54:304–313.

Menzel, M.A., D.M. Krishon, T.C. Carter, and J. Laerm. 1999. Notes on tree roost characteristics of the northern yellow bat (*Lasiurus intermedius*), the Seminole bat (*L. seminolus*), the evening bat (*Nycticeius humeralis*), and the eastern pipistrelle (*Pipistrellus subflavus*). Florida Scientist 62:185–193.

Mirowsky, K.M. 1997. Bats in palms: precarious habitat. Bats 15:3–6.

Mumford, R.E. 1973. Natural history of the red bat (*Lasiurus borealis*) in Indiana. Periodicum Biologorum 75:155–158.

Mumford, R.E., and D.A. Zimmerman. 1963. The southern yellow bat in New Mexico. Journal of Mammalogy 44:417–418.

Neill, W.T. 1952. Hoary bat in a squirrel's nest. Journal of Mammalogy 33:113.

O'Shea, T.J., and M.A. Bogan, eds. 2003. Monitoring trends in bat populations of the United States and territories: problems and prospects: U.S. Geological Survey Information and Technology Report ITR-2003-0003.

Pierson, E.D. 1998. Tall trees, deep holes, and scarred landscapes: conservation biology of North American bats, pp. 309–325, *in* Bat biology and conservation (T.H. Kunz and P.A. Racey, eds.). Smithsonian Institution Press, Washington, D.C.

Sandel, J.K., G.R. Benatar, K.M. Burke, C.W. Walker, T.E. Lacher Jr., and R.L. Honeycutt. 2001. Use and selection of winter hibernacula by the eastern pipistrelle (*Pipistrellus subflavus*) in Texas. Journal of Mammalogy 82:173–178.

Saugey, D.A, B.G. Crump, R.L. Vaughn, and G.A. Heidt. 1998. Notes on the natural history of *Lasiurus borealis* in Arkansas. Journal of the Arkansas Academy of Science 52:92–98.

Sherman, H.B. 1945. The Florida yellow bat, *Dasypterus floridanus*. Proceedings of the Florida Academy of Science 1:102–128.

Shulz, M., and D. Hannah. 1998. Relative abundance, diet and roost selection of the tube-nosed insect bat, *Murina florium,* on the Atherton Tablelands, Australia. Wildlife Research 25:261–271.

Shump, K.A., Jr., and A.U. Shump. 1980. Comparative insulation in vespertilionids bats. Comparative Biochemistry and Physiology 66A:351–354.

———. 1982a. *Lasiurus cinereus.* Mammalian Species 185:1–5.

———. 1982b. *Lasiurus borealis.* Mammalian Species 183:1–6.

Snow, T.K. 1996. Western red bat. Arizona Wildlife Views 39:5.

Spencer, S.G., P.C. Choucair, and B.R. Chapman. 1988. Northward expansion of the southern yellow bat, *Lasiurus ega,* in Texas. Southwestern Naturalist 33:493.

U.S. Fish and Wildlife Service. 1998. Recovery plan for the Hawaiian hoary bat. U.S. Fish and Wildlife Service, Portland, OR.

Veilleux, J.P., and S.L. Veilleux. 2004. Intra-annual and interannual fidelity to summer roost areas by female eastern pipistrelles, *Pipistrellus subflavus*. American Midland Naturalist 152:196–200.

Veilleux, J.P., J.O. Whitaker Jr., and S.L. Veilleux. 2003. Tree-roosting ecology of reproductive female eastern pipistrelles, *Pipistrellus subflavus*, in Indiana. Journal of Mammalogy 84:1068–1075.

———. 2004. Reproductive stage influences roost use by tree roosting female eastern pipistrelles, *Pipistrellus subflavus*. Ecoscience 11:249–256.

Vonhof, M.J., and R.M.R. Barclay. 1996. Roost-site selection and roosting ecology of forest-dwelling bats in southern British Columbia. Canadian Journal of Zoology 74:1797–1805.

Webster, W.D., J.K. Jones Jr., and R.J. Baker. 1980. *Lasiurus intermedius*. Mammalian Species 132:1–3.

Weyandt, S.E., T.E. Lee, and J.C. Patton. 2001. Noteworthy record of the western yellow bat, *Lasiurus xanthinus* (Chiroptera: Vespertilionidae), and a report on the bats of Eagle Nest Canyon, Val Verde County, Texas. Texas Journal of Science 53:289–292.

Whitaker, J.O., Jr. 1998. Life history and roost switching in six summer colonies of eastern pipistrelles in buildings. Journal of Mammalogy 79:651–659.

Whitaker, J.O., Jr., R.K. Rose, and T.M. Padgett. 1997. Food of the red bat *Lasiurus borealis* in winter in the Great Dismal Swamp, North Carolina and Virginia. American Midland Naturalist 137:408–411.

Wilkins, K.T. 1987. *Lasiurus seminolus*. Mammalian Species 280:1–5.

Williams, M. 1989. Americans and their forests: a historic geography. Cambridge University Press, New York, NY.

Willis, C.K.R. 2003. Physiological ecology of roost selection in female, forest-living big brown bats (*Eptesicus fuscus*) and hoary bats (*Lasiurus cinereus*). Ph.D. dissertation, University of Regina, Regina, Saskatchewan.

Willis, C.K.R., and R.M. Brigham. 2005. Physiological and ecological aspects of roost selection by reproductive female hoary bats (*Lasiurus cinereus*). Journal of Mammalogy 86:85–94.

Wilson, D.E., and S. Ruff, eds. 1999. The complete book of North American mammals. Smithsonian Institution Press, Washington, D.C.

Winchell, J.M., and T.H. Kunz. 1996. Day-roosting activity budgets of the eastern pipistrelle bat, *Pipistrellus subflavus* (Chiroptera: Vespertilionidae). Canadian Journal of Zoology 74:431–441.

FORAGING ECOLOGY OF BATS IN FORESTS

Michael J. Lacki, Sybill K. Amelon, and Michael D. Baker

4

Bats have a greater diversity of behavior, diet, and morphology than any other mammalian order. As the primary predators of nocturnal insects, bats play a significant role in all forested ecosystems (Fenton 2003). Despite the importance of bats in forests, the information on foraging behavior for many species in North America is limited. This dearth of information exists because bats are difficult to study and many techniques and methods, developed to assess habitat use and behavior in other mammalian species, are simply not suitable for the study of nocturnal flying mammals (Kunz 1988a).

Optimal foraging theory helps explain foraging behavior of animals using mathematical constructs in the context of evolutionary theory (Stephens and Krebs 1986). Optimal foraging theory assumes that "fitness" (Ehrlich and Holm 1963) is strictly an increasing function of foraging efficiency and that natural selection would, in the long term, lead to the evolution of behaviors that maximize the efficiency at which animals acquire food. A widely accepted measure of efficiency is the rate at which energy is obtained relative to time spent in a particular foraging habitat (Charnov 1976). Presumably, bats allocate considerable time and effort to foraging and likely react to changes in available prey with different strategies for acquiring food (Barclay 1985; Freeman 1979). Experiments and field observations show that bats can discriminate among various habitats and items of food and adjust their behavior to utilize, more-or-less optimally, these varying resources (Barclay 1991; Barclay and Brigham 1994; Norberg 1994).

For temperate-zone bats, the need for resources occurs during three critical life-history periods: maternity season, migration, and hibernation. During the maternity and migratory periods, and during winter for some foliage-roosting species, resources include roosting, foraging, drinking, and commuting habitats. It is important to understand not only the habitats that bats use for foraging, but also the components of the area that are significant from the perspective of energy loss or gain by bats (Murray and Kurta 2004). Presently, few studies have examined the influence of required habitats and the spatial arrangement of these habitats on energy balance in bats (Anthony et al. 1981; Salcedo et al. 1995), and none has examined survivorship in forest-dwelling bats.

In this chapter, we present an overview of the existing information on the foraging behavior of bats in forests, with an emphasis on North American species, and integrate this information into the context of conservation and management of forest-dwelling bats in North America. We emphasize ecomorphology and echolocation, ontogeny and energetics, feeding ecology and diets of insectivorous bats, and the use of foraging habitat by bats in forests, with attention paid to advancements in field methodology and procedures for data analysis. We close with recommendations for further study aimed at improving both our understanding of the foraging ecology of bats in forests and our capability for managing habitat to ensure long-term conservation of these species.

ECOMORPHOLOGY AND ECHOLOCATION
ECOMORPHOLOGY

Ecomorphology is the study of morphology within the context of its biological role (Karr and James 1975; Swartz and Norberg 1998). Ecomorphology is an area of evolutionary biology separate and distinct from functional morphology in that explanations are hypothetical-deductive as opposed to historical-narrative (Bock 1994). In other words, functional morphology provides measures of physical form that represent adaptations in response to environmental cues over evolutionary time. Ecomorphology incorporates the results of functional morphology into analyses of how physical forms of living organisms help species adapt to their natural environments (Bock 1990, 1994; Winkler 1988). Bats are excellent subjects for ecomorphological studies because of the array of morphological forms they possess in association with their unique mammalian adaptation for nocturnal foraging by flight (Arita and Fenton 1997).

A suite of morphological factors influences foraging behavior in insectivorous bats, including body mass and the size and shape of the skull, jaws, and wings. Ecomorphological studies show that the structure of insectivorous bat assemblages is influenced by the correlation between diet and morphology, with morphologically similar species feeding on similar foods (Findley and Black 1983). Among sympatric lasiurines, for example, large bats with large jaws are capable of eating a wider range of prey sizes than small bats, resulting in a broader feeding niche for these bats when food resources are plentiful, although such differences are less apparent when availability of insects is low (Hickey et al. 1996).

Within North American *Myotis*, large-bodied species with long ears tend to forage near and within vegetation, and those that forage over water have shorter ears and smaller bodies (Fenton and Bogdanowicz 2002). Further, species of *Myotis* intermediate in these morphological features alternate between foraging habitats (Fenton and Bogdanowicz 2002). Head size also is reflected in the foraging strategy of some species, with species that use continuous flight to capture small airborne prey having proportionally shorter heads and tooth rows than bats that forage from perches (sally foragers) and consume larger and often nonairborne prey (Fenton 1989). Morphological studies also demonstrate that species that feed on hard-bodied insects tend to have fewer but larger teeth, wider dentaries, and more fully developed cranial crests than bats that feed on soft-bodied insects, such as dipterans and moths (Freeman 1981). For example, there are three sympatric species of *Myotis* in the Midwest that are similar in body size, skull length, and wingspan. The northern myotis (*Myotis septentrionalis*) has a longer maxillary tooth row than the Indiana myotis (*M. sodalis*), and the little brown myotis (*M. lucifugus*) shows greater variation in this morphometric trait than the other two species (Lee and McCracken 2004). Diets of the three species correspond to these structural differences such that the northern myotis feeds more frequently on large beetles, the Indiana myotis eats primarily soft-bodied moths, and the little brown myotis exhibits the most diverse diet of the three species (Lee and McCracken 2004).

The size and shape of wings determine flight speed and maneuverability (Norberg 1981; Norberg and Rayner 1987; Stockwell 2001), which, in turn, affect where a bat is likely to forage. Differences in wing morphology facilitate partitioning of available foraging habitats among species (Crome and Richards 1988; Heller and von Helversen 1989; Kingston et al. 2000). The importance of wingspan, wing area, and body mass for flight has been recognized for a long time (Farney and Fleharty 1969), with shape of the wing tip, in particular, affecting differences in maneuverability among species (Findley et al. 1972). Shape of the wing tip is correlated with aspect ratio, which is the length of the wingspan squared divided by the surface area of the wing (Aldridge and Rautenbach 1987; Arita and Fenton 1997). This measure, coupled with wing loading, which is the mass of the bat divided by its total wing area, is used to evaluate flight and habitat use in insectivorous bats (Aldridge and Rautenbach 1987; Chruszcz and Barclay 2003; Fenton 1990; Fenton and Bell 1979; Patriquin and Barclay 2003; Saunders and Barclay 1992). Bats with lower wing loadings, whether accompanied by small body size or not, exhibit greater maneuverability and are more capable of exploiting foraging space that is cluttered with vegetation (Aldridge 1986, 1987). Length of wings is directly related to flight speed in bats (Hayward and Davis 1964), and this, coupled with the need for an increase in minimum velocity as the mass of the body increases (Pennycuick 1975), suggests that some large bats fly at faster speeds than small species, but large bats are less maneuverable and thus more likely to forage in open, uncluttered habitats. The mean and variance of flight speeds should differ among bats with similar wing morphologies when pronounced differences in body size exist (Salcedo et al. 1995).

Assemblages of bats inhabiting North American forests typically consist of several morphologically similar species (Saunders and Barclay 1992). Most species possess either a moderate aspect ratio and low wing loading or a low aspect ratio and a moderate wing loading (table 4.1; Norberg and Rayner 1987). In North American forests, the hoary bat (*Lasiurus cinereus*) is the lone species with a high aspect ratio, and only a few species (hoary bat; eastern red bat, *L. borealis;* and evening bat, *Nycticeius humeralis*) possess high wing loadings. Thus, most species in North American forests have bodies designed for feeding in forest canopies or near the clutter of vegetation.

Table 4.1. Feeding strategy, aspect ratio, wing loading, and foraging habitats of North American species of forest-dwelling bats

Species	Feeding strategy*	Aspect ratio†	Wing loading†	Foraging habitats‡
Myotis californicus	AH	low	low	forest canopies
Myotis evotis	AH, GL	moderate	low	riparian forest, forest edge
Myotis keenii	AH, GL(?)	low	moderate	forest canopies, forest clearings
Myotis lucifugus	AH, GL(?)	low	moderate	riparian forest, forest edge
Myotis septentrionalis	AH, GL	low(?)	moderate(?)	intact forest, forest edge
Myotis sodalis	AH, GL(?)	low	moderate	riparian forest, upland forest
Myotis thysanodes	AH, GL(?)	moderate	low	riparian forest
Myotis volans	AH	low	low	forest canopies, forest clearings
Lasiurus borealis	AH	moderate	high	riparian forest, forest gaps
Lasiurus cinereus	AH	high	high	clearcut, forest gaps
Lasionycteris noctivagans	AH	moderate	low	clearcut, riparian forest
Pipistrellus subflavus	AH	moderate	low	riparian forest, forest gaps
Eptesicus fuscus	AH	moderate	low	riparian forest, forest gaps
Nycticeius humeralis	AH	low	high	forest gaps
Corynorhinus rafinesquii	GL, AH(?)	moderate	low	upland forest
Corynorhinus townsendii	GL, AH(?)	low	moderate	riparian forest, forest clearings
Antrozous pallidus	GL	moderate	low	forest clearings

*Feeding strategies are aerial hawking (AH) and gleaning (GL) and are presented in order of importance to a species. *Source:* Modified from Barclay and Brigham (1991).
† *Source:* Norberg and Rayner (1987)
‡ *Source:* Based on data in Adam et al. (1994), Barclay (1991), Burford and Lacki (1995), Clark et al. (1993), Fellers and Pierson (2002), Hobson and Holland (1995), Hogberg et al. (2002), Hurst and Lacki (1999), Menzel et al. (2001, 2002), Nagorsen and Brigham (1995), Owen et al. (2004), Patriquin and Barclay (2003), Waldien and Hayes (2001), and Wethington et al. (1996).

Forests vary in structure and composition with age, productivity, and history of disturbance; therefore, the habitats they offer to bats differ in the amount of available clutter (Hayes and Loeb, chap. 8 in this volume). Using multivariate analyses of wing morphology, Norberg (1994) demonstrated that gleaners, or bats that take insect prey from the surface of objects, cluster in different areas of multivariate space than aerial hawkers, or bats that capture insects in flight. The morphological attributes that establish these clusters have been interpreted as having an adaptive basis. Saunders and Barclay (1992) suggested that large differences in morphology may restrict individual species to different foraging habitats, but that small morphological differences influence the actual prey that are available to bats within a habitat. These and other investigators have ascribed a greater importance to plasticity in foraging behavior of morphologically similar individuals, whether members of the same or different species, than to strictly species-specific differences in morphology (Barclay 1991; Chruszcz and Barclay 2003; Fenton 1990; Saunders and Barclay 1992). For instance, both long-legged myotis (*Myotis volans*) and little brown myotis have low to moderate wing loading and should be maneuverable enough to exploit cluttered habitats (table 4.1), but both species also forage in open spaces (Fenton and Bell 1979; Saunders and Barclay 1992). Likewise, although long-eared myotis (*M. evotis*) have broad wings that are energetically expensive, they are capable of prolonged foraging bouts in open habitats that allow profitable aerial hawking (Chruszcz and Barclay 2003).

Furlonger et al. (1987) demonstrated significant associations in the foraging behavior of bats in Ontario among wing design, foraging activity, and habitat features, but no species they examined was restricted to foraging in a single habitat. The California myotis (*Myotis californicus*) and western small-footed myotis (*M. ciliolabrum*) differ ecologically (Black 1974; Constantine 1998; Woodsworth 1981), in skull morphology (Gannon et al. 2001) and in structure of echolocation calls (Fenton and Bell 1981; Gannon et al. 2001; O'Farrell et al. 1999a), yet these species are not different in the sequence divergence of mitochondrial DNA (Rodriguez and Ammerman 2004). This suggests that plasticity in foraging behavior may facilitate development of morphological differences without measurable genetic divergence.

ECHOLOCATION

Echolocation is as important as wing morphology in influencing the ability of bats to forage in cluttered versus open foraging space (Aldridge and Rautenbach 1987; Norberg 1994; Norberg and Rayner 1987). Most echolocation signals consist of constant frequency (CF), quasi-constant frequency (QCF), or frequency-modulated (FM) components, with many species using combinations of these components. Bats produce calls that vary in absolute frequency, range of frequencies (bandwidth), harmonic structure, duration, and intensity (Fenton 1990; Neuweiler 1989). In general, high-intensity signals produced in the larynx interfere with detection of low-intensity echoes of the same frequency by the ear, so many bats temporally separate the two processes, resulting in a cycling of outgoing pulses of sound and intervening listening periods (Fenton et al. 1995; Kalko and Schnitzler 1993). All bats in North American forests have low-duty cycles, meaning these bats produce sound while echolocating less than 20% of the time and use the other 80% for listening (Arita and Fenton 1997; Bogdanowicz et al. 1999).

The structure of search-phase calls, or signals emitted by bats when attempting to locate prey, is correlated with type of habitat and foraging strategy (Schnitzler and Kalko 1998). Species that capture prey in open areas with no obstacles experience different conditions relative to echolocation than do bats foraging in forested sites with large amounts of clutter. In general, the structure of echolocation signals is species specific, but the signal of each species varies according to the task encountered (Schnitzler and Kalko 1998). Broadband calls at high frequency are better suited to cluttered locations (Simmons and Stein 1980), whereas narrowband calls of low frequency are more suited to open sites (Neuweiler 1983). Thus, one would predict that species with short maneuverable wings and low wing loadings, which perform well in cluttered habitats, should also use broadband calls comprised of high frequencies. In general, this is true; species that typically forage in clutter ("narrow space bats," sensu Schnitzler and Kalko 1998) use calls that have low intensity, short duration, and high peak frequency, and include a wide range of frequencies, whereas open-space foragers have calls that are more intense, longer in duration, and lower in frequency (Fenton 1990). Just as large species of bats are capable of eating prey that are unavailable to smaller bats, small bats adapted to clutter can readily access insects in both cluttered and open habitats (Fenton 1990). Studies do not agree, however, on whether morphology (Bogdanowicz et al. 1999) or the structure of echolocation calls (Barclay and Brigham 1991; Swartz et al. 2003) limits the prey base that is "truly" available to bats in forests.

Many insects possess their own ecomorphological adaptations and use echolocation pulses from a bat as a warning system (Jones and Rydell 2003; Roeder 1969; Treat 1955). Hearing has evolved in the Lepidoptera multiple times, including members of the Sphingoidea, Noctuoidea, Geometroidea, and Pyraloidea (Waters 2003). The tympanic organ of some families of moths, such as the Noctuidae and Arctiidae, permit these insects to detect signals of echolocating bats, resulting in evasive maneuvers by the moths that vary according to the relative distance between moth and bat (Achyara and Fenton 1992; Roeder 1964, 1967). Arctiid moths also produce clicks that are audible to bats. These clicks are designed presumably to avoid predation by bats either by startling approaching bats (Bates and Fenton 1990), jamming echolocation signals (Fullard et al. 1979), or serving as aposematic signals, the latter because arctiid moths are distasteful to bats (Dunning 1968). Noctuid moths can detect the echolocation pulse of foraging bats from roughly ten times the distance that bats detect an object that is the size of the moth (Surlykke et al. 1999), which suggests the need for bats to evolve effective countermeasures. One such measure is that many moth-eating bats produce broadband echolocation calls at either high (>55 kHz) or low (<20 kHz) frequencies, because these signals are inaudible to most tympanate moths (Fullard 1987; Fullard and Dawson 1997; Rydell and Arlettaz 1994; Surlykke 1988).

Structure of echolocation calls varies among bats inhabiting North American forests and not all species fit neatly into the extremes of open- versus clutter-adapted foraging strategies (Fullard et al. 1991). Gleaning bats, for example, are species that detect, locate, and capture at least some of their prey from vegetation or other surfaces (Faure et al. 1993). Gleaners are similar in body design to species that forage in cluttered habitats (Fenton 1990), but their strategy of prey capture relies less on echolocation and more on sight (Bell 1985) or prey-generated sounds (Faure and Barclay 1992). Although generally considered uncommon (Faure et al. 1993), species capable of gleaning at least some of the time are well represented in communities of bats inhabiting North American forests (table 4.1) (Burford et al. 1999).

ONTOGENY AND ENERGETICS
ONTOGENY AND FLIGHT

Juveniles of many animal species use resources differently as juveniles than they do as adults, exhibiting patterns consistent with what ecologists have termed the ontogenetic niche (Polis 1984; Werner and Gilliam

1984), in which juveniles act as "ecological" species distinct from adults

relative to the use of available resources in the environment. Juvenile bats fit this paradigm, especially in terms of their feeding behavior relative to adult bats. For example, juvenile big brown bats (*Eptesicus fuscus*) rely more on soft-bodied insects and eat a broader array of insects than do adults (Hamilton and Barclay 1998), and the dietary composition of juvenile little brown myotis shifts seasonally as young bats develop (Adams 1996). Juvenile little brown myotis demonstrate a more random diet than adults (Anthony and Kunz 1977), with juveniles of this species spending less time foraging in cluttered habitats, especially individuals with small wings (Adams 1996, 1997). Young *Myotis* take about 20 days after birth to develop the wing morphology that they will use as adults (Adams 1992; Powers et al. 1991), and even longer to develop their foraging skills (Kunz 1974; Kunz and Anthony 1996). Thus, habitats near day-roosting sites must provide the insects required by young bats during their final stage of somatic development. Success of juveniles may be facilitated by shifts in the foraging behavior of adults, with adult females foraging farther from the maternity roost once young become volant (Anthony and Kunz 1977) or feeding in more cluttered habitats that are temporarily unavailable to inexperienced young (Adams 1997).

ENERGETICS AND FORAGING BEHAVIOR

Kurta et al. (1989) estimated energetic requirements for little brown myotis during pregnancy (33.7 kJ d^{-1}) and lactation (41.3 kJ d^{-1}) and predicted that the amount of insects needed to support these energetic demands is 5.5 g d^{-1} and 6.7 g d^{-1}, respectively. Flight can be ten times as energetically expensive as roosting (Barclay 1989; Kurta et al. 1987). Consequently, foraging accounts for 61–66% of daily energetic requirements for reproductive female little brown myotis, resulting in lower amounts of energy allocated to tissue and milk production relative to other mammals (Kurta et al. 1989). Size of foraging areas used by lactating, female little brown myotis decreases by 51% compared with that of pregnant females, although the duration of foraging is similar between pregnant and lactating bats (Henry et al. 2002). Presumably, milk production and the time spent nursing young constrain the length of foraging distances of adult females of this species (Henry et al. 2002). Female eastern red and hoary bats that hunt in concentrations of insects attracted to streetlights sustain more positive energy budgets due to reduced foraging times when feeding within areas of high prey density, and this behavior is likely important during periods that are more energetically demanding such as pregnancy and lactation (Salcedo et al. 1995).

Nightly variation in the time spent foraging by bats is related to reproductive condition, prey density, and ambient temperature (Anthony et al. 1981). Energy loss by bats is minimized by long periods of night roosting sandwiched between short foraging bouts, especially when temperature and prey density are low (Anthony et al. 1981; Ormsbee et al., chap. 5 in this volume). Analyses of reproductive rates across 103 species of bats suggest that females in temperate regions adjust reproduction based on their physical condition, availability of prey, and weather (Barclay et al. 2004) with the latter two factors likely having an impact on the time spent foraging by reproductive female bats. Use of intermittent foraging bouts by bats can result in 25% less energetic expenditure than maintaining prolonged or continuous foraging activity (Bonaccorso and McNab 2003). Available data, however, suggest most species of insectivorous bats forage for most of the night with only brief periods of night roosting (Audet 1990; Entwistle et al. 1996; Krull et al. 1991; Murray and Kurta 2004; O'Donnell 2002).

DIET OF NORTH AMERICAN BATS

All species of bats inhabiting North American forests are insectivorous, and diet has been detailed at multiple locations across the continent for several species. Most species of forest-dwelling bats demonstrate, in at least one study, moderate dietary selection by relying on at least one order of insects to provide more than 40% of the volume of their diet (table 4.2). Localized selection of prey by some species of bats, not typical of their diet elsewhere in the distribution, suggests an ability to switch and fixate on an especially abundant or easily captured prey. The northern myotis and eastern pipistrelle (*Pipistrellus subflavus*) show no evidence for selection of any order of insects in any study, and perhaps these species are closest to what might be considered "foraging opportunists" among bats in North American forests. The California myotis, Rafinesque's big-eared bat (*Corynorhinus rafinesquii*), and Townsend's big-eared bat (*C. townsendii*), in contrast, demonstrate high selection (>80% volume in the diet) for a single order (Lepidoptera), with the two species of big-eared bats doing so across all studies for which data are available. We suggest that big-eared bats are the best examples of "foraging specialists" among bats in North American forests. The big brown bat is also highly selective, with Coleoptera forming more than 80% of the diet by volume in most studies. However, the ability of big brown bats to switch to other taxa, such as Trichoptera and Hemiptera, indicates a species with more flexibility in dietary choice than expected for a true specialist.

Although specialization occurs, all species of bats in North American forests feed at some point on multiple orders of insects (table 4.2), with local diets often comprised of a range of insects that changes seasonally, presumably because of shifts in availability of different types of insects (Murray and Kurta 2002; Whitaker 1972; Whitaker et al. 1981a). In one study alone (Whitaker et al. 1977), the taxonomic orders of prey recorded in the diet of bats in western Oregon include Diptera (flies), Isoptera (termites), Trichoptera (caddis flies), Lepidoptera (moths), Coleoptera (beetles), Hymenoptera (wasps and ants), Homoptera (leaf hoppers), Hemiptera (bugs), Orthoptera (crickets), Neuroptera (lacewings), Chilopoda (centipedes), and Arachnida (harvestmen and spiders), with the majority of bats sampled consuming multiple orders of insects. Other studies demonstrate variation in the diet of bats by season (Brack and LaVal 1985; Whitaker and Clem 1992), year (Adams 1997; Kurta and Whitaker 1998), age class (Adams 1996; Hamilton and Barclay 1998), ge-

Table 4.2. Evidence for dietary specialization on insect orders by forest-dwelling bats in North America, along with the maximum number of orders of insects and other arthropods reported eaten by species of bat

Species	Number of orders of insects	Lepidoptera	Coleoptera	Diptera	Homoptera	Other
Myotis californicus	10	>80	p	p	p	
Myotis evotis	12	>40	p	p	p	
Myotis lucifugus	9	p	p	>40	p	
Myotis septentrionalis	9	p	p	p	p	
Myotis sodalis	12	>80	p	>40	p	>40 Trichoptera
Myotis volans	6	>40	p	p	p	
Lasiurus borealis	9	>40	p	p	p	
Lasiurus cinereus	7	>40	p	p	p	>40 Odonata
Lasionycteris noctivagans	11	>40	p	>40	p	
Pipistrellus subflavus	7	p	p	p	p	
Eptesicus fuscus	10	p	>80	p	p	>40 Trichoptera; Hemiptera
Nycticeius humeralis	9	p	>40	p	p	
Corynorhinus rafinesquii	7	>80	p	p	p	
Corynorhinus townsendii	7	>80	p	p	p	
Antrozous pallidus	11	p	>40	p	–	

Sources: Barclay 1985; Belwood and Fenton 1976; Brack and LaVal 1985; Brigham 1990; Brigham and Saunders 1990; Dalton et al. 1986; Easterla and Whitaker 1972; Hamilton and Barclay 1998; Hurst and Lacki 1997; Johnston and Fenton 2001; Kurta and Whitaker 1998; Leslie and Clark 2002; Murray and Kurta 2002; Rolseth et al. 1994; Sample and Whitmore 1993; Whitaker 1972, 1995, 2004; Whitaker and Clem 1992; and Whitaker et al. 1977, 1981a, 1981b.

Note: Data for specialization are orders of insects reported at least once as >80% volume in the diet, at least once as >40% volume in the diet, or always <40% volume (p, or present) in the diet; all data are based on sample sizes of ≥10 fecal pellets per species of bat per study.

ographic location (Belwood and Fenton 1976), and time of night (Lee and McCracken 2001; Whitaker et al. 1996). This variability has implications for management and conservation of bats in forests, because maintaining an adequate diversity of insects will likely require a variety of habitats including habitat for both aquatic and terrestrial insects.

Small prey are consumed, at least occasionally, by almost all North American species of bats (Griffith and Gates 1985; Whitaker 1972; Whitaker et al. 1977), but for most bats, there appears to be an upper limit to the size of prey that is eaten, presumably because of an inability to capture or handle large items in flight (Burford and Lacki 1998). Additionally, large insects are digested by bats more efficiently than small prey, because small insects cannot be manipulated as effectively as large ones and the structures of small insects comprised largely of chitin, that is, legs and wings, are eaten along with the digestible parts (Barclay et al. 1991). Collectively, these patterns suggest that some bats may attempt to maximize energy intake by selecting prey within a limited range of body size (Belwood and Fullard 1984; Swift et al. 1985).

IDENTIFYING DIETS OF BATS

In this section, we comment on specific methodological issues relevant to interpretation of data on the feeding patterns of bats in forests and, thus, how biologists study and perceive these bats as predators of insects. Data on foods of insectivorous bats can be acquired from four sources: stomach contents, feces, culled parts, and direct observations of feeding bats (Whitaker 1988); the latter approach has limited utility in the study of bats in North American forests and will not be commented on here.

Analysis of stomach contents is only appropriate under specialized circumstances (Whitaker et al. 1981b), because killing bats for data on diet is unacceptable when a nonintrusive method of similar accuracy—fecal analysis—can be used instead (Whitaker 1988). Insectivorous bats are capable of cutting a chitinous exoskeleton into small pieces (Freeman 1979), and thus, whether material is obtained by means of stomach contents or feces, the task of identifying insects from chewed parts is equally challenging. Fecal analysis has the advantage of samples being easily acquired, either directly from captured bats or via collections beneath roosting sites. Fecal samples can be obtained throughout the maternity season to permit tracking of seasonal (Brack and LaVal 1985; Whitaker and Clem 1992) and annual (Adams 1997; Kurta and Whitaker 1998) shifts in diet. Furthermore, feces from individual bats can be assigned to sex and age classes for additional comparison (Adams 1996; Hamilton and Barclay 1998). Because of these advantages, fecal analysis remains the most frequently used method in dietary studies of insectivorous bats.

The major limitation of fecal analysis is the taxonomic level to which insect fragments can be assigned (Whitaker 1988). Many published studies identify prey only to the ordinal level (Brack and LaVal 1985; Brigham 1990; Brigham and Saunders 1990; Hamilton and Barclay 1998; Leslie and Clark 2002), although resolution to family has been achieved for many orders of insects (Whitaker 2004; Whitaker et al. 1977, 1981a, 1981b), especially Diptera and Coleoptera (Belwood and Fenton 1976; Kurta and Whitaker 1998; Lacki et al. 1995; Whitaker 1995; Whitaker and Clem 1992). Improvements in the resolution of scale to which parts of insects in feces can be identified would significantly increase our understanding of predator-prey relationships between bats and insects. Pioneering work using amplified species-specific DNA sequences of insect prey in bat feces suggests that new advancements in this area are forthcoming (McCracken et al. 2004).

Data from fecal studies are typically presented as "percent volume," the percentage by volume of each taxon in the sample, and "percent frequency," the percentage of bats sampled that eat a particular taxon (Whitaker 1988). Percent frequency indicates how widely used a certain type of prey is across all bats in a population or a community, whereas percent volume more accurately reflects the contribution of a particular taxon to the total diet. We recommend that both measures be calculated because each provides useful information. For example, a taxon that appears in only some fecal samples from a population of bats (low-percent frequency), yet represents almost the entire sample when it does occur, may be just as important to the bats as a taxon that is found throughout many samples but always in small amounts (low-percent volume).

Identification of diets through culled parts has been used primarily with bats that feed on moths, although studies addressing culled pieces of other taxa exist (Bell 1982; Johnston and Fenton 2001; LaVal and LaVal 1980b). An advantage to using culled parts is that prey often can be identified to species, especially if these remnants are obtained soon after being discarded by the feeding bat to prevent decomposition and breakdown of identifiable features of the culled remains. Biologists have even been able to determine sex of prey from culled insect parts (Acharya 1995), although such studies are rare. Discarded parts of insects can be collected throughout the maternity season in feeding areas or at feeding roosts away from the maternity site (Ormsbee et al., chap. 5 in this volume), permitting determination of seasonal and annual variation in diet (Lacki and LaDeur 2001; LaVal and LaVal 1980a). However, studies that identify prey from discarded wings of moths are probably biased toward macrolepidopterans, because microlepidopterans are likely eaten in flight and not carried to a roost for later consumption (Burford and Lacki 1998; Sample and Whitmore 1993). Dietary studies of insectivorous bats should be based on more than one method, if possible, to provide a more comprehensive picture of what the bats are eating (Whitaker 1988).

FIELD STUDIES OF INSECT PREY

A number of traps are available for sampling insects (Kunz 1988b). For convenience, we organize these into three categories: those that capture insects passively, such as malaise traps and sticky traps (Belwood and Fenton 1976; Brigham and Saunders 1990); those that use mechanical aids to sample insects, including rotary and suction traps (Brigham 1990; Brigham and Saunders 1990); and those that use an attractant to lure insects to the trap, such as light traps (Burford et al. 1999; Tibbels and Kurta 2003). All traps have biases and limitations, not the least of which is that all measure the availability of insects to the researcher, and none measures actual availability of insects to bats; thus, data from traps must be treated as an index of the availability of insects to bats (Whitaker 1994).

Selection of the type of trap can be specific to the species of bat being studied. For example, pitfall traps are well suited for capturing large, ground-dwelling arthropods (Kunz 1988b), and this approach works well when studying the diet of gleaning bats that feed off the ground (Bauerova 1978). Studies on big brown bats have used a combination of trap types to capture the range of insects that occur in the diet of this species, including whirligig, sticky, and malaise traps (Brigham 1990; Brigham and Saunders 1990), and black-light traps have been used with success to describe the prey base available to bats that feed extensively on moths (Burford et al. 1999; Grindal and Brigham 1998; Tibbels and Kurta 2003). Regardless, just as use of multiple methods can enhance resolution of the diets of bats (Whitaker 1988), use of multiple trap types should result in a clearer representation of what insects are available for the bats to eat (Brigham and Saunders 1990; Kunz 1988b).

Most studies measuring insects that are available to bats do so in the context of a narrowly defined experiment (Brigham 1990; Brigham and Saunders 1990). These studies are helpful in elucidating the biology of specific species of bats or the behavior of insects as it relates to how bats feed (Acharya 1995; Yack 1988). Such studies, however, cannot identify broad-scale patterns in the distribution and abundance of the community of insects or how temporal and geographic patterns in abundance of insects influence the use of foraging habitat by bats.

Bats make multiple choices in determining where and when to feed each night (Whitaker 1994), and the first choice is finding the desired feeding area (fig. 4.1). This is a landscape-scale decision based on the localized abundance and distribution of insects. The second choice takes place when bats begin capturing prey from among the insects that are present at the feeding site. Both decisions are fundamentally important to understand, but almost all studies of feeding in insectivorous bats have targeted the second choice. Whitaker (1994) suggested that the first decision is probably more important to bats, because it places bats where the insects actually are and where the likelihood of capturing prey is greater.

COUPLING DIET AND HABITAT STUDIES
OF FOREST-DWELLING BATS

Data on abundance of insects in forests rarely has been integrated with information on use of forested habitats by bats (Burford et al. 1999; Grindal and Brigham 1998; Hayes and Loeb, chap. 8 in this volume; Tibbels and Kurta 2003). Thus, we know little of how abundant insects are within specific habitats, how forest bats make use of these insects and habitats, or how these patterns change on a nightly, seasonal, or annual basis. Consequently, the opportunity for progress in these areas is immense. One potential avenue of research is to place insect traps repeatedly within different habitats to index the abundance of insects by habitat and time (Burford et al. 1999). Traps can also be set at different heights to examine vertical stratification of potential prey (Kunz 1988b). When combined, data on dietary choice and habitat use by bats could be used to test hypotheses concerning patterns in activity of bats among forested habitats or among different types of structure within forested habitats. Such studies could facilitate an understanding of the relationship between silvicultural practices used in North American forests and habitat quality of bats, by measuring activity of bats simultaneous with presence of insect prey before and after implementing silvicultural treatments (Tibbels and Kurta 2003). Whitaker (1994) suggested that studies be designed so that habitats are sampled for insects in proportion to the availability of habitats in the landscape and, when possible, that distance from known colony sites be factored into the determination of what habitats are actually available to the bats, with habitats closer to a roosting site being considered more available than habitats further from the roost.

Some bats may be able to transfer information on the quality and location of food patches to other members of a colony through observation and by following each other in flight (Gaudet and Fenton 1984; Wilkinson 1992). This possibility suggests that young bats may learn the location of feeding areas by following adults (Brigham 1990) and that knowledge of historic or primary feeding areas is sustained over time within the collective memories of a colony of bats. Arlettaz (1996) demonstrated a distinction between primary (traditional) and secondary

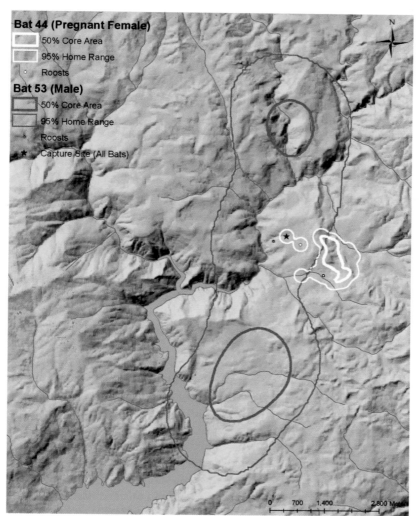

Figure 4.1.
Home ranges (95% kernel), core areas (50% kernel), and location of roosts for a pregnant female and an adult male long-legged myotis (*Myotis volans*) in north-central Idaho (J. Johnson, unpubl. data). The home range of the adult male is substantially larger and possesses two core areas, one along a steep montane ravine and another in low-elevation habitat near a reservoir. The pregnant female has a much smaller home range situated in upland habitat that includes a single core area.

Legend (on map):

Bat 44 (Pregnant Female)
- 50% Core Area
- 95% Home Range
- Roosts

Bat 53 (Male)
- 50% Core Area
- 95% Home Range
- Roosts
- ★ Capture Site (All Bats)

Scale: 0 — 700 — 1,400 — 2,800 Meters

are the number of personnel needed, the minimal ability to distinguish among individually tagged bats, the short time for which chemiluminescent tags are visible, and the difficulty in seeing tagged bats that are flying within a cluttered forest (Fenton et al. 1977; LaVal et al. 1977). Reflective tape, typically attached to wing bands, has many of the same disadvantages, and in addition, seeing the tape requires strong illumination, which may or may not affect the bat's behavior.

RADIOTELEMETRY

Improvements in technology, especially the production of small transmitters, are leading to more studies using radiotelemetry to evaluate activity and habitat use of bats in North American forests, including species of *Myotis,* all of which weigh less than 14 g (fig. 4.2). Adding mass of any type to a bat potentially affects flight performance, because aerodynamic theory suggests that many parameters of flight, such as power requirements, wing-beat frequency, and turning radius are proportional to mass in some way (Aldridge and Brigham 1988; Hughes and Rayner 1991; Webb et al. 1992). Bats are, of course, adapted to accommodate large increases in body mass over the short term, such as a 20% increase in mass after feeding (Gould 1955; Kunz 1974; Kunz et al. 1995), and over the long-term, such as 30% increases in mass associated with pregnancy (Kurta and Kunz 1987) and deposition of fat for hibernation (Kunz et al. 1998). However, the added mass of a radiotransmitter may compound the problem of these natural gains in mass for a bat, perhaps interfering with normal behavior or foraging success, and ultimately having an impact on the survival of the individual.

Thus, a continuing question for biologists using radiotelemetry is how much additional mass is too much? Caccamise and Hedin (1985) arbitrarily recommended that transmitters not exceed 5% of body mass for a bird. Shortly thereafter, the guideline was extrapolated to studies on bats (Aldridge and Brigham 1988). In the past 15 years, the "5% rule" has evolved from a simple recommendation to minimize alteration of behavior of radiotracked animals to a "commandment," which, if violated, is evidence for cruelty toward the bat. Unfortunately, no laboratory data or theoretical equations have suggested that 5% is any better than 3% or 7% or some other proportion, and evidence from the field is equivocal.

Figure 4.2.
An adult, male long-legged myotis (*Myotis volans*) with transmitter.
Photo by M. Baker

Neubaum et al. (2005) recently demonstrated that big brown bats carrying transmitters that weighed less than 5% of body mass show normal reproductive activity and body mass during the year following radiotracking, and they touted their study as a test of the 5% rule; unfortunately, all their transmitters weighed less than 5% of the animal's mass, so one cannot say whether or not exceeding 5% would have been detrimental. Kurta and Murray (2002) radiotracked Indiana myotis using transmitters that represented 8% of body mass and these bats also showed normal reproductive activity and body mass when recaptured in subsequent summers. Although there is no empirical basis for the 5% rule, we believe that it remains a useful guideline to follow until further data are available. Additionally, we suggest that even 5% of mass may be too much for some studies, especially those involving newly volant young, near-term pregnant females, and fattened individuals immediately prior to hibernation.

Regardless of the size of transmitters, studies examining use of foraging habitat and foraging patterns in bats with radiotelemetry largely have been ground-based approaches, as opposed to tracking from aircraft. Success of ground-based telemetry depends on a range of factors including the number of personnel and tracking units available (Menzel et al. 2001), the presence of roads across the landscape (Dobkin et al. 1995), and inclement weather that can inhibit tracking or cause early failure of transmitters (Adam et al. 1994). Other factors affecting success are the ruggedness of the terrain and errors caused by signal bounce (Kufeld et al. 1987), error due to movement of bats as data are collected (Schmutz and White 1990), and autocorrelation or lack of independence in the data due to the short time between recorded observations (Hansteen et al. 1997; Otis and White 1999). Additionally, errors that are inherent in the equipment being used (White and Garrott 1990) or caused by variation in the skill of field personnel (Mills and Knowlton 1989) can affect the quality of data that are obtained. Consequently, the exact location of an animal is often not known with certainty, and the location of the animal is assigned to an area or error polygon (White and Garrott 1990). The size of the polygon is determined by trials that measure errors associated with collection of the data in the field (Hurst and Lacki 1999; Wethington et al. 1996); unfortunately, most studies do not acknowledge or document the extent of their error.

Other approaches using radiotelemetry, such as satellite tracking with global positioning systems (Keating et al. 1991) or aerial searches (White and Garrott 1990), are not feasible or have had little use to date with forest-dwelling bats. Satellite telemetry potentially is very useful, especially for species that cover large distances each night, but this technique requires transmitters that currently are too heavy for insectivorous bats to carry. Although aerial telemetry is feasible with current-sized transmitters, its use in North America mostly has been limited to finding diurnal roosts because of the danger associated with low-altitude flight at night. Nevertheless, nocturnal aerial telemetry has been used to locate radiotagged Townsend's big-eared bats (Dobkin et al. 1995) and Indiana myotis (A. Hicks, pers. comm.), and multiple species of *Myotis* have been tracked using light tags and a helicopter in Missouri (LaVal et al. 1977).

ACOUSTIC MONITORING

The development of acoustic-monitoring devices has arguably changed the landscape of investigation into the ecology of bats in forests more than any other technique (Weller and Zabel 2002). Early attempts to evaluate habitat use by bats with ultrasonic detectors were constrained by the inability of equipment to distinguish among species or species groups; thus, the data represented only an index of activity of the entire community of bats (Thomas 1988). Since Thomas's (1988) seminal work, many studies have used ultrasonic detectors for identification and survey of bats in forests (Ahlén and Baagøe 1999; Burford and Lacki 1995; Grindal 1999; Grindal et al. 1999; Grindal and Brigham 1998; Hogberg et al. 2002; Humes et al. 1999; Menzel et al. 2002; Owen et al. 2004; Patriquin and Barclay 2003; Tibbels and Kurta 2003; Zimmerman and Glanz 2000). Using search-phase calls, both qualitative identification from visual inspection of call structure (Humes et al. 1999; O'Farrell et al. 1999b) and quantitative identification using statistical algorithms (Betts 1998; Britzke and Murray 2000) have been used successfully to study forest-dwelling bats.

There are several types of instruments for converting ultrasonic calls of bats into audible sound or a recording for analysis (Parsons et al. 2000). The most common types of ultrasonic detectors are heterodyne, frequency division, and time expansion, and each has associated advantages and disadvantages. Heterodyne detectors are the simplest and least expensive. These detectors operate by sensing a narrow (often adjustable) range of frequencies (ca. 3–5 kHz) and producing a low-quality, analog output. These detectors superimpose an internally generated pure tone with the incoming ultrasonic signal and subsequently filter the sound to identify the difference between the two frequencies. Because only a narrow range of frequencies is sampled at once, considerable information concerning a call, such as minimum frequency or presence of harmonics, often is not available, yet these detectors are very sensitive to the tuned frequency over relatively long distances. Heterodyne detectors have major limitations: only bats that produce signals within the selected bandwidth will be detected and, although recordings of heterodyne signals can be made, recordings cannot be used with sound-analysis software.

In general, frequency-division detectors are moderately priced detectors that operate simultaneously over a broad range of frequencies. Frequency-division detectors electronically divide the incoming ultrasonic frequency by a factor of 10 or 16, for example, and the result is an output audible to humans. Calls are heard and recorded, if desired, in real time. These detectors are less sensitive to a specific frequency than heterodyne units and are unable to detect signals at as great a distance, but different species emitting calls with different frequencies can be detected and recorded simultaneously because there is no preselected bandwidth. Information regarding the time and frequency characteristics of the calls is retained, and in some models, information from harmonics can also be evaluated. The output of a frequency-division detector can be analyzed based on zero-crossing periods, and the result is a display of the frequency-time structure of the call. These detectors can be configured to operate manually or remotely, thus allowing multiple sites to be monitored concurrently, a feature that is especially important when the number of personnel is limited. Characteristics of the calls may be quantified using sound-analysis software and may subsequently be analyzed to identify species or species groups.

Time-expansion detectors also operate over a broad range of frequencies but are usually the most expensive. These detectors digitally record the ultrasonic signal and subsequently replay it at a reduced speed, and, therefore, not in real time. The recordings are high quality and retain all information from the calls including harmonics and variations in amplitude. These detectors also can record manually or remotely but require large amounts of digital storage capacity. Because these units do not operate in real time, some calls may be missed during downloading. Time-expansion units are usually more sensitive than other broadband detectors and retain more aural characteristics for analysis (Fenton et al. 2001).

An obvious advantage of using ultrasonic detection is that bats do not need to be handled, thereby eliminating potential stress and injury to the animal. Unfortunately, the intensity of echolocation calls decreases with increasing distance because of geometric, atmospheric, and environmental attenuation of sound (Schnitzler and Kalko 1998); therefore, the distance that a bat can be detected is influenced by environmental conditions, intensity of its call, and amount of clutter in the habitat. Low-frequency sounds, with their correspondingly large wavelength, are less affected by attenuation and thus carry farther than high-frequency sounds with short wavelengths.

Because of variation in call structure and the influence of local habitats on echolocating bats, as well as potential geographic variation, a library of known calls by species and habitat is essential for identification of unknown calls using ultrasonic detectors (O'Farrell et al. 1999a). With experience, researchers can use reference calls and flight behavior to identify many species of bats (Fenton 1970; Fenton and Bell 1981). Implicit in these studies, however, is that echolocation calls can be correctly assigned to a species or species group, but not all researchers agree on how reliable the present technology is in achieving the needed resolution, especially for differentiating among species of *Myotis* (Barclay 1999; Biscardi et al. 2004; Burnett et al. 2004; Hayes 2000).

Additional problems potentially exist with the use of bat detectors, including the assumption that the area sampled by a detector is representative of the habitat under study (Hayes 2000). Weller and Zabel (2002) demonstrated that placement of detectors influenced the amount and quality of data obtained, with instruments placed higher in the canopy or oriented toward less clutter recording higher levels of activity than detectors close to the ground or oriented toward more dense vegetation. Another limitation of acoustic monitoring is related to the intensity of vocalizations produced by different species. Some species, such as various *Corynorhinus,* emit sounds of low intensity and are less likely than other species to be detected by acoustic equipment, and consequently, such species are probably underrepresented in most acoustic-monitoring studies (O'Farrell and Gannon 1999). Furthermore, identification of individual bats is not possible, in general, making it unclear as to how many bats comprise a given suite of calls (Hayes 2000). For example, the same bat passing back and forth 12 times in a night cannot be differentiated from two, four, or six individuals of the same species, each passing the same reference point multiple times. Consequently, no indication of population size or density can be achieved with acoustic monitoring, and the data are only an index of activity, albeit to the species level in some cases (Hayes and Loeb, chap. 8 in this volume; Weller, chap. 10 in this volume). Hayes (1997) and Broders (2003) also demonstrated that the activity of bats varies considerably among nights at the same site, creating potential for bias unless sufficient replication occurs. Fortunately, the increasing affordability of acoustic equipment is making this limitation less of a constraint.

SIZE, SHAPE, AND OVERLAP OF FORAGING AREAS

Many studies attempt to measure spatial use of habitat by bats using radiotelemetry, thereby providing information on size of area used, associations with habitats near roosting sites, and the distance traveled (Adam et al. 1994; Audet 1990; Clark et al. 1993; Robinson and Stebbings 1997). In North America, most studies of spatial use have examined big-eared bats (Adam et al. 1994; Clark et al. 1993; Dobkin et al. 1995; Fellers and Pierson 2002; Hurst and Lacki 1999) or Indiana myotis (Butchkoski and Hassinger 2002; Gumbert et al. 2002; Hobson and Holland 1995; Kiser and Elliott 1996; Murray and Kurta 2004). Advancement in this field is somewhat hindered by a lack of consistency in terminology and approach, rendering direct comparison among studies difficult. In general, data are reported as foraging areas (Krull et al. 1991) or as estimates of home range (Robinson and Stebbings 1997), neither of which should be compared with the other because they have different meanings. Home range, as originally described by Burt (1943), implies the entire area used by an animal in meeting its life requisites. Such an estimate should include roosting sites, foraging sites, and drinking sites, as well as the corridors on the landscape used to travel among these locations (fig. 4.1). In turn, a foraging area should only refer to where a bat concentrates its feeding activity, such as the vicinity of a streetlight, along a stream channel, a clearing in the forest, or a topographically defined watershed in a forested landscape. Bats also may feed when traveling along corridors to and from feeding and roosting sites (Verboom and Spoelstra 1999); therefore, these areas could be included in estimates of the size of foraging areas. Estimates calculated in this way, however, are more comparable to home range and should be interpreted as such.

Numerous quantitative methods exist for evaluating spatial data from radiotelemetric studies, including parametric estimators, such as bivariate ellipses (Jennrich and Turner 1969; Koeppl et al. 1975), and nonparametric estimators, including minimum and maximum convex polygons (Hayne 1949; Mohr 1947), harmonic means (Dixon and Chapman 1980), and Fourier series estimates (Anderson 1982). All methods have strengths and weaknesses (table 4.3). Bivariate ellipses require data to be normally distributed (Jennrich and Turner 1969; Koeppl et al. 1975), a condition that often is not met when using data collected from foraging bats. Harmonic mean and Fourier series estimators are less constrained by properties of the data but require larger samples than other methods for estimates to be reliable (Anderson 1982; Dixon and Chapman 1980). Opportunities for obtaining large sets of data for individual bats are limited, however, because the small transmitters that are used generally remain attached to a bat for less than 10 days (Barclay and Kurta, chap. 2 in this volume). Furthermore, the difficulty in maintaining contact with foraging bats, especially in forested terrain, makes achieving large samples even more unlikely. Most biologists, therefore, rely on the more simple nonparametric approaches, especially estimates based on minimum convex polygons (Clark et al. 1993; Robinson and Stebbings 1997). Unfortunately, these methods consistently overestimate the area used by bats, because habitat that is never visited often is included in the estimate (Hayne 1949; Mohr 1947).

Development of software and methods to calculate estimates and to display the shape and appearance of foraging areas and home ranges has kept pace with the evolution of quantitative methods discussed above (Kie et al. 1996; Lawson and Rodgers 1997). Kernel methods are most

Table 4.3. Attributes of various home-range and foraging-area estimators

| Parameter | Home-range estimators | | | | |
	Minimum convex polygon	Bivariate normal	Harmonic mean	Fixed kernel	Adaptive kernel
Type of estimator	nonparametric	parametric	nonparametric	nonparametric	nonparametric
Minimum sample	70	20	70	30	30
Bias with small samples	high	low	high	moderate	moderate
Assumptions of the distribution	none	normal	none	none	none
Based on utilization distributions	no	no	yes	yes	yes
Sensitivity to outliers	high	high	low	low	low
Bias at inner contour	NA	good	poor	moderate	good
Bias at outer contour	NA	poor	poor	good	moderate
Delineates core areas	no	no	yes	yes	yes
Sensitive to autocorrelation	low	high	high	low	low

Sources: Anderson 1982; Dixon and Chapman 1980; Hayne 1949; Jennrich and Turner 1969; Kie et al. 1996; Koeppl et al. 1975; Lawson and Rodgers 1997; Mohr 1947; Samuel and Garton 1985; Schoener 1981; Smith 1983; Swihart and Slade 1985; White and Garrott 1990; and Worton 1989.

commonly used for generating estimates of area use by bats (Menzel et al. 2001, 2005; Waldien and Hayes 2001). Kernel methods rely on the utilization distributions associated with data on a bat's position and compare those with random samples of locations to derive an estimate of the area that is used; such methods are arguably superior to traditional convex-polygon techniques (Seaman et al. 1998; Worton 1989). Kernel methods also provide estimates of area based on polygons that represent 100%, 95%, or even smaller percentages, that is, core areas, of the distribution of foraging positions that are obtained (fig. 4.1). Previous research on bats focused on 95%-use distributions (Hurst and Lacki 1999; Menzel et al. 2001), presumably to eliminate outliers in the data that otherwise would be included in the 100%-use distribution and that might upwardly bias the estimates of area. Seaman et al. (1999) demonstrated that 30 locations per animal are sufficient to generate an estimate of area using kernel methods, although they recommended obtaining more than 50 locations per animal when possible.

Despite the emphasis placed on the size of areas used by forest-dwelling bats, less attention has been paid to shape of these areas, even though visual representations are sometimes provided (fig. 4.1) (Clark et al. 1993; Menzel et al. 2001). Hutchinson and Lacki (1999) showed that foraging areas of eastern red bats in Kentucky are elongated along a single axis and associated with a preferred stream corridor. Wilhide et al. (1998b) demonstrated that foraging habitat used by Ozark big-eared bats (*Corynorhinus townsendii ingens*) in Arkansas also occurred along stream drainages.

Overlap in areas of use among individual bats of the same species has received limited attention. Estimates of overlap are available for long-tailed bats (*Chalinolobus tuberculatus*) in New Zealand (31–68%; O'Donnell 2001) and Rafinesque's big-eared bat in Kentucky (34–86%; Hurst and Lacki 1999). Overlap of foraging areas in greater mouse-eared bats occurred in West Germany, even though more than one male was never found using the same foraging area simultaneously (Audet 1990). Conversely, Murray and Kurta (2004) reported two female Indiana myotis occasionally foraging in the same area at the same time, although members of a colony generally foraged apart. A composite home range was identified for colonies of serotine bats (*Eptesicus serotinus*) in England, and different colonies overlapped in their use of available foraging habitat (Robinson and Stebbings 1997). This latter finding has potential implications for managing bats in xeric forests, where colonies likely are distributed across landscapes and where drinking and feeding sites may be limited and irregularly spaced.

DISTANCE TRAVELED

A common metric in foraging studies is an estimate of distance traveled, including straight-line distances to and among foraging, drinking, capture, and roosting sites (Fellers and Pierson 2002; Waldien and Hayes 2001). Typically the maximum distance traveled by bats is reported

(Adam et al. 1994; Dobkin et al. 1995). Data on distance traveled are valuable because they can be easily obtained and spatially interpreted, and they can be used for conservation purposes, regardless of whether bats are radiotagged, light tagged, or mist netted. Because only a subsample of a colony of bats is ever marked or captured, reported values for maximum distance traveled from roosting sites to foraging areas may be underestimates of the true value for a particular colony. Further, Murray and Kurta (2004) demonstrated conclusively that Indiana myotis travel nonlinear paths between roosting and foraging sites and, therefore, use much longer flight paths than the maximum straight-line distances between roosting and foraging sites would suggest. Nevertheless, we still believe that straight-line estimates of maximum distance traveled are useful in evaluating conservation needs of bat colonies and can be a valuable tool for indexing habitat requirements, especially when management and conservation decisions must be made quickly and when data on local habitat use by bats would take too long to collect and analyze.

EVALUATING DATA ON HABITAT USE

Selection of resources is a hierarchical process of behavioral choices by bats that results in differential use of habitats (Block and Brennan 1993). Numerous methods have been developed for analyzing data on the use of resources. These include univariate methods based on classifying data, such as chi-square goodness-of-fit tests (Neu et al. 1974) or rank-order tests (Friedman 1937; Johnson 1980; Quade 1979), as well as multivariate methods, including logistic regression (Thomasma et al. 1991), polytomous logistic regression (North and Reynolds 1995), and discrete choice models (Cooper and Millspaugh 1999). Each method has strengths and weaknesses that range from appropriate identification of the experimental unit to accuracy of point estimates (Alldredge and Ratti 1986, 1992).

Data on bats in forests have been analyzed traditionally using classification-based methods that place locations of an animal into distinct categories of habitat for analysis (Conner et al. 2003). Frequency of locations of an animal among categories is then either rank-ordered by availability of habitat for testing or compared among categories scaled against the proportional availability of habitats in the landscape. Habitats are classified as preferred (selected), used at random, or avoided. Implicit in this approach is that the area of inference, or area of habitat available to the animal, is known and can be quantified. Among available techniques, the method of Neu et al. (1974) has been used most frequently in studies of North American bats (Adam et al. 1994; Clark et al. 1993). Unfortunately, most studies examining use of foraging habitat by bats have interpreted habitat use directly from the distribution of locations without considering availability of habitats on the landscape (Miller et al. 2003).

Recently, biologists have begun to use distance-based methods for analysis of habitat-use data (Aebischer et al. 1993; Conner et al. 2003), including studies on bats (Menzel et al. 2005; Sparks et al. 2005). Distance-based methods use distances between locations of an animal and habitats in the analysis and are superior to classification-based methods for evaluating the importance of features of the landscape, such as riparian corridors or ponds (Conner et al. 2003). Other methods of analysis that account for the decline in likelihood of use of habitats situated farther from the roosting site, such as that which occurs with central-place foragers (Rosenberg and McKelvey 1999), are appropriate for use with bats that roost in a specific location, such as a cave, a cluster of snags, or an individual tree. An important caveat with using any method for assessing resource use in bats, however, is that inferences concerning optimum habitat or survival requirements may not be directly revealed by resource-use analysis, because resources selected by bats in a particular study may not always be those preferred in another area. For instance, if less preferred resources are the only ones available, selection studies are likely to show some of them to be important.

Data from studies based on ultrasonic detectors also provide insight into habitat use by bats. Ultrasonic detectors can be placed at "known" locations in the habitat and can be maneuvered in space and time to measure variation in activity among habitats and seasons (Weller and Zabel 2002). Studies based on ultrasonic detectors, for instance, demonstrate the importance of landmarks and corridors for commuting and foraging bats (Verboom et al. 1999; Walsh and Harris 1996) and have measured the relative importance of different vertical strata to bats foraging in closed-canopy forest (Kalcounis et al. 1999). As Miller et al. (2003) pointed out, studies based on the use of ultrasonic detectors are "Design 1 studies" (sensu Thomas and Taylor 1990) and are limited in scope, because in Design 1 studies individual animals are not identified and inferences on habitat selection can only be made at the level of the population. The ability to identify calls of individual bats would greatly increase the value of ultrasonic detectors in evaluating habitat use of bats.

FORAGING HABITAT OF NORTH AMERICAN BATS: WHAT WE DO KNOW

Historically, the value of foraging habitat to the conservation of bats has been viewed as secondary to availability and location of roosting sites (Kunz 1982, 2002). This premise works well when developing conservation plans for species that roost in permanent structures, such as caves and mines, because these roosting sites are more easily protected and managed for in the long term than are ephemeral roosts, such as snags and live trees, or foraging habitats. Vaughan et al. (1997) indicate that explanations for differences in the level of activity of bats among sites often are confounded by an inability to distinguish between the importance of proximity to the roosting site and quality of a particular foraging habitat, suggesting the need for landscape-scale studies that evaluate quality of foraging habitat at both sites used and not used by the bats while feeding. In the following sections, we discuss existing data sets that are available to help us begin to draw a picture of what forest bats are doing or likely will do in selecting habitats in which to feed.

In general, the distribution of bats with elevation suggests that reproductively active females use forests at lower elevations, presumably to minimize thermoregulatory costs and increase foraging efficiency (Brack et al. 2002; Cryan et al. 2000). Ford et al. (2002) demonstrated female-biased sex ratios in the eastern red bat in regions with high ambient temperatures in summer, and implicated altitudinal and other geographic influences in determining the distribution of adult females. Thomas (1988) found male-biased sex ratios for bats at higher elevations in forests of Douglas-fir (*Pseudotsuga menziesii*) in Washington and Oregon, and suggested that freedom from rearing pups permits adult males to enter torpor more often than females and to live in locations where the weather is frequently inclement and food resources are scarce. Similarly, Grindal et al. (1999) postulated that juvenile bats should inhabit high-elevation forests once young bats are fully weaned, because they too could enter torpor more often and reduce their overall energetic expenditures. Grindal et al. (1999), however, countered this suggestion by speculating that use of lower elevations by juveniles could also be advantageous by permitting greater access to food and allowing a more rapid accumulation of fat reserves necessary to survive winter hibernation. Clearly, more data are needed on the distribution and use of habitats by all sex and age classes of forest-dwelling bats across elevational gradients. Knowledge of such differences will be important for setting aside foraging and roosting habitats in topographically complex forested landscapes.

SIZE OF FORAGING AREAS

We surveyed published estimates of size of foraging areas to evaluate inter- and intraspecific patterns among North American bats (table 4.4). Size of foraging areas of males varies considerably, ranging from lows of 24–25 ha for Rafinesque's (Menzel et al. 2001) and Virginia big-eared bats (*Corynorhinus townsendii virginianus*) (Adam et al. 1994), respectively, to 3,026 ha for an Indiana myotis (Rommé et al. 2002). Each of these estimates was obtained in late summer or fall. The size of foraging areas of male bats typically varies more than that of females of the same species, despite female bats having been studied more often than males.

Estimates of the size of foraging areas of female bats vary from 10 ha for an Ozark big-eared bat (Wethington et al. 1996) to 727 ha for a lactating female of the same species (Clark et al. 1993). Furthermore, multiple sites of concentrated feeding activity occur within foraging areas of Ozark big-eared bats (Clark et al. 1993; Wethington et al. 1996). The size of foraging areas used by female eastern red bats increases from pregnancy to lactation to postlactation in the forested mountains of Kentucky (Hutchinson and Lacki 1999). However, the opposite pattern occurs in little brown myotis foraging in the St. Lawrence River estuary, with a 51% decrease in the area used by adult females from pregnancy to lactation (Henry et al. 2002). Studies of big-eared bats have demonstrated variation among subspecies in the relative size of foraging areas during preg-

Table 4.4. Range of values reported for size of foraging areas (ha) and length of commuting distances (km) for species of North American forest-dwelling bats

Species	Size of foraging areas (ha)			Length of commuting distances (km)		
	No. studies	Minimum	Maximum	No. studies	Minimum	Maximum
Myotis evotis	1	–	38	1	–	0.52
Myotis lucifugus	1	18	30	1	1.7	2.6
Myotis sodalis	5	39	3,026	7	0.5	10
Lasiurus borealis	1	113	925	1	–	7.4
Corynorhinus rafinesquii	2	24	260	1	–	1.22
Corynorhinus townsendii	3	10	727	6	2.5	24

Sources: Adam et al. 1994; Butchkoski and Hassinger 2002; Clark et al. 1993; Dobkin et al. 1995; Fellers and Pierson 2002; Gumbert et al. 2002; Henry et al. 2002; Hobson and Holland 1995; Hurst and Lacki 1999; Hutchinson and Lacki 1999; Kiser and Elliott 1996; Kurta et al. 2002; LaVal et al. 1977; Menzel et al. 2001; Murray and Kurta 2004; Rommé et al. 2002; Waldien and Hayes 2001; Wethington et al. 1996; and Wilhide et al. 1998b.

nancy, lactation, and postlactation. Female Virginia big-eared bats increase their foraging area in August with the onset of postlactation (Adam et al. 1994), whereas female Ozark big-eared bats increase their foraging areas during lactation, in midsummer, only to reduce the size of their foraging areas during postlactation (Clark et al. 1993).

The paucity of estimates of foraging areas for forest-dwelling bats clearly limits generalizations of foraging area requirements for these species, especially across differing stages of reproduction. One commonality among studies, however, is that foraging areas usually encompass a body of open water or a riparian corridor (Waldien and Hayes 2001; Wilhide et al. 1998b), suggesting that proximity to water may be critically important to the choice of foraging areas for many species (Carter et al. 2002).

COMMUTING DISTANCES

Values reported for commuting distances appear to be strongly affected by the methods used to obtain them (table 4.4). For example, the greatest distance reported, 24 km for Townsend's big-eared bat (Dobkin et al. 1995), was obtained with the aid of fixed-wing aircraft. When ground-based telemetry is involved, commuting distances decrease to maximums of 10.3 km for Indiana myotis (Rommé et al. 2002) and 8.4 km for Virginia big-eared bats (Adam et al. 1994). Regardless, the shortest commuting distances, 0.52 km for long-eared myotis (Waldien and Hayes 2001) and 1.22 km for Rafinesque's big-eared bat (Hurst and Lacki 1999), if used to delimit the radius of nondisturbance zones surrounding maternity sites, would still enclose a sizeable area of land and ensure adequate foraging habitat for newly volant young of these species (Hurst and Lacki 1999).

THE SIGNIFICANCE OF PATCH SIZE

The role of patch (i.e., a block of habitat distinct from the surrounding landscape matrix) size in the ecology of bats in North America has received little study. The importance of larger patches for the conservation of wildlife, in general, is now dogma founded largely on the principles of island biogeography (MacArthur and Wilson 1967). Debate continues, though, over criteria that should be used in selecting the location of patches to receive protective measures (Margules and Pressey 2000; Pressey et al. 1993). Studies of bats in tropical systems suggest that large habitat patches are effective for the protection of abundant and common species of bats, but may fall short of expectations in protecting habitat of rare species (Andelman and Willig 2002).

The extent to which forest-dwelling bats become limited by a shortage of quality foraging habitat as the size of forest patch declines is unclear. Bats use small patches (<5 ha) of old-growth redwood (*Sequoia sempervirens*) stands in California, in part because of their proximity to riparian corridors that the bats use for foraging (Zielinski and Gellman 1999). Studies in Sweden (de Jong 1995), however, demonstrate that forest patches of less than 20 ha are seldom used and support fewer populations and species of bats than larger patches, indicating a negative effect due to forest fragmentation. Bats in arid ecosystems must seek out localized, high-density patches of prey across landscapes where insects are irregularly distributed, both spatially and temporally (Bell 1980). One might expect a similar response for bats inhabiting xeric forests located in the rain shadow of large mountain ranges, where water and insects are likely to be scarce and patchily distributed.

The importance of patch size in the use of nonforested habitat by bats has been examined for a number of species (fig. 4.3). Some bats avoid large clearings or clearcuts when foraging (de Jong 1995; Murray and Kurta 2004; Patriquin and Barclay 2003; Perdue and Steventon 1996), whereas bats in other locations exhibit the opposite response (Burford and Lacki 1995; Erickson and West 1996; Owen et al. 2004; Patriquin and Barclay 2003; Tibbels and Kurta 2003). The size of the openings, however, was not always reported, so it is difficult to assess whether differential use of openings by foraging bats was a function of the size of the clearing or the clearcut studied. Different species likely will show varying abilities and willingness to cross or make use of open terrain (Law et al. 1999; Schulze et al. 2000), so variation in numerical response among species should be anticipated with increasing levels of forest fragmentation.

HABITAT STRUCTURE WITHIN FOREST PATCHES

Existing studies permit an examination of bat activity in stands of varying age (Krusic et al. 1996; Thomas 1988) and subject to different silvicultural practices (Erickson and West 1996; Hayes and Adam 1996; Tibbels and Kurta 2003). Activity is high in older forests (Crampton and Barclay 1996; Krusic et al. 1996) and often very low in sapling and regenerating stands (Burford and Lacki 1995; Erickson and West 1996). The latter response presumably is due to greater clutter associated with regenerating stands and, thus, a reduced ability of aerial-foraging bats to capture insects at these sites. Patterns of use by bats of forests in intermediate seral stages are also inconsistent across studies (Tibbels and Kurta 2003; Patriquin and Barclay 2003). Presently, the number of stud-

a

b

Figure 4.3.
Forested landscapes fragmented with timber harvests in north-central Idaho (a) and the east Cascades in Washington (b). Photo b shows a roost used by long-legged myotis (*Myotis volans*) situated at the edge of a timber harvest.
Photos by M. Baker

ies completed is insufficient to cover the full range of forest types used by bats, so minimal inference can be drawn on the relative foraging success of bats in hardwood and conifer forests. Limited data suggest, however, that hardwood forests may be associated with higher levels of bat activity than conifer forests (de Jong 1995; Kalcounis et al. 1999).

A few studies have measured the response of bats to specific silvicultural practices. For example, thinning apparently reduces clutter in forests, facilitating foraging by bats (Erickson and West 1996; Humes et al. 1999), and may be an effective technique for generating foraging habitat in sapling and pole-sized stands. However, there was no increase in bat activity in stands of red pine (*Pinus resinosa*) that were 7 years postthinning compared with unthinned stands, although activity in both types of stands was very low relative to small openings nearby (Tibbels and Kurta 2003). Group-selection harvests are associated with high levels of bat activity, in particular, in the newly created gaps in the canopy, compared with forests where no harvesting occurs (Menzel et al. 2002). In this study, gaps were created only in bottomland forest, and it remains to be seen if the same response would take place after creating canopy gaps in forests at higher elevations. Unfortunately, little other information is available on how bats respond to man-made alterations in the structure and composition of forests.

HABITAT STRATIFICATION

Design of the body and structure of echolocation calls affects how different species of bats access cluttered environments (Aldridge and Rautenbach 1987; Crome and Richards 1988). This premise holds for bats both within forest canopies (Kalcounis et al. 1999) and along forest edges (Grindal 1996; Hogberg et al. 2002). Vertical stratification is especially important in tropical forests and helps support the complex bat communities inhabiting these ecosystems (Bernard 2001; Hodgkison et al. 2003; Kalko and Handley 2001). In turn, bats in boreal forests have demonstrated use of both the canopy and the area above the canopy for foraging (Kalcounis et al. 1999). Another element of habitat structure that may be important to the foraging behavior of bats is the configuration of edge habitats adjacent to patches of forest. Edges exist in a variety of forms, from sharply defined to gradual, depending on the manner in which local habitat is disturbed or created, and this difference in habitat complexity might affect foraging behavior among bat species of different morphology and echolocation call structure (Aldridge and Rautenbach 1987; Crome and Richards 1988).

THE IMPORTANCE OF LANDSCAPE FEATURES

Particular landscape features appear crucial to the ecology of forest-dwelling bats for foraging and commuting, including high-elevation ponds (fig. 4.4) (Lacki 1996; Wilhide et al. 1998a), riparian systems with well-developed bank-side vegetation (Hayes and Adam 1996; Verboom et al. 1999; Warren et al. 2000), trails or roads within intact forest (Palmeirim and Etheridge 1985; Zimmerman and Glanz 2000), and sharply defined boundaries, including edges of forests (Grindal 1996; Hogberg et al. 2002) and cliffs and canyon walls (Adam et al. 1994; Caire et al. 1984). Landscape structure accounts for more than half the variation in distribution of bats in montane ecosystems of Switzerland (Jaberg and Guisan 2001), suggesting that data on spatial distribution of forest resources may be used to design approaches in habitat management that facilitate foraging success of bats in forests. Attempts to build models that

a

b

Figure 4.4.
High-elevation ponds used by multiple species of bats in the east Cascades in south-central Oregon: (a) cattle pond and mist net and (b) water storage pond used in fighting fires.
Photos by M. Baker

predict foraging activity of bats in the Pacific Northwest using landscape features, however, have been largely unsuccessful (Erickson and West 2003).

FINAL REMARKS

Declining populations of bats are a concern throughout North America (Kunz and Fenton 2003; Kunz and Racey 1998; O'Shea and Bogan 2003), yet efforts at conservation have been hampered by insufficient knowledge of the factors influencing populations of bats, their habitat requirements, and their roles in different forested ecosystems. To develop guidelines for integrating habitat needs of bats at strategic and operational planning levels and to provide for conservation strategies regionally or rangewide, roosting and foraging habitats and their interspersion must be addressed at multiple spatial and temporal scales during each critical life-history period for all species of bats (Fenton 2003). Loss or modification of roosting and foraging habitat, or their juxtaposition, may affect crucial life-history parameters, such as birth rate or adult survivorship.

Land-use change is one of several factors contributing to recent declines in populations of bats (Crampton and Barclay 1996; Stebbings 1988). Reduction in the number of large trees and snags within forests may affect roost availability and may increase the distance between sites used for roosting, foraging, and drinking (Erickson and West 1996). The increased energetic cost of longer commuting distances between roosting and foraging sites may be particularly important to pregnant and lactating females and, therefore, to reproductive success (Kunz 1987). As available habitats change from a primarily forested condition to a heavily fragmented matrix of forest (fig. 4.3), species of bats that avoid openings when commuting or foraging may be restricted in their access to more distant resources (Limpens and Kapteyn 1991; Murray and Kurta 2004; Winhold et al. 2005). It is reasonable to assume that changes in composition, structure, and distribution of forested habitats have already altered the prey base for bats throughout North American forests.

Much progress still needs to be made in understanding how forest-dwelling bats locate and acquire food resources from among those that are available. In heavily forested landscapes, such as the Southeast and the Pacific Northwest, land managers historically have selected older stands for harvesting, thereby producing landscapes increasingly dominated by younger and structurally less complex even-aged stands. The ramifications of these changes in stand age and structure remain unclear. Younger forests may support fewer roosting sites (Crampton and Barclay 1998; Thomas 1988), although more data are needed to confirm this hypothesis. The implication of geographic-scale changes in forest structure for the prey base and foraging habitat of bats is largely unknown.

Technological advances in the past decade have enhanced research and conservation efforts (Barclay and Brigham 1996), but there continues to be a lack of methodologies and consistent, repeatable approaches to research that provide the information needed to address how changes in available habitats for bats translates into change in population size or density (O'Shea and Bogan 2003). An ultimate goal of foraging studies should be to enhance the long-term conservation of the species under study. To accomplish this goal, future efforts will need to integrate information on roosting and foraging components and estimate the role of each in sustaining populations of bats at multiple spatial scales (Turner et al. 1995). Additionally, careful consideration of goals, experimental design, methodologies, underlying assumptions of selected methods, sample sizes, demographics, and appropriate analyses will contribute to our understanding of "why" bats use particular resources not just "that" they use them (Anderson 2001; Millspaugh and Marzluff 2001).

Unfortunately, we can never know the "true" habitat conditions that existed for bats prior to human settlement, because pristine or baseline conditions are now limited to remote regions of large wilderness areas. Further, landscapes are always in flux, and patterns in the distribution of resources for bats also are changing over time (Duchamp et al., chap. 9 in this volume; Law and Dickman 1998). Nevertheless, we can make comparisons across time among forested landscapes that vary in extent of fragmentation or habitat modification, to assess the degree of response by bats, if any, associated with increased levels of change in habitat. At this scale, well-designed field experiments of foraging behavior are now possible through the use of ultrasonic detectors, insect traps, and radio-telemetry equipment. Tropical species of bats already have experienced a reduction in species richness with increased habitat fragmentation (Estrada et al. 1993), and it appears that a threshold can be crossed whereby sufficient habitat is lost from the landscape and some species are eliminated. We should anticipate the same for bats in North American forests and respond with well-informed management policies that are based on sound field data and that work to sustain the foraging and roosting habitat of these species.

LITERATURE CITED

Acharya, L. 1995. Sex-biased predation on moths by insectivorous bats. Animal Behaviour 49:1461–1468.

Acharya, L., and M.B. Fenton. 1992. Echolocation behaviour of vespertilionid bats (*Lasiurus cinereus* and *Lasiurus borealis*) attacking airborne targets including arctiid moths. Canadian Journal of Zoology 70:1292–1298.

Adam, M.D., M.J. Lacki, and T.G. Barnes. 1994. Foraging areas and habitat use of the Virginia big-eared bat in Kentucky. Journal of Wildlife Management 58:462–469.

Adams, R.A. 1992. Stages of development and sequence of bone formation in the little brown bat, *Myotis lucifugus*. Journal of Mammalogy 73:160–167.

———. 1996. Size-specific resource use in juvenile little brown bats, *Myotis lucifugus* (Chiroptera: Vespertilionidae): is there an ontogenetic shift? Canadian Journal of Zoology 74:1204–1210.

———. 1997. Onset of volancy and foraging patterns of juvenile little brown bats, *Myotis lucifugus*. Journal of Mammalogy 78:239–246.

Aebischer, N.J., P.A. Robertson, and R.E. Kenward. 1993. Compositional analysis of habitat use from animal radio-tracking data. Ecology 74:1313–1325.

Ahlén, I., and H.J. Baagøe. 1999. Use of ultrasound detectors for bat studies in Europe: experiences from field identification, surveys and monitoring. Acta Chiropterologica 1:137–150.

Aldridge, H.D.J.N. 1986. Maneuverability and ecological segregation in the little brown bat (*Myotis lucifugus*) and yuma (*M. yumanensis*) bat (Chiroptera: Vespertilionidae). Canadian Journal of Zoology 64:1878–1882.

———. 1987. Turning flight of bats. Journal of Experimental Biology 128:419–425.

Aldridge, H.D.J.N., and R.M. Brigham. 1988. Load carrying and maneuverability in an insectivorous bat: a test of the 5% "rule" of radio-telemetry. Journal of Mammalogy 69:379–382.

Aldridge, H.D.J.N., and I.L. Rautenbach. 1987. Morphology, echolocation, and resource partitioning in insectivorous bats. Journal of Animal Ecology 56:763–778.

Alldredge, J.R., and J.T. Ratti. 1986. Comparison of some statistical techniques for analysis of resource selection. Journal of Wildlife Management 50:157–165.

———. 1992. Further comparison of some statistical techniques for analysis of resource selection. Journal of Wildlife Management 56:1–9.

Andelman, S.J., and M.R. Willig. 2002. Alternative configurations of conservation reserves for Paraguayan bats: considerations of spatial scale. Conservation Biology 16:1352–1363.

Anderson, D.J. 1982. The home range: a new nonparametric estimation technique. Ecology 63:103–112.

Anderson, D.R. 2001. The need to get the basics right in wildlife field studies. Wildlife Society Bulletin 29:1294–1297.

Anthony, E.L.P., and T.H. Kunz. 1977. Feeding strategies of the little brown bat, *Myotis lucifugus,* in southern New Hampshire. Ecology 58:775–786.

Anthony, E.L.P., H.M. Stack, and T.H. Kunz.1981. Night roosting and the nocturnal time budget of the little brown bat, *Myotis lucifugus:* effects of reproductive status, prey density, and environmental conditions. Oecologia 51:151–156.

Arita, H.T., and M.B. Fenton. 1997. Flight and echolocation in the ecology and evolution of bats. Trends in Ecology and Evolution 12:53–58.

Arlettaz, R. 1996. Feeding behaviour and foraging strategy of free-living mouse-eared bats, *Myotis myotis* and *Myotis blythii.* Animal Behaviour 51:1–11.

Arlettaz, R., G. Dändliker, E. Kasybekov, J. Pillet, S. Rybin, and J. Zima. 1995. Feeding habits of the long-eared desert bat, *Otonycteris hemprichi* (Chiroptera: Vespertilionidae). Journal of Mammalogy 76:873–876.

Audet, D. 1990. Foraging behavior and habitat use by a gleaning bat, *Myotis myotis* (Chiroptera: Vespertilionidae). Journal of Mammalogy 71:420–427.

Barclay, R.M.R. 1985. Long-versus short-range foraging strategies of hoary (*Lasiurus cinereus*) and silver-haired (*Lasionycteris noctivagans*) bats and the consequences for prey selection. Canadian Journal of Zoology 63:2507–2515.

———. 1989. The effect of reproductive condition on the foraging behavior of female hoary bats, *Lasiurus cinereus.* Behavioral Ecology and Sociobiology 24:31–37.

———. 1991. Population structure of temperate zone insectivorous bats in relation to foraging behavior and energy demand. Journal of Animal Ecology 60:165–178.

———. 1999. Bats are not birds: a cautionary note on using echolocation calls to identify bats: a comment. Journal of Mammalogy 80:290–296.

Barclay, R.M.R., and R.M. Brigham. 1991. Prey detection, dietary niche breadth, and body size in bats: why are aerial insectivores so small? American Naturalist 137:693–703.

———. 1994. Constraints on optimal foraging: a field test of prey discrimination by echolocating insectivorous bats. Animal Behaviour 48:1013–1021.

———, eds. 1996. Bats and forests symposium. Victoria, British Columbia, Canada. Research Branch, British Columbia Ministry of Forests, Victoria, BC.

Barclay, R.M.R., M.A. Dolan, and A. Dyck. 1991. The digestive efficiency of insectivorous bats. Canadian Journal of Zoology 69:1853–1856.

Barclay, R.M.R., J. Ulmer, C.J.A. MacKenzie, M.S. Thompson, L. Olsen, J. McCool, E. Cropley, and G. Poll. 2004. Variation in the reproductive rate of bats. Canadian Journal of Zoology 82:688–693.

Bates, D.L., and M.B. Fenton. 1990. Aposematism or startle? Predators learn their responses to the defences of prey. Canadian Journal of Zoology 68:49–52.

Bauerova, Z. 1978. Contribution to the trophic ecology of *Myotis myotis.* Folia Zoologica 27:305–316.

Bell, G.P. 1980. Habitat use and response to patches of prey by desert insectivorous bats. Canadian Journal of Zoology 58:1876–1883.

———. 1982. Behavioral and ecological aspects of gleaning by a desert insectivorous bat. Behavioral Ecology and Sociobiology 10:217–223.

———. 1985. The sensory basis of prey location by the California leaf-nosed bat *Macrotus californicus* (Chiroptera: Phyllostomidae). Behavioral Ecology and Sociobiology 16:343–347.

Belwood, J.J., and M.B. Fenton. 1976. Variation in the diet of *Myotis lucifugus* (Chiroptera: Vespertilionidae). Canadian Journal of Zoology 54:1674–1678.

Belwood, J.J., and J.H. Fullard. 1984. Echolocation and foraging behaviour in the Hawaiian hoary bat, *Lasiurus cinereus semotus*. Canadian Journal of Zoology 62:2113–2120.

Bernard, E. 2001. Vertical stratification of bat communities in primary forests of Central Amazon, Brazil. Journal of Tropical Ecology 17:115–126.

Betts, B.J. 1998. Effect of inter-individual variation in echolocation calls on identification of big brown and silver-haired bats. Journal of Wildlife Management 62:1003–1010.

Biscardi, S., J. Orprecio, M.B. Fenton, A. Tsoar, and J.M. Ratclliffe. 2004. Data, sample sizes and statistics affect the recognition of species of bats by their echolocation calls. Acta Chiropterologica 6:347–363.

Black. H.L. 1974. A north temperate bat community: structure and prey populations. Journal of Mammalogy 55:138–157.

Block, W.M., and L.A. Brennan. 1993. The habitat concept in ornithology: theory and applications, pp. 35–91, *in* Current ornithology (D.M. Power, ed.), vol. 11. Plenum Press, New York.

Bock, W.J. 1990. From Biologische Anatomie to Ecomorphology: Proceedings of the Third International Congress on Vertebrate Morphology. Netherlands Journal of Zoology 40:254–277.

———. 1994. Concepts and methods in ecomorphology. Journal of Biosciences 19:403–413.

Bogdanowicz W., M.B. Fenton, and K. Daleszcyk. 1999. The relationships between echolocation calls, morphology and diet in insectivorous bats. Journal of Zoology (London) 247:381–393.

Bonaccorso, F.J., and B.K. McNab. 2003. Standard energetics of leaf-nosed bats (Hipposideridae): its relationship to intermittent and protracted foraging tactics in bats and birds. Journal of Comparative Physiology B 173:43–53.

Brack, V., Jr., and R.K. LaVal. 1985. Food habits of the Indiana bat in Missouri. Journal of Mammalogy 66:308–315.

Brack, V., Jr., C.W. Stihler, R.J. Reynolds, and C. Butchkoski. 2002. Effect of climate and elevation on distribution and abundance in the mideastern United States, pp. 221–225, *in* The Indiana bat: biology and management of an endangered species (A. Kurta and J. Kennedy, eds.). Bat Conservation International, Austin, TX.

Brigham, R.M. 1990. Prey selection by big brown bats (*Eptesicus fuscus*) and common nighthawks (*Chordeiles minor*). American Midland Naturalist 124:73–80.

Brigham, R.M., H.D.J.N. Aldridge, and R.L. Mackey. 1992. Variation in habitat use and prey selection by Yuma bats, *Myotis yumanensis*. Journal of Mammalogy 73:640–645.

Brigham, R.M., and M.B. Saunders. 1990. The diet of big brown bats (*Eptesicus fuscus*) in relation to insect availability in southern Alberta, Canada. Northwest Science 64:7–10.

Britzke, E.R., and K.L. Murray. 2000. A quantitative method for the selection of identifiable search-phase calls using the Anabat II detector system. Bat Research News 41:33–36.

Broders, H.G. 2003. Another quantitative measure of bat species activity and sam-

pling intensity considerations for the design of ultrasonic monitoring studies. Acta Chiropterologica 5:235–241.

Burford, L.S., and M.J. Lacki. 1995. Habitat use by *Corynorhinus townsendii virginianus* in the Daniel Boone National Forest. American Midland Naturalist
134:340–345.

———. 1998. Moths consumed by *Corynorhinus townsendii virginianus* in eastern
Kentucky. American Midland Naturalist 139:141–146.

Burford, L.S., M.J. Lacki, and C.V. Covell Jr. 1999. Occurrence of moths among
habitats in a mixed mesophytic forest: implications for management of forest bats. Forest Science 45:323–332.

Burnett, S.C., M.B. Fenton, K.A. Kazial, W.M. Masters, and G.F. McCracken. 2004.
Variation in echolocation: notes from a workshop. Bat Research News
45:187–197.

Burt, W.H. 1943. Territoriality and home range concepts as applied to mammals.
Journal of Mammalogy 24:346–352.

Butchkoski, C.M., and J.D. Hassinger. 2002. Ecology of a maternity colony roosting
in a building, pp. 130–142, *in* The Indiana bat: biology and management of
an endangered species (A. Kurta and J. Kennedy, eds.). Bat Conservation International, Austin, TX.

Caccamise, D.F., and R.S. Heddin. 1985. An aerodynamic basis for selecting transmitter loads in birds. Wilson Bulletin 97:306–318.

Caire, W., J.F. Smith, S. McGuire, and M.A. Royce. 1984. Early foraging behavior of
insectivorous bats in western Oklahoma. Journal of Mammalogy 65:319–
324.

Carter, T.C., S.K. Carroll, J.E. Hofmann, J.E. Gardner, and G.A. Feldhamer. 2002.
Landscape analysis of roosting habitat in Illinois, pp. 160–164, *in* The Indi-
ana bat: biology and management of an endangered species (A. Kurta and
J. Kennedy, eds.). Bat Conservation International, Austin, TX.

Charnov, E.L. 1976. Optimal foraging, the marginal value theorem. Theoretical
Population Biology 9:129–136.

Chruszcz, B.J., and R.M.R. Barclay. 2003. Prolonged foraging bouts of a solitary/
gleaning bat, *Myotis evotis.* Canadian Journal of Zoology 81:823–826.

Clark, B.S., D.M. Leslie Jr., and T.S. Carter. 1993. Foraging activity of adult female
Ozark big-eared bats (*Plecotus townsendii ingens*) in summer. Journal of
Mammalogy 74:422–427.

Conner, L.M., M.D. Smith, and L.W. Burger. 2003. A comparison of distance-based
and classification-based analyses of habitat use. Ecology 84:526–531.

Constantine, D.G. 1998. An overlooked external character to differentiate *Myotis
californicus* and *Myotis ciliolabrum* (Vespertilionidae). Journal of Mammalogy 79:624–630.

Cooper A.B., and J.J. Millspaugh. 1999. The application of discrete choice models to
wildlife resource selection studies. Ecology 80:566–575.

Crampton, L.H., and R.M.R. Barclay. 1996. Habitat selection by bats in fragmented
and unfragmented aspen mixedwood stands of different ages, pp. 238–259,
in Bats and forests symposium (R.M.R. Barclay and R.M. Brigham, eds.).
Research Branch, British Columbia Ministry of Forests, Victoria, BC.

———. 1998. Selection of roosting and foraging habitat by bats in different-aged
aspen mixedwood stands. Conservation Biology 12:1347–1358.

Crome, F.H.J., and G.C. Richards. 1988. Bats and gaps: microchiropteran community structure in a Queensland rain forest. Ecology 69:1960–1969.

Cryan, P.M., M.A. Bogan, and J.S. Altenbach. 2000. Effect of elevation on distribution of female bats in the Black Hills, South Dakota. Journal of Mammalogy
81:719–725.

Dalton, V.M., V. Brack Jr., and P.M. McTeer. 1986. Food habits of the big-eared bat,
Plecotus townsendii virginianus, in Virginia. Virginia Journal of Science
37:248–254.

de Jong, J. 1995. Habitat use and species richness of bats in a patchy landscape. Acta Theriologica 40:237–248.

Dixon, K.R., and J.A. Chapman. 1980. Harmonic mean measure of animal activity areas. Ecology 61:1040–1044.

Dobkin, D.S., R.D. Gettinger, and M.G. Gerdes. 1995. Springtime movements, roost use, and foraging activity of Townsend's big-eared bat (*Plecotus townsendii*) in central Oregon. Great Basin Naturalist 55:315–321.

Dunning, D.C. 1968. Warning sounds of moths. Zeitschrift für Tierpsychologie 25:129–138.

Easterla, D.A., and J.O. Whitaker Jr. 1972. Food habits of some bats from Big Bend National Park, Texas. Journal of Mammalogy 53:887–890.

Ehrlich, P.R., and R.W. Holm. 1963. The process of evolution. McGraw-Hill Book Company, New York.

Entwistle, A.C., P.A. Racey, and J.R. Speakman. 1996. Habitat exploitation by a gleaning bat, *Plecotus auritus*. Philosophical Transactions of the Royal Society of London B 351:921–931.

Erickson, J.L., and S.D. West. 1996. Managed forests in the Western Cascades: the effects of seral stage on bat habitat use patterns, pp. 215–227, *in* Bats and forests symposium (R.M.R. Barclay and R.M. Brigham, eds.). Research Branch, British Columbia Ministry of Forests, Victoria, BC.

———. 2003. Associations of bats with local structure and landscape features of forested stands in western Oregon and Washington. Biological Conservation 109:95–102.

Estrada, A., R. Coates-Estrada, and D. Meritt. 1993. Bat species richness and abundance in tropical rain forest fragments and in agricultural habitats at Los Tuxtlas, Mexico. Ecography 16:309–318.

Farney, J., and E.D. Fleharty. 1969. Aspect ratio, loading, wing span, and membrane areas of bats. Journal of Mammalogy 50:362–367.

Faure, P.A., and R.M.R. Barclay. 1992. The sensory basis of prey detection by the long-eared bat, *Myotis evotis*, and the consequences for prey selection. Animal Behaviour 44:31–39.

Faure, P.A., J.H. Fullard, and J.W. Dawson. 1993. The gleaning attacks of the northern long-eared bat, *Myotis septentrionalis*, are relatively inaudible to moths. Journal of Experimental Biology 178:173–189.

Fellers, G.M., and E.D. Pierson. 2002. Habitat use and foraging behavior of Townsend's big-eared bat (*Corynorhinus townsendii*) in coastal California. Journal of Mammalogy 83:167–177.

Fenton, M.B. 1970. A technique for monitoring bat activity with results obtained from different environments in southern Ontario. Canadian Journal of Zoology 48:847–851.

———. 1989. Head size and the foraging behaviour of animal-eating bats. Canadian Journal of Zoology 67:2029–2035.

———. 1990. The foraging behaviour and ecology of animal-eating bats. Canadian Journal of Zoology 68:411–422.

———. 2003. Science and the conservation of bats: where to next? Wildlife Society Bulletin 31:6–15.

Fenton, M.B., D. Audet, M.K. Obrist, and J. Rydell. 1995. Signal strength, timing, and self-deafening: the evolution of echolocation in bats. Paleobiology 21:229–242.

Fenton, M.B., and G.P. Bell. 1979. Echolocation and behavior in four species of *Myotis* (Chiroptera: Vespertilionidae). Canadian Journal of Zoology 57:1271–1277.

———. 1981. Recognition of species of insectivorous bats by their echolocation calls. Journal of Mammalogy 62:233–243.

Fenton, M.B., and W. Bogdanowicz. 2002. Relationships between external morphology and foraging behaviour: bats in the genus *Myotis*. Canadian Journal of Zoology 80:1004–1013.

Fenton, M.B., S. Bouchard, M.J. Vonhof, and J. Zogouris. 2001. Time-expansion and zero-crossing period meter systems present significantly different views of echolocating bats. Journal of Mammalogy 82:721–727.

Fenton, M.B., N.G.H. Boyle, T.M. Harrison, and D.J. Oxley. 1977. Activity patterns, habitat use, and prey selection by some African insectivorous bats. Biotropica 9:73–85.

Findley, J.S., and H. Black. 1983. Morphological and dietary structuring of a Zambian insectivorous bat community. Ecology 64:625–630.

Findley, J.S., E.H. Studier, and D.E. Wilson. 1972. Morphologic properties of bat wings. Journal of Mammalogy 53:429–444.

Ford, W.M., M.A. Menzel, J.M. Menzel, and D.J. Welch. 2002. Influence of summer temperature on sex ratios in eastern red bats (*Lasiurus borealis*). American Midland Naturalist 147:179–184.

Freeman, P.W. 1979. Specialized insectivory: beetle-eating and moth-eating molossid bats. Journal of Mammalogy 60:467–479.

———. 1981. Correspondence of food habits and morphology in insectivorous bats. Journal of Mammalogy 62:166–173.

Friedman, M. 1937. The use of ranks to avoid the assumption of normality implicit in the analysis of variance. Journal of the American Statistical Association 32:675–701.

Fullard, J.H. 1987. Sensory ecology and neuroethology of moths and bats: interactions in a global perspective, pp. 244–272, *in* Recent advances in the study of bats (M.B. Fenton, P.A. Racey, and J.M.V. Rayner, eds.). Cambridge University Press, Cambridge, MA.

Fullard, J.H., and J.W. Dawson. 1997. The echolocation calls of the spotted bat (*Euderma maculatum*) are relatively inaudible to moths. Journal of Experimental Biology 200:129–137.

Fullard, J.H., M.B. Fenton, and J.A. Simmons. 1979. Jamming bat echolocation: the clicks of arctiid moths. Canadian Journal of Zoology 57:647–649.

Fullard, J.H., C. Koehler, A. Surlykke, and N.L. McKenzie. 1991. Echolocation ecology and flight morphology of insectivorous bats (Chiroptera) in southwestern Australia. Australian Journal of Zoology 39:427–438.

Furlonger, C. L., H.J. Dewar, and M.B. Fenton. 1987. Habitat use by foraging insectivorous bats. Canadian Journal of Zoology 65:284–288.

Gannon, W.L., R.E. Sherwin, T.N. de Carvalho, and M.J. O'Farrell. 2001. Pinnae and echolocation call differences between *Myotis californicus* and *M. ciliolabrum* (Chiroptera: Vespertilionidae). Acta Chiropterologica 3:77–91.

Gaudet, C.L., and M.B. Fenton. 1984. Observational learning in three species of insectivorous bats (Chiroptera). Animal Behaviour 32:385–388.

Gould, E. 1955. The feeding efficiency of insectivorous bats. Journal of Mammalogy 36:399–407.

Griffith, L.A., and J. E. Gates. 1985. Food habits of cave-dwelling bats in the central Appalachians. Journal of Mammalogy 66:451–460.

Grindal, S.D. 1996. Habitat use by bats in fragmented forests, pp. 260–272, *in* Bats and forests symposium (R.M.R. Barclay and R.M. Brigham, eds.). Research Branch, British Columbia Ministry of Forests, Victoria, BC.

———. 1999. Habitat use by bats, *Myotis* spp., in western Newfoundland. Canadian Field-Naturalist 113:258–263.

Grindal, S.D., and R.M. Brigham. 1998. Short-term effects of small-scale habitat disturbance on activity by insectivorous bats. Journal of Wildlife Management 62:996–1003.

Grindal, S.D., J.L. Morissette, and R.M. Brigham. 1999. Concentration of bat activity in riparian habitats over an elevational gradient. Canadian Journal of Zoology 77:972–977.

Gumbert, M.W., J.M. O'Keefe, and J.R. MacGregor. 2002. Roost fidelity in Kentucky, pp. 143–152, *in* The Indiana bat: biology and management of an endangered species (A. Kurta and J. Kennedy, eds.). Bat Conservation International, Austin, TX.

Hamilton, I.M., and R.M.R. Barclay. 1998. Diets of juvenile, yearling, and adult big brown bats (*Eptesicus fuscus*) in southeastern Alberta. Journal of Mammalogy 79:764–771.

Hansteen, T.L., H.P. Andreassen, and R.A. Ims. 1997. Effects of spatiotemporal scale on autocorrelation and home range estimators. Journal of Wildlife Management 61:280–290.

Hart, J.A., G.L. Kirkland Jr., and S.C. Grossman. 1993. Relative abundance and habitat use by tree bats, *Lasiurus* spp., in southcentral Pennsylvania. Canadian Field-Naturalist 107:208–212.

Hayes, J.P. 1997. Temporal variation in activity of bats and the design of echolocation-monitoring studies. Journal of Mammalogy 78:514–524.

———. 2000. Assumptions and practical considerations in the design and interpretation of echolocation-monitoring studies. Acta Chiropterologica 2:225–236.

Hayes, J.P., and M.D. Adam. 1996. The influence of logging riparian areas on habitat utilization by bats in western Oregon, pp. 228–237, *in* Bats and forests symposium (R.M.R. Barclay and R.M. Brigham, eds.). Research Branch, British Columbia Ministry of Forests, Victoria, BC.

Hayne, D.W. 1949. Calculation of size of home range. Journal of Mammalogy 30:1–18.

Hayward, B.J., and R.P. Davis. 1964. Flight speed in western bats. Journal of Mammalogy 45:236–242.

Heller, K., and O. von Helversen. 1989. Resource partitioning of sonar frequency bands in rhinolophoid bats. Oecologia 80:178–186.

Henry, M., D.W. Thomas, R. Vaudry, and M. Carrier. 2002. Foraging distances and home range of pregnant and lactating little brown bats (*Myotis lucifugus*). Journal of Mammalogy 83:767–774.

Hickey, M.B.C., L. Acharya, and S. Pennington. 1996. Resource partitioning by two species of vespertilionid bats (*Lasiurus cinereus* and *Lasiurus borealis*) feeding around street lights. Journal of Mammalogy 77:325–334.

Hobson, C.S., and J.N. Holland. 1995. Post-hibernation movement and foraging habitat of a male Indiana bat, *Myotis sodalis* (Chiroptera: Vespertilionidae) in Western Virginia. Brimleyana 23:95–101.

Hodgkison, R., S.T. Balding, Z. Akbar, and T.H. Kunz. 2003. Roosting ecology and social organization of the spotted-winged fruit bat, *Balionycteris maculata* (Chiroptera: Pteropodidae), in a Malaysian lowland dipterocarp forest. Journal of Tropical Ecology 19:667–676.

Hodgkison, R., S.T. Balding, A. Zubaid, and T.H. Kunz. 2001. Vertical stratification of Malaysian fruit bats. Bat Research News 42:101.

Hogberg, L.K., K.J. Patriquin, and R.M.R. Barclay. 2002. Use by bats of patches of residual trees in logged areas of the boreal forest. American Midland Naturalist 148:282–288.

Hughes, P.M., and J.M.V. Rayner. 1991. Addition of artificial loads to long-eared bats *Plecotus auritus:* handicapping flight performance. Journal of Experimental Biology 161:285–298.

Humes, M.L., J.P. Hayes, and M.W. Collopy. 1999. Bat activity in thinned, unthinned, and old-growth forests in western Oregon. Journal of Wildlife Management 63:553–561.

Hurst, T.E., and M.J. Lacki. 1997. Food habits of Rafinesque's big-eared bat in southeastern Kentucky. Journal of Mammalogy 78:525–528.

———. 1999. Roost selection, population size and habitat use by a colony of

Rafinesque's big-eared bats (*Corynorhinus rafinesquii*). American Midland Naturalist 142:363–371.

Hutchinson, J.T., and M.J. Lacki. 1999. Foraging behavior and habitat use of red bats in mixed mesophytic forests of the Cumberland Plateau, Kentucky, pp. 171–177, *in* Proceedings, 12th central hardwood forest conference (J.W. Stringer and D.L. Loftis, eds.). USDA Forest Service, Southern Research Station, General Technical Report SRS-24. Asheville, NC.

Jaberg, C., and A. Guisan. 2001. Modelling the distribution of bats in relation to landscape structure in a temperate mountain environment. Journal of Applied Ecology 38:1169–1181.

Jennrich, R.I., and F.B. Turner. 1969. Measurement of non-circular home range. Journal of Theoretical Biology 22:227–237.

Johnson, D.H. 1980. The comparison of usage and availability measurements for evaluating resource preference. Ecology 61:65–71.

Johnston, D.S., and M.B. Fenton. 2001. Individual and population-level variability in diets of pallid bats (*Antrozous pallidus*). Journal of Mammalogy 82:362–373.

Jones, G., and J. Rydell. 2003. Attack and defense: interactions between echolocating bats and their insect prey, pp. 301–345, *in* Bat ecology (T.H. Kunz and M.B. Fenton, eds.). University of Chicago Press, Chicago, IL.

Jung, T.S., I.D. Thompson, R.D. Titman, and A.P. Applejohn. 1999. Habitat selection by forest bats in relation to mixed-wood stand types and structure in central Ontario. Journal of Wildlife Management 63:1306–1319.

Kalcounis, M.C., K.A. Hobson, R.M. Brigham, and K.R. Hecker. 1999. Bat activity in the boreal forest: importance of stand type and vertical strata. Journal of Mammalogy 80:673–682.

Kalko, E.K.V., and C.O. Handley Jr. 2001. Neotropical bats in the canopy: diversity, community structure, and implications for conservation. Plant Ecology 153:319–333.

Kalko, E.K.V., and H.U. Schnitzler. 1993. Plasticity in echolocation signals of European pipistrelle bats in search flight: implications for habitat use and prey detection. Behavioral Ecology and Sociobiology 33:415–428.

Karr, J.R., and F.C. James. 1975. Eco-morphological configurations and convergent evolution of species and communities, pp. 258–291, *in* Ecology and evolution of communities (M. L. Cody and J. M. Diamond, eds.). Harvard University Press, Cambridge, MA.

Keating, K.A., W.G. Brewster, and C.H. Key. 1991. Satellite telemetry: performance of animal-tracking systems. Journal of Wildlife Management 55:160–171.

Kie, J.G., J.A. Baldwin, and C.J. Evans. 1996. CALHOME: a program for estimating animal home ranges. Wildlife Society Bulletin 24:342–344.

Kingston, T., G. Jones, A. Zubaid, and T.H. Kunz. 2000. Resource partitioning in rhinolophoid bats revisited. Oecologia 124:332–342.

Kiser, J.D., and C.L. Elliott. 1996. Foraging habitat, food habits, and roost tree characteristics of the Indiana bat (*Myotis sodalis*) during autumn in Jackson County, Kentucky. Unpublished Report. Kentucky Department of Fish and Wildlife Resources, Frankfort, KY.

Koeppl, J.W., N.A. Slade, and R.S. Hoffman. 1975. A bivariate home range model with possible application to ethological data analysis. Journal of Mammalogy 56:81–90.

Krull, D., A. Schumm, W. Metzner, and G. Neuweiler. 1991. Foraging areas and foraging behavior in the notch-eared bat, *Myotis emarginatus* (Vespertilionidae). Behavioral Ecology and Sociobiology 28:247–253.

Krusic, R.A., M. Yamasaki, C.D. Neefus, and P.J. Pekins. 1996. Bat habitat use in White Mountain National Forest. Journal of Wildlife Management 60:625–631.

Kufeld, R.C., D.C. Bowden, and J.M. Siperek Jr. 1987. Evaluation of a telemetry system for measuring habitat usage in mountainous terrain. Northwest Science 61:249–256.

Kunz, T.H. 1974. Feeding ecology of a temperate insectivorous bat (*Myotis velifer*). Ecology 55:693–711.

———. 1982. Roosting ecology of bats, pp. 1–55, *in* Ecology of bats (T.H. Kunz, ed.). Plenum Press, New York.

———. 1987. Post-natal growth and energetics of suckling bats, pp. 396–420, *in* Recent advances in the study of bats (M.B. Fenton, P.A. Racey, and J.M.V. Rayner, eds.). Cambridge University Press, Cambridge, UK.

———. 1988a. Ecological and behavioral methods for the study of bats. Smithsonian Institution Press, Washington, DC.

———. 1988b. Methods of assessing the availability of prey to insectivorous bats, pp. 191–210, *in* Ecological and behavioral methods for the study of bats. (T.H. Kunz, ed.). Smithsonian Institution Press, Washington, D.C.

———. 2002. Roosting ecology and population genetic structure of an old-world tent making bat, *Cynopterus sphinx*. Bat Research News 43:53.

———. 2003. Censusing bats: challenges, solutions, and sampling biases, pp. 9–19, *in* Monitoring trends in bat populations of the United States and territories: problems and prospects (T.J. O'Shea and M.A. Bogan, eds.). U.S. Geological Survey Information and Technology Report ITR-2003-0003.

Kunz, T.H., and E.L.P. Anthony. 1996. Variation in nightly emergence behavior in the little brown bat, *Myotis lucifugus* (Chiroptera: Vespertilionidae), pp. 225–236, *in* Contributions in mammalogy: a memorial volume honoring J. Knox Jones, Jr. (H.H. Genoways and R.J. Baker, eds.). Texas Tech University Press, Lubbock, TX.

Kunz, T.H., and M.B. Fenton, eds. 2003. Bat ecology. University of Chicago Press, Chicago, IL.

Kunz, T.H., and A. Kurta. 1988. Capture methods and holding devices, pp. 1–29, *in* Ecological and behavioral methods for the study of bats (T.H. Kunz, ed.). Smithsonian Institution Press, Washington, DC.

Kunz, T.H., and P.A. Racey, eds. 1998. Bat biology and conservation. Smithsonian Institution Press, Washington, DC.

Kunz, T.H., D.W. Thomas, G.C. Richards, C.D. Tidemann, E.D. Pierson, and P.A. Racey. 1996. Observational techniques for bats, pp. 105–114, *in* Measuring and monitoring biological diversity: standard methods for mammals (D.E. Wilson, F.R. Cole, J.D. Nichols, R. Rudran, and M.S. Foster, eds.). Smithsonian Institution Press, Washington, DC.

Kunz, T.H., J.O. Whitaker Jr., and M.D. Wadonoli. 1995. Dietary energetics of the insectivorous Mexican free-tailed bat (*Tadarida brasiliensis*) during pregnancy and lactation. Oecologia 101:407–415.

Kunz, T.H., J.A. Wrazen, and C., D. Burnett. 1998. Changes in body mass and fat reserves in pre-hibernating little brown bats (*Myotis lucifugus*). Ecoscience 5:8–17.

Kurta, A. 1982. Flight patterns of *Eptesicus fuscus* and *Myotis lucifugus* over a stream. Journal of Mammalogy 63:335–337.

Kurta, A., G.P. Bell, K.A. Nagy, and T.H. Kunz. 1989. Water balance of free-ranging little brown bats (*Myotis lucifugus*) during pregnancy and lactation. Canadian Journal of Zoology 67:2468–2472.

Kurta, A., K.A. Johnson, and T.H. Kunz. 1987. Oxygen consumption and body temperature of female little brown bats (*Myotis lucifugus*) under simulated roost conditions. Physiological Zoology 60:386–397.

Kurta, A., and T.H. Kunz. 1987. Size of bats at birth and maternal investment during pregnancy. Symposia of the Zoological Society of London 57:79–107.

Kurta, A., T.H. Kunz, and K.A. Nagy. 1990. Energetics and water flux of free-ranging big brown bats (*Eptesicus fuscus*) during pregnancy and lactation. Journal of Mammalogy 71:59–65.

Kurta, A., and S.W. Murray. 2002. Philopatry and migration of banded Indiana bats (*Myotis sodalis*) and effects of radio transmitters. Journal of Mammalogy 83:585–589.

Kurta, A., S.W. Murray, and D.H. Miller. 2002. Roost selection and movements across the summer landscape, pp. 118–129, *in* The Indiana bat: biology and management of an endangered species (A. Kurta and J. Kennedy, eds.). Bat Conservation International, Austin, TX.

Kurta, A., and J.O. Whitaker Jr. 1998. Diet of the endangered Indiana bat (*Myotis sodalis*) on the northern edge of its range. American Midland Naturalist 140:280–286.

Lacki, M.J. 1996. The role of research in conserving bats on managed forests, pp. 39–48, *in* Bats and forests symposium (R.M.R. Barclay and R.M. Brigham, eds.). Research Branch, British Columbia Ministry of Forests, Victoria, BC.

Lacki, M.J., and T.A. Bookhout. 1983. A survey of bats in Wayne National Forest, Ohio. Ohio Journal of Science 83:45–50.

Lacki, M.J., L.S. Burford, and J.O. Whitaker Jr. 1995. Food habits of gray bats in Kentucky. Journal of Mammalogy 76:1256–1259.

Lacki, M.J., and K.M. LaDeur. 2001. Seasonal use of lepidopteran prey by Rafinesque's big-eared bats (*Corynorhinus rafinesquii*). American Midland Naturalist 145:213–217.

LaVal, R.K., R.L. Clawson, M.L. LaVal, and W. Caire. 1977. Foraging behavior and nocturnal activity patterns of Missouri bats, with emphasis on the endangered species *Myotis grisescens* and *Myotis sodalis*. Journal of Mammalogy 58:592–599.

LaVal, R.K., and M.L. LaVal. 1980a. Prey selection by a neotropical foliage-gleaning bat, *Micronycteris megalotis*. Journal of Mammalogy 61:324–327.

———. 1980b. Prey selection by the slit-faced bat *Nycteris thebaica* (Chiroptera: Nycteridae) in Natal, South Africa. Biotropica 12:241–246.

Law, B.S., J. Anderson, and M. Chidel. 1999. Bat communities in a fragmented landscape on the south-west slopes of New South Wales, Australia. Biological Conservation 88:333–345.

Law, B.S., and C.R. Dickman. 1998. The use of habitat mosaics by terrestrial vertebrate fauna: implications for conservation and management. Biodiversity and Conservation 7:323–333.

Lawson, E.J.G., and A.R. Rodgers. 1997. Differences in home-range size computed in commonly used software programs. Wildlife Society Bulletin 25:721–729.

Lee, Y.-F., and G.F. McCracken. 2001. Timing and variation in the emergence and return of Mexican free-tailed bats, *Tadarida brasiliensis mexicana*. Zoological Studies 40:309–316.

———. 2004. Flight activity and food habits of three species of *Myotis* bats (Chiroptera: Vespertilionidae) in sympatry. Zoological Studies 43:589–597.

Leslie, D.M., Jr., and B.S. Clark. 2002. Feeding habits of the endangered Ozark big-eared bat (*Corynorhinus townsendii ingens*) relative to prey abundance. Acta Chiropterologica 4:173–182.

Limpens, H.J.G.A., and C. Kapteyn. 1991. Bats, their behavior and linear landscape elements. Myotis 29:63–71.

MacArthur, R.H., and E.O. Wilson. 1967. The theory of island biogeography. Monographs in population biology, no. 1. Princeton University Press, Princeton, NJ.

Margules, C.R., and R.L. Pressey. 2000. Systematic conservation planning. Nature 405:243–253.

McCracken, G.F., V.A. Brown, M. Eldridge, Y. Lee, and S. Vege. 2004. Fecal DNA analysis to identify species of insects in the diets of bats. Bat Research News 45:240–241.

Menzel, J.M., W.M. Ford, M.A. Menzel, T.C. Carter, J.E. Gardner, J.D. Garner, and J.E. Hofmann. 2005. Summer habitat use and home-range analysis of the endangered Indiana bat. Journal of Wildlife Management 69:430–436.

Menzel, M.A., T.C. Carter, J.M. Menzel, W.M. Ford, and B.R. Chapman. 2002. Effects of group selection silviculture in bottomland hardwoods on the spatial activity patterns of bats. Forest Ecology and Management 162:209–218.

Menzel, M.A., J.M. Menzel, W.M. Ford, J.W. Edwards, T.C. Carter, J.B. Churchill, and J.C. Kilgo. 2001. Home range and habitat use of male Rafinesque's big-eared bats (*Corynorhinus rafinesquii*). American Midland Naturalist 145:402–408.

Miller, D.A., E.B. Arnett, and M.J. Lacki. 2003. Habitat management for forest-roosting bats of North America: a critical review of habitat studies. Wildlife Society Bulletin 31:30–44.

Mills, L.S., and F.F. Knowlton. 1989. Observer performance in known and blind radio-telemetry accuracy tests. Journal of Wildlife Management 53:340–342.

Millspaugh, J.J., and J.M. Marzluff, eds. 2001. Radiotelemetry and animal populations. Academic Press, San Diego, CA.

Mohr, C.O. 1947. Table of equivalent populations of North American small mammals. American Midland Naturalist 37:223–249.

Murray, S.W., and A. Kurta. 2002. Spatial and temporal variation in diet, pp. 182–192, *in* The Indiana bat: biology and management of an endangered species (A. Kurta and J. Kennedy, eds.). Bat Conservation International, Austin, TX.

———. 2004. Nocturnal activity of the endangered Indiana bat (*Myotis sodalis*). Journal of Zoology (London) 262:197–206.

Nagorsen, D.W., and R.M. Brigham. 1995. Bats of British Columbia. University of British Columbia Press, Vancouver, BC.

Neu, C.W., C.R. Byers, and J.M. Peek. 1974. A technique for analysis of utilization-availability data. Journal of Wildlife Management 38:541–545.

Neubaum, D.J., M.A. Neubaum, L.E. Ellison, and T.J. O'Shea. 2005. Survival and condition of big brown bats (*Eptesicus fuscus*) after radiotagging. Journal of Mammalogy 86:95–98.

Neuweiler, G. 1983. Echolocation and adaptivity to ecological constraints, pp. 280–302, *in* Neuroethology and behavioural physiology (F. Huber and H. Markl, eds.). Springer-Verlag, Berlin, Germany.

———. 1989. Foraging ecology of and audition in echolocating bats. Trends in Ecology and Evolution 4:160–166.

Norberg, U.M. 1981. Flight morphology and the ecological niche. Symposia of the Zoological Society of London 48:173–197.

———. 1994. Wing design, flight performance, and habitat use in bats, pp. 205–239, *in* Ecological morphology: integrative organismal biology (P.C. Wainwright and S.M. Reilly, eds.). University of Chicago Press, Chicago, IL.

Norberg, U.M., and J.M.V. Rayner. 1987. Ecological morphology and flight in bats (Mammalia; Chiroptera): wing adaptations, flight performance, foraging strategy and echolocation. Philosophical Transactions of the Royal Society of London B 316:335–427.

North, M.P., and J.H. Reynolds. 1996. Microhabitat analysis using radiotelemetry locations and polytomous logistic regression. Journal of Wildlife Management 60:639–653.

O'Donnell, C.F.J. 2001. Home range and use of space by *Chalinolobus tuberculatus*, a temperate rainforest bat from New Zealand. Journal of Zoology (London) 253:253–264.

————. 2002. Influence of sex and reproductive status on nocturnal activity of long-tailed bats (*Chalinolobus tuberculatus*). Journal of Mammalogy 83:794–803.

O'Farrell, M.J., C. Corben, W.L. Gannon, and B.W. Miller. 1999a. Confronting the dogma: a reply. Journal of Mammalogy 80:297–302.

O'Farrell, M.J., and W.L. Gannon. 1999. A comparison of acoustic versus capture techniques for the inventory of bats. Journal of Mammalogy 80:24–30.

O'Farrell, M.J., B.W. Miller, and W.L. Gannon. 1999b. Qualitative identification of free-flying bats using the Anabat detector. Journal of Mammalogy 80:11–23.

O'Shea, T.J., and M.A. Bogan, eds. 2003. Monitoring trends in bat populations of the United States and territories: problems and prospects. U.S. Geological Survey Information and Technology Report ITR-2003-0003.

Otis, D.L., and G.C. White. 1999. Autocorrelation of location estimates and the analysis of radiotracking data. Journal of Wildlife Management 63:1039–1044.

Owen, S.F., M.A. Menzel, J.W. Edwards, W.M. Ford, J.M. Menzel, B.R. Chapman, P.B. Wood, and K.V. Miller. 2004. Bat activity in harvested and intact forest stands in the Allegheny Mountains. Northern Journal of Applied Forestry 21:154–159.

Palmeirim, J., and K. Etheridge. 1985. The influence of man-made trails on foraging by tropical frugivorous bats. Biotropica 17:82–83.

Parsons, S., A.M. Boonman, and M.K. Obrist. 2000. Advantages and disadvantages of techniques for transforming and analyzing Chiropteran echolocation calls. Journal of Mammalogy 81:927–938.

Patriquin, K.J., and R.M.R. Barclay. 2003. Foraging by bats in cleared, thinned and unharvested boreal forest. Journal of Applied Ecology 40:646–657.

Pennycuick, C.J. 1975. Mechanics of flight, pp. 1–75, *in* Avian biology (D. Farner and J.R. King, eds.). Academic Press, London, U.K.

Perdue, M., and J.D. Steventon. 1996. Partial cutting and bats: a pilot study, pp. 273–276, *in* Bats and forests symposium (R.M.R. Barclay and R.M. Brigham, eds.). Research Branch, British Columbia Ministry of Forests, Victoria, BC.

Perkins, J.M., and S.P. Cross. 1988. Differential use of some coniferous forest habitats by hoary and silver-haired bats in Oregon. The Murrelet 69:21–24.

Polis, G.A. 1984. Age structure component of niche width and intraspecific resource partitioning: can age groups function as ecological species. The American Naturalist 123:541–564.

Powers, L.V., S.C. Kandarian, and T.H. Kunz. 1991. Ontogeny of flight in the little brown bat, *Myotis lucifugus:* behavior, morphology, and muscle histochemistry. Journal of Comparative Physiology A 168:675–685.

Pressey, R.L., C.J. Humphries, C.R. Margules, R.I. Vane-Wright, and P.H. Williams. 1993. Beyond opportunism: key principles for systematic reserve selection. Trends in Ecology and Evolution 8:124–128.

Quade, D. 1979. Using weighted rankings in the analysis of complete blocks with additive block effects. Journal of the American Statistical Association 74:680–683.

Robinson, M.F., and R.E. Stebbings. 1997. Home range and habitat use by the serotine bat, *Eptesicus serotinus,* in England. Journal of Zoology (London) 243:117–136.

Rodriguez, R.M., and L.K. Ammerman. 2004. Mitochondral DNA divergence does not reflect morphological differences between *Myotis californicus* and *Myotis ciliolabrum.* Journal of Mammalogy 85:842–851

Roeder, K.D. 1964. Aspects of the noctuid tympanic nerve response having significance in the avoidance of bats. Journal of Insect Physiology 10:529–546.

————. 1967. Turning tendency of moths exposed to ultrasound while in stationary flight. Journal of Insect Physiology 13:873–888.

————. 1969. Nerve cells and insect behaviour. Harvard University Press, Cambridge, MA.

Rolseth, S.L., C.E. Koehler, and R.M.R. Barclay. 1994. Differences in the diets of juvenile and adult hoary bats, *Lasiurus cinereus*. Journal of Mammalogy 75:394–398.

Rommé, R.C., A.B. Henry, R.A. King, T. Glueck, and K. Tyrell. 2002. Home range near hibernacula in spring and autumn, pp. 153–158, *in* The Indiana bat: biology and management of an endangered species (A. Kurta and J. Kennedy, eds.). Bat Conservation International, Austin, TX.

Rosenberg, D.K., and K.S. McKelvey. 1999. Estimation of habitat selection for central-place foraging animals. Journal of Wildlife Management 63:1028–1038.

Rydell, J., and R. Arlettaz. 1994. Low-frequency echolocation enables the bat *Tadarida teniotis* to feed on tympanate insects. Proceedings of the Royal Society of London B 257:175–178.

Salcedo, H.D., M.B. Fenton, M.B. Hickey, and R.W. Blake. 1995. Energetic consequences of flight speeds of foraging red and hoary bats (*Lasiurus borealis* and *Lasiurus cinereus;* Chiroptera: Vespertilionidae). Journal of Experimental Biology 198:2245–2251.

Sample, B.E., and R.C. Whitmore. 1993. Food habits of the endangered Virginia big-eared bat in West Virginia. Journal of Mammalogy 74:428–435.

Samuel, M.D., and E.O. Garton. 1985. Home range: a weighted normal estimate and tests of underlying assumptions. Journal of Wildlife Management 49:513–519.

Saunders, M.B., and R.M.R. Barclay. 1992. Ecomorphology of insectivorous bats: a test of predictions using two morphologically similar species. Ecology 73:1335–1345.

Schmutz, J.A., and G.C. White. 1990. Error in telemetry studies: effects of animal movement on triangulation. Journal of Wildlife Management 54:506–510.

Schnitzler, H.U., and E.K.V. Kalko. 1998. How echolocating bats search and find food, pp. 183–196, *in* Bat biology and conservation (T.H. Kunz and P.A. Racey, eds.). Smithsonian Institution Press, Washington, DC.

Schoener, T.W. 1981. An empirically based estimate of home range. Journal of Theoretical Biology 20:281–325.

Schulze, M.D., N.E. Seavy, and D.F. Whitacre. 2000. A comparison of phyllostomid bat assemblages in undisturbed Neotropical forest and in forest fragments of a slash-and-burn farming mosaic in Petin, Guatemala. Biotropica 32:174–184.

Seaman, D.E., B. Griffith, and R.A. Powell. 1998. KERNELHR: a program for estimating animal home ranges. Wildlife Society Bulletin 26:95–100.

Seaman, D.E., J.J. Millspaugh, B.J. Kernohan, G.C. Brundige, K.J. Raedeke, and R.A. Gitzen. 1999. Effects of sample size on kernel home range estimates. Journal of Wildlife Management 63:739–747.

Simmons, J.A., and R.A. Stein. 1980. Acoustic imaging in bat sonar: echolocation signals and the evolution of echolocation. Journal of Comparative Physiology 135:61–84.

Smith, W.P. 1983. A bivariate normal test for elliptical home-range models: biological implications and recommendations. Journal of Wildlife Management 47:613–619.

Sparks, D.W., C.M. Ritzi, J.E. Duchamp, and J.O. Whitaker Jr. 2005. Foraging habitat of Indiana bat (*Myotis sodalis*) at an urban-rural interface. Journal of Mammalogy 86:713–718.

Stebbings, R. 1988. The conservation of European bats. Christopher Helm, London, U.K.

Stephens, D.W., and J.R. Krebs. 1986. Foraging theory. Princeton University Press, Princeton, NJ.

Stockwell, E.F. 2001. Morphology and flight manoeuvrability in New World leaf-nosed bats (Chiroptera: Phyllostomidae). Journal of Zoology (London) 254:505–514.

Surlykke A. 1988. Interaction between echolocating bats and their prey, pp. 551–566, *in* Animal sonar: processes and performance. (P.E. Nachtigall and P.W.B. Moore, eds.). NATO ASI Series A, Life Sciences, vol. 156. Plenum Press, New York.

Surlykke, A., M. Filskov, J.H. Fullard, and E. Forrest. 1999. Auditory relationships to size in noctuid moths: bigger is better. Naturwissenschaften 86:238–241.

Swartz, S.M., P.W. Freeman, and E.F. Stockwell. 2003. Ecomorphology of bats: comparative and experimental approaches relating structural design to ecology, pp. 257–300, *in* Bat ecology (T.H. Kunz and M.B. Fenton, eds.). The University of Chicago Press, Chicago, IL.

Swartz, S.M., and U.M. Norberg. 1998. Part two, functional morphology, pp. 91–92, *in* Bat biology and conservation (T.H. Kunz and P.A. Racey, eds.). Smithsonian Institution Press, Washington, D.C.

Swift, S.M., P.A. Racey, and M.I. Avery. 1985. Feeding ecology of *Pipistrellus pipistrellus* (Chiroptera: Vespertilionidae) during pregnancy and lactation. II. Diet. Journal of Animal Ecology 54:217–225.

Swihart, R.K., and N.A. Slade. 1985. Testing for independence of observations in animal movements. Ecology 66:1176–1184.

Thomas, D.L., and E.J. Taylor. 1990. Study designs and tests for comparing resource use and availability. Journal of Wildlife Management 54:322–330.

Thomas, D.W. 1988. The distribution of bats in different ages of Douglas-fir forests. Journal of Wildlife Management 52:619–626.

Thomasma, L.E., T.D. Drummer, and R.O. Peterson. 1991. Testing the habitat suitability index model for the fisher. Wildlife Society Bulletin 19:291–297.

Tibbels, A.E., and A. Kurta. 2003. Bat activity is low in thinned and unthinned stands of red pine. Canadian Journal of Forest Research 33:2436–2442.

Treat, A.E. 1955. The response to sound in certain Lepidoptera. Annals of the Entomological Society of America 48:272–284.

Turner, M.G., G.J. Arthaud, R.T. Engstrom, S.J. Hejl, J. Liu, S. Loeb, and K. McKelvey. 1995. Usefulness of spatially explicit population models in land management. Ecological Applications 5:12–16.

Vaughan, N., G. Jones, and S. Harris. 1997. Habitat use by bats (Chiroptera) assessed by means of a broad-band acoustic method. Journal of Applied Ecology 34:716–730.

Verboom, B., A.M. Boonman, and H.J.G.A. Limpens. 1999. Acoustic perception of landscape elements by the pond bat (*Myotis dasycneme*). Journal of Zoology (London) 248:59–66.

Verboom, B., and K. Spoelstra. 1999. Effects of food abundance and wind on the use of tree lines by an insectivorous bat, *Pipistrellus pipistrellus*. Canadian Journal of Zoology 77:1393–1401.

Waldien, D.L., and J.P. Hayes. 2001. Activity areas of female long-eared myotis in coniferous forests in western Oregon. Northwest Science 75:307–314.

Walsh, A.L., and S. Harris. 1996. Factors determining the abundance of vespertilionid bats in Britain: geographical, land class and local habitat relationships. Journal of Applied Ecology 33:519–529.

Warren, R.D., D.A. Waters, J.D. Alringham, and D.J. Bullock. 2000. The distribution of Daubenton's bats (*Myotis daubentonii*) and pipistrelle bats (*Pipistrellus pipistrellus*) (Vespertilionidae) in relation to small-scale variation in riverine habitat. Biological Conservation 92:85–91.

Waters, D.A. 2003. Bats and moths: what is there left to learn? Physiological Entomology 28:237–250.

Webb, P.I., J.R. Speakman, and P.A. Racey. 1992. Inter- and intra-individual variation in wing loading and body mass in female pipistrelle bats: theoretical implications for flight performance. Journal of Zoology (London) 228:669–673.

Weller, T.J., and C.J. Zabel. 2002. Variation in bat detections due to detector orientation in a forest. Wildlife Society Bulletin 30:922–930.

Werner, E.E., and J.F. Gilliam. 1984. The ontogenetic niche and species interactions in size-structured populations. Annual Review of Ecology and Systematics 15:393–425.

Wethington, T.A., D.M. Leslie Jr., M.S. Gregory, and M.K. Wethington. 1996. Pre-hibernation habitat use and foraging activity by endangered Ozark big-eared bats (*Plecotus townsendii ingens*). American Midland Naturalist 135:218–230.

Whitaker, J.O., Jr. 1972. Food habits of bats from Indiana. Canadian Journal of Zoology 50:877–883.

———. 1988. Food habits analysis of insectivorous bats, pp. 171–189, *in* Ecological and behavioral methods for the study of bats (T.H. Kunz, ed.). Smithsonian Institution Press, Washington, DC.

———. 1994. Food availability and opportunistic versus selective feeding in insectivorous bats. Bat Research News 35:75–77.

———. 1995. Food of the big brown bat *Eptesicus fuscus* from maternity colonies in Indiana and Illinois. American Midland Naturalist 134:346–360.

———. 2004. Prey selection in a temperate zone insectivorous bat community. Journal of Mammalogy 85:460–469.

Whitaker, J.O., Jr., and P. Clem. 1992. Food of the evening bat *Nycticeius humeralis* from Indiana. American Midland Naturalist 127:211–214.

Whitaker, J.O., Jr., C. Maser, and S.P. Cross. 1981a. Food habits of eastern Oregon bats, based on stomach and scat analyses. Northwest Science 55:281–292.

———. 1981b. Foods of Oregon silver-haired bats, *Lasionycteris noctivagans*. Northwest Science 55:75–77.

Whitaker, J.O., Jr., C. Maser, and L.E. Keller. 1977. Food habits of bats of western Oregon. Northwest Science 51:46–55.

Whitaker, J.O., Jr., C. Neefus, and T.H. Kunz. 1996. Dietary variation in the Mexican free-tailed bat (*Tadarida brasiliensis mexicana*). Journal of Mammalogy 77:716–724.

White, G.C., and R.A. Garrott. 1990. Analysis of wildlife radio-tracking data. Academic Press, New York.

Wilhide, J.D., M.J. Harvey, V.R. McDaniel, and V.E. Hoffman. 1998a. Highland pond utilization by bats in the Ozark National Forest, Arkansas. Journal of the Arkansas Academy of Science 52:110–112.

Wilhide, J.D., V.R. McDaniel, M.J. Harvey, and D.R. White. 1998b. Telemetric observations of foraging Ozark big-eared bats in Arkansas. Journal of the Arkansas Academy of Science 52:113–116.

Wilkinson, G.S. 1992. Information transfer at evening bat colonies. Animal Behaviour 44:501–518.

Winhold, L., E. Hough, and A. Kurta. 2005. Long-term fidelity of tree-roosting bats to a home area. Bat Research News 46:9–10.

Winkler, H. 1988. An examination of concepts and methods in ecomorphology, pp. 2246–2253, *in* Acta XIX Congress of International Ornithology (H. Ouellet, ed.). National Museum of Natural Sciences, Ottawa, ON.

Woodsworth, G.C. 1981. Spatial partitioning by two species of sympatric bats, *Myotis californicus* and *Myotis leibii*. M.S. thesis, Carleton University, Ottawa, ON.

Worton, B.J. 1989. Kernel methods for estimating the utilization distribution in home-range studies. Ecology 70:164–168.

Yack, J.E. 1988. Seasonal partitioning of atympanate moths in relation to bat activity. Canadian Journal of Zoology 66:753–755.

Zielinski, W.J., and S.T. Gellman. 1999. Bat use of remnant old-growth redwood stands. Conservation Biology 13:160–167.

Zimmerman, G.S., and W.E. Glanz. 2000. Habitat use by bats in eastern Maine. Journal of Wildlife Management 64:1032–1040.

IMPORTANCE OF NIGHT ROOSTS TO THE ECOLOGY OF BATS

5

Patricia C. Ormsbee, James D. Kiser, and Stuart I. Perlmeter

A succinct definition of night roosting is "anytime a bat stops flying at night" (S. Cross, pers. comm.). Discerning the role of this behavior in the lives of bats seems fundamental to understanding the life history of these volant, nocturnal organisms, but the difficulty in studying bats roosting at night, especially in forested areas, has discouraged rigorous studies focused solely on night roosting. One does not have to delve far into the literature on bats to realize that the word "roost" often is an accepted synonym for "day roost," and as a matter of practicality, most information on night roosting has been collected incidental to studies of day roosting in bats (Kunz 1982; Kunz and Lumsden 2003).

Much of the information that does exist on night roosting reflects surveillance of individual roost sites and how and when members of a colony use these sites (Anthony et al. 1981; Catto et al. 1995; Swift 1980). Less common are studies using radiotelemetry that focus on individuals and monitor roosting habits relative to nightly activities and areas of use by bats (Audet 1990; Krull et al. 1991; Murray and Kurta 2004). Most studies of night roosting have focused on reproductively active females, whereas data for males and nonreproductive females, and for bats from seasons other than summer, are rare.

By default, night roosting includes the tending of young at a maternity roost between foraging bouts, nights when bats elect not to emerge from a day roost because of inclement weather, and the period each night when bats return to the day roost before dawn. Although we acknowledge that these activities constitute night roosting in its strictest sense, our coverage of these behaviors in this chapter is limited. Instead, we emphasize night roosting of bats at sites away from day roosts. Our focus is on forest-dwelling bats in North America, i.e., species that day or night roost in trees or forage in woodlands, although we occasionally include pertinent material on species of bats living in temperate areas of other continents. Additionally, we emphasize studies of bats that used radiotelemetry because of the range in behaviors of individual bats that can be discerned using this approach.

In this chapter, we review the functions of night roosts, summarize night-roosting structures and behavior, propose factors that influence night roosting, and describe fidelity to night roosts. We also discuss night roosting and its relationship to the conservation of bats in North America and suggest future research. As Hayes (2003) points out, when we investigate similarities among species in search of underlying patterns, we must remember that each species has a unique natural history and set of behaviors, and decisions by resource managers and conservationists should reflect such uniqueness. This is certainly an admonition worth heeding when evaluating and acting upon information that is available for night roosting.

FUNCTIONS OF NIGHT ROOSTS AND NIGHT ROOSTING

Numerous physiological, ecological, and behavioral functions have been proposed to explain the behavior of night roosting, ranging from digestion of food, avoidance of predators, and information transfer. There is, however, immense variation in night-roosting behavior, both within and among species of bats, which makes it difficult to identify consistent patterns or even to discern the exact function of a specific night-roosting event. Nevertheless, reviewing proposed functions of night roosts and night roosting provides a biological foundation for our discussion and may help formulate future hypotheses.

RESTING

Resting seems the most obvious function of night roosting (Kunz 1982; Wilkinson 1992). Nevertheless, it is difficult to identify, let alone quantify, specific resting behaviors, because rest from flight inevitably occurs whenever a bat begins night roosting. Hence, even though a bat might night roost to socialize or to investigate a potential day roost, it also is simultaneously resting from the rigors of flight.

There are two general forms of rest, one of which is behavioral and the other physiological. Some inactivity (i.e., night roosting) may occur because the animal has nothing better to do and is equivalent to the "laziness" of Herbers (1981). This could be a bat that has satisfied its foraging requirements for the night and retires to a night roost, biding its time until it makes a predawn move to a suitable day roost. In this case, resting would be a simple choice between "continue flying for no particular reason" or "not continue flying."

In contrast, physiological rest is induced by fatigue, which may be caused by a buildup of lactate, inadequate amounts of muscle glucose, or overheating (Hill et al. 2004). Overheating is unlikely to induce fatigue in temperate zone bats, because these animals typically have reduced rectal temperatures when flying in the cool of the night, suggesting peripheral vasoconstriction and a need to conserve, not eliminate, core body heat (Kurta 1986; O'Farrell and Bradley 1977; Thomas 1987). Although lactate increases to high levels whenever an animal engages in prolonged supramaximal exercise (Hill et al. 2004), we assume that bats generally avoid this problem by having a wing shape that minimizes the required power for flight and by flying near their velocities for either minimum power or minimum cost of transport (Norberg 1987; Thomas 1987). Prolonged, high-intensity exercise below supramaximal levels can deplete glucose in muscles, forcing these cells to rely on fuel obtained from the liver. Fatigue can set in should this outside supply of energy not be delivered quickly enough. Although such physiological fatigue is plausible, no study has yet examined circulating levels of glucose or lactate in bats that are foraging, so the role of these physiological processes in inducing fatigue and eventually forcing the animal to rest remains unknown. If night roosting does occur in bats to recover physiologically from bouts of activity, the frequency and duration of rest probably is dictated by several factors including reproductive status of the bat (e.g., flight is presumably more difficult for females in advanced pregnancy; Kunz 1974), physical fitness of individuals, and intraspecific and interspecific differences in physiology and morphology (e.g., types of muscle fibers; Hermanson 1998).

PROMOTING DIGESTION

Many aerial insectivores are capable of consuming large amounts of prey in a short time, occasionally in excess of 20% of their body mass (Anthony and Kunz 1977; Gould 1955; Kunz 1974; Kunz et al. 1995), and some bats may night roost only because they need to make room in their gastrointestinal tracts before foraging again (Barclay 1982). Further, blood flow to the large flight muscles probably is maximized during foraging at the expense of nonlocomotor organs, such as those of the digestive tract (Thomas 1987). Night-roosting bats are likely inactive, and blood flow to the digestive tract, pancreas, and liver presumably is enhanced, speeding digestion, absorption, and processing of food when compared with periods of flight.

RETREATING FROM INCLEMENT WEATHER

Cool or wet weather may influence decisions by bats as to whether they should night roost (Anthony et al. 1981; Brigham 1991; Butchkoski and Hassinger 2002; Clem 1992; Kunz 1973, 1974; Mayrberger and Mayrberger 2002; Rydell 1989). For example, the lesser noctule (*Nyctalus leisleri*) in England never night roosted away from the day roost except during periods of rain (Shiel et al. 1999). Increased night roosting may be more likely if the roost is warmer than ambient conditions, protected from wind and rain, and has a structure that allows clustering by bats to reduce metabolic expenditure, because these attributes minimize the energetic cost of roosting and maximize the energetic difference between foraging and roosting (Barclay 1982; Brigham 1991; Clem 1992; Perlmeter 1995).

RETREATING FROM PREDATORS

Nocturnal predation on bats appears uncommon (Tuttle and Stevenson 1982), and we found no observation of a North American species of bat retreating to a night roost to specifically avoid predators. Certain night-roosting structures, however, may provide protection from predation (Pierson et al. 1996), and bats may avoid a night roost that is occupied by a potential predator (Renison 2003). Although night roosting in colonies may indicate a herd or flock strategy that reduces the risk of predation through increased vigilance (Lingle 2001; Pulliam 1973; Roverud and Chappell 1991), this effect simply makes the night-roosting experience less dangerous, and it is not likely to be the cause of night roosting per se.

FEEDING

Feeding roosts are sites that are used by bats to dismantle and consume food. Although use of feeding roosts is common among frugivorous species in tropical areas (Heithaus and Fleming 1978), feeding roosts are used by some predominantly insectivorous species in North America, especially species that glean large prey as part of their foraging repertoire. A male pallid bat (*Antrozous pallidus*), for example, used a large cherry tree (*Prunus serotinus*) as a feeding perch and culled parts of insects (scarab beetles) were found beneath six other trees within 30 m of the feeding perch (Rambaldini and Brigham 2004). Similarly, pallid bats in Arizona consistently used cliffs and shallow caves as night roosts to consume their prey (O'Shea and Vaughan 1977), and the Virginia big-eared bat (*Corynorhinus townsendii virginianus*) repeatedly used rock shelters as feeding roosts in Kentucky (Lacki et al. 1993). Culled parts of insects, such as moth wings, can be used by biologists as an indication of whether a cave, bridge, or other site is used as a feeding roost (Lewis 1994).

SCOUTING

Many forest-dwelling bats use several different trees for day roosting in a single season (Barclay and Kurta, chap. 2 in this volume; Carter and Menzel, chap. 3 in this volume), but it is largely unknown how and when a bat decides that an individual tree is suitable as a day roost. Murray and Kurta (2004) recently noted that both day roosts and night roosts of the Indiana myotis (*Myotis sodalis*) were located in trees within the foraging areas used by the colony. These authors speculated that one function of brief periods (10–30 min) of night roosting by Indiana myotis was possibly to evaluate a particular tree as a future day roost. Likewise, R. M. R. Barclay (pers. comm.) noted that female hoary bats (*Lasiurus cinereus*) occasionally night roosted briefly in an individual tree when commuting between foraging areas and maternity sites, and a day or two later the mother would move her pups to that tree for day roosting. In Germany, Bechstein's bats (*Myotis bechsteini*) briefly night roosted in newly erected bat boxes days or weeks before the boxes actually were used as day roosts (Kerth and Reckardt 2003). Forest-dwelling bats regularly switch day roosts and scouting for future diurnal roosts may be a common nocturnal behavior.

INFORMATION TRANSFER

Although some bats night roost as individuals (Murray and Kurta 2004), communal night roosting is very common and the presence of multiple individuals creates the possibility of night roosts acting as "information centers," where knowledge is conveyed from one bat to another. Indeed, information transfer often is proposed as a factor leading to the evolution of communal behavior in areas where resources (e.g., food and roosts) are patchily and unpredictably distributed (Richner and Heeb

1996), and there is some evidence of this phenomenon in temperate zone bats. For example, Wilkinson (1992; but see Kerth et al. 2001) demonstrated the existence of information transfer in evening bats (*Nycticeius humeralis*) in Missouri by showing that individuals followed each other to roosting and feeding sites. Similarly, Kerth and Reckardt (2003) showed that female Bechstein's bats exchanged information about new roosts (bat boxes). Although these studies did not specifically identify a role for night roosts in information transfer, there is no reason to believe that information transfer occurs only in day roosts.

SWARMING AND MATING

Swarming is a complex nocturnal activity, primarily involving species of *Myotis* and *Pipistrellus* in North America, that takes place at mines or caves in late summer and early fall (Davis and Hitchcock 1965; Fenton 1969; Hall and Brenner 1968; Humphrey and Cope 1976; Zembal and Gall 1980). At this time, few bats roost deep in a cave or mine during the day, but up to thousands may visit the entrance of the site at night, with the exact number of bats varying greatly depending on the specific locality and time of year. Some bats just fly through the underground roost site (Taylor 2002; Thomas et al. 1979), whereas many bats roost for at least short periods. No study of swarming in bats has yet to monitor the roosting behavior of radiotagged individuals.

There are two proposed functions of swarming, the first of which is speculative and the second of which is known. First, swarming may be a way of introducing young-of-the-year to potential hibernation sites. Second, mating occurs during swarming, in particular, in late August and September (Fenton 1969). Caves or mines that are used by only a small number of bats as night roosts for resting, feeding, or protection in early summer, may become sites used for mating in late summer or hibernation in winter. This shift in function of a particular roost is evidenced by an increase in number of bats that visit each night, an increase in the ratio of males to females over time, and the buildup of a torpid population of bats during the day (Fenton 1969; Thomas et al. 1979).

Adult males often are present at other types of night-roosting structures, such as buildings or bridges, along with adult females (Butchkoski and Hassinger 2002; Perlmeter 1996). Although presence of both sexes at night roosts during most of the year probably is a passive response to desirable roosting conditions, males that roost with females in late summer could increase their potential for reproductive success if females become sexually receptive soon after weaning their pups (Perlmeter 1996; Shiel et al. 1999). Regardless, for many species of North American bats, biologists still have little knowledge of how or when mating activity takes place, and we are unaware of any observation of copulation involving bats at night roosts, other than swarming sites.

OTHER SOCIAL INTERACTIONS

Many authors speculate that night roosting has a social function, but most reports are restricted to simple descriptions of apparent social interactions or group behaviors (Lewis 1994; Renison 2003). For instance, the pattern of departures and arrivals at night roosts by long-legged myotis (*Myotis volans*) and little brown myotis (*M. lucifugus*) seem to be coordinated events that indicate social cohesion (Perlmeter 1995), and flying pallid bats apparently locate night-roosting clusters by vocally communicating with bats already in the roost (O'Shea and Vaughan 1977). Night roosts may serve as focal points for social interaction and group cohesion of bats, as do day roosts, especially when night-roosting sites are predictable in space and time (Lewis 1994, 1995; Perlmeter 1995; Willis and Brigham 2004). Studies are needed, however, that provide field tests of specific hypotheses relating to the social functions of night roosting by bats.

CONSERVATION OF ENERGY

Bats may cease flying and begin night roosting specifically to save energy (e.g., when foraging is unproductive), or the proximal reason for night roosting may not be directly related to energy conservation at all (e.g., social interactions). Nevertheless, maintaining a normothermic body temperature can be costly for bats because they have a large surface-to-mass ratio that results in high mass-specific thermal conductance and the potential for rapid heat loss (McNab 1982). Consequently, if bats spend prolonged periods roosting at night (table 5.1), it is reasonable to assume that they select a roost site or roosting strategies that minimize energetic expenditures while in the night roost.

Although use of torpor is a valuable strategy for energy conservation, an array of physiological processes, including digestion, fetal and juvenile growth, milk production, and sperm production, are enhanced at normal body temperatures through Q_{10} effects and optimization of enzyme activity (Humphries et al. 2003; Kurta and Kunz 1988; Racey 1973; Wilde et al. 1999). Consequently, many bats, especially reproductive females and young, favor minimizing the energetic expense of thermoregulation

Table 5.1. Duration of individual bouts of night roosting, as determined with radiotelemetry, for bats in North America

Species	Duration (min)	Type of structure	Location	Source
Antrozous pallidus	180–480	cave	Arizona	O'Shea and Vaughan 1977
Corynorhinus townsendii	90–120	building, bridge	West Virginia	Stihler 1994
Euderma maculatum	0		British Columbia	Wai-Ping and Fenton 1989
Euderma maculatum	120–360	tree	Arizona	Rabe et al. 1998
Euderma maculatum	60–180	unknown in forest	Arizona	C. Chambers, pers. comm.
Macrotus californicus	30–360	tree	Arizona	Dalton 2001
Myotis evotis	<20	bridge	Oregon	Waldien and Hayes 2001
Myotis septentrionalis	60	tree	South Dakota	Swier 2003
Myotis sodalis	10–30	tree	Michigan	Murray and Kurta 2004
Myotis sodalis	10–15	building, tree	Pennsylvania	Butchkoski and Hassinger 2002
Myotis yumanensis	<10	undetermined	Washington	G. Falxa, pers. comm.

by behavioral rather than physiological means (torpor). Strategies for conserving energy at night roosts are similar to those used by bats during the day and include clustering, roosting in sites that entrap air warmed by the body, and use of structures with higher than ambient temperatures (Anthony et al. 1981; Barclay 1989; Burnett and August 1981; Herreid 1967; Kunz 1974; Kurta 1985; Kurta et al. 1989; Perlmeter 1996; Pierson et al. 1996).

NIGHT-ROOSTING STRUCTURES

Bats night roost in an array of man-made and natural structures, and there is diversity both within and between species in use of night roosts (table 5.2). Night-roosting sites used by bats include relatively permanent structures such as buildings, bridges, caves, and mines, as well as ephemeral sites such as tree cavities and foliage (Kunz 1982). Despite the prevalence of ephemeral roosting structures in forested landscapes of North America, night roosting by bats in relatively permanent sites has been studied more thoroughly for two reasons. Permanent structures used for night roosting by bats often are found more easily by researchers because they are more accessible to humans, and these structures can accommodate a larger number of bats than are usually supported by roosts comprised of cracks and crevices in dead trees (snags) or spaces behind bark.

Table 5.2. Night-roosting structures reported for bats in North America

Species	Structures	Sources
Antrozous pallidus	bridge, building, cave, rock crevice, rock overhang, tree	Chapman et al. 1994; Hermanson and O'Shea 1983; Lewis 1994; O'Shea and Vaughan 1977; Rambaldini and Brigham 2004; Renison 2003
Corynorhinus rafinesquii	bridge, building, rock shelter, tree	Clark et al. 1998; M. Gumbert, pers. comm.; J. McGregor, J. Kiser, and L. Meade, pers. comm.; A. Trousdale, pers. comm.
Corynorhinus townsendii townsendii	bridge, building, cave, mine, unknown in forest stand	Cross 1998; Cross et al. 1998; Fellers and Pierson 2002; Perlmeter 1996; Pierson et al. 1996; Pierson et al. 1999
Corynorhinus townsendii virginianus	building, rock shelter, tree	Adam et al. 1994; Lacki et al. 1993; Stihler 1994, 1995; J. McGregor, J. Kiser and L. Meade, pers. comm.
Eptesicus fuscus	bridge, building, cave, mine, rock crevice, rock shelter, tree	Adam and Hayes 2000; Barclay 1982; Brigham 1991; Cross 1977; Cross and Waldien 1995; Kiser et al. 2002; J. MacGregor and H. Bryan, pers. comm.; L. Meade, pers. comm.; Nagorsen and Brigham 1993; Perlmeter 1996; Pierson et al. 1996; Renison 2003; Swier 2003
Euderma maculatum	tree	Rabe et al. 1998; Siders and Steffensen 1998
Lasionycteris noctivigans	bridge, tree	P. Ormsbee, unpubl. data; D. Waldien, pers. comm.

(*continued*)

Table 5.2. Continued

Species	Structures	Sources
Lasiurus blossevillii	unknown in forest stand	Pierson et al. 2002
Lasiurus borealis	unknown in forest stand	S. Amelon, pers. comm.
Lasiurus cinereus	tree	Barclay 1989
Macrotus californicus	building, cave, mine, rock feature, tree	Bell et al. 1986; Dalton 2001
Myotis austroriparius	cistern	Sherman 2004
Myotis californicus	bridge, cave, mine	Adam and Hayes 2000; Albright 1959; Cross and Waldien 1995; Perlmeter 1996
Myotis evotis	bridge, building, cave, mine, tree	Adam and Hayes 2000; Albright 1959; Nagorsen and Brigham 1993; Perlmeter 1996; Pierson et al. 1996; Purdy 2002; Tigner and Aney 1994; Waldien and Hayes 2001
Myotis grisescens	bridge	S. Amelon, pers. comm.; Johnson et al. 2002
Myotis leibii	bridge, building, cave, rail tunnel, rock shelter	J. Adams, pers. comm.; Brack et al. 2006 ; Davis et al. 1965; M. Gumbert, pers. comm.; J. Kiser and J. Mac-Gregor, pers. comm.
Myotis lucifugus	bridge, building, cave, mine, tree	Adam and Hayes 2000; Albright 1959; Anthony et al. 1981; Barclay 1982; Fenton and Barclay 1980; Pearl and Fenton 1996; Perlmeter 1996; Pierson et al. 1996; Swier 2003; Thomas et al. 1979; Zembal and Gall 1980
Myotis septentrionalis	bridge, building, cave, mine, tree	S. Amelon, pers. comm.; Kiser et al. 2002; J. Kiser and M. Gumbert, unpubl. data; Sparks et al. 2000; Swier 2003; Tigner 2001
Myotis sodalis	bat box, bridge, building, cave, tree	Butchkoski and Hassinger 2002; Kiser et al. 2002; J. Kiser and M. Gumbert, unpubl.data; Murray and Kurta 2004
Myotis thysanodes	bridge, building, cave, mine, rock crevice, tree	Adam and Hayes 2000; Albright 1959; Nagorsen and Brigham 1993; O'Farrell and Studier 1980; Purdy 2002
Myotis volans	bridge, cave, mine, tree	Adam and Hayes 2000; Albright 1959; Nagorsen and Brigham 1993; Ormsbee 1996; Perlmeter 1996; Pierson et al. 1996; Zembal and Gall 1980
Myotis yumanensis	bat box, bridge, building, cave, mine, tree	Adam and Hayes 2000; Albright 1959; Betts 1997; Falxa 2004; Nagorsen and Brigham 1993; Perlmeter 1996; Pierson et al. 1996
Nycticeius humeralis	building	Clem 1992
Pipistrellus subflavus	bridge	Kiser et al. 2002

BUILDINGS

Human interactions with bats roosting in buildings generally happen more frequently than at other sites, and consequently, reports of bats night roosting in buildings are common (table 5.1). Some species, such as Yuma myotis (*Myotis yumanensis*) and big brown bat (*Eptesicus fuscus*)

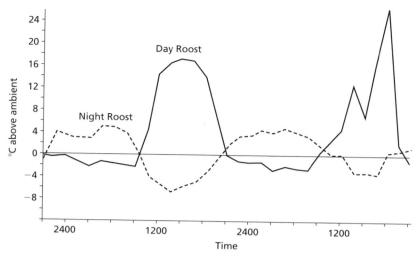

Figure 5.1.
Temperature differences of day- and night-roosting locations, compared with ambient temperatures, measured in and adjacent to a building used by little brown myotis (*Myotis lucifugus*) in Ontario (Barclay 1982).
From Journal of Mammalogy, Alliance Communications Group, Allen Press, Inc.

actually occur more frequently in buildings than in natural roosts (Kunz and Reynolds 2003). Garages, barns, and other outbuildings are frequently used by bats for night roosting, and within houses, bats commonly night roost in attics, under porch awnings, behind chimneys, and under or inside eaves (Kunz 1982).

Although buildings may serve as replacements for natural roosts in human-dominated landscapes (Brigham 1991; Kunz and Reynolds 2003), these structures often are used even in remote forested regions, and in many cases, buildings simply may provide the warmest, safest option for night roosting. Buildings located in forests are typically stable over time, often inaccessible to predators, and frequently uninhabited by humans at night. Additionally, roosting sites within buildings typically provide a range of temperatures, protection from predators and the elements, and roosting habitat that structurally mimics natural roosts (i.e., crevices and cavities).

The same building occasionally functions as both a day and a night roost, with bats using separate locations in the structure during different times, presumably due to variation in thermal characteristics (Anthony et al. 1981; Barclay 1982). Daytime temperatures taken at locations used for day roosting by female little brown myotis in southeastern Ontario were warmer than daytime temperatures of sites in the same building that were used for night roosting (Barclay 1982). This pattern was reversed at night, so that nighttime temperatures in locations used for night roosting were warmer than nighttime temperatures of day-roosting sites (Barclay 1982) (fig. 5.1). Further, the mortises of a barn were as much as 6°C warmer than nighttime ambient temperatures in Ontario, even when bats were not present (Barclay 1982), and entrapment of body heat resulted in occupied roost temperatures that were up to 20°C warmer than ambient at a similar night-roosting site in New Hampshire (Anthony et al. 1981). Use of different locations within the same building for day and night roosting may be analogous to bats roosting in different trees in different parts of a forest under different weather conditions (Willis and Brigham 2004).

Many forest-dwelling bats night roost under bridges (15 of 23 species; table 5.1), even in forested habitat, and up to six species of bats have been found under the same bridge (Pierson et al. 1996). The widespread use of bridges may reflect the high suitability of these structures as night roosts. However, it also may reflect an abundance of surveys directed specifically at investigating these potential roosts, because bridges often are more accessible, more convenient to inspect for evidence of night roosting (e.g., culled insect parts or feces), and less numerous across the landscape than are snags, hollow trees, or buildings. Like buildings, though, bridges are stable, protected sites that provide a range of potential roosting temperatures and are uninhabited by humans at night. Additionally, bridges are conveniently located over the waterways that bats use for drinking, commuting, and foraging; thus, accessing these sites may require little time and energy (Adam and Hayes 2000; Keeley and Tuttle 1999; Perlmeter 1996; Pierson et al. 1996; Renison 2003).

Certain types of bridges in specific locations seem preferred as night roosts by bats (fig. 5.2). Bridges used as night roosts in North America most often are made of concrete and have concave chambers under the bridges in which the bats roost (Adam and Hayes 2000; Keeley and Tuttle 1999; Kiser et al. 2002; Perlmeter 1995, 1996; Pierson et al. 1996; Renison 2003). During the day, the concrete absorbs solar radiation, and at night, the bridge cools more slowly than the surrounding air. This thermal inertia results in air temperatures in chambers beneath the bridge that are higher (often by 10°C or more) and more stable than external ambient conditions (Kiser et al. 2002; Pierson et al. 1996). Large bridges tend to have higher temperatures and more night-roosting activity than small bridges (Adam and Hayes 2000; Perlmeter 1996; Renison 2003), and bridges that are not shaded by nearby trees also may be warmer and yield more bat activity (Kiser et al. 2002). Other variables such as elevation, topography, and orientation of the bridge may influence temperature of specific roosting sites (Perlmeter 1996; Renison 2003). Even though structural and thermal characteristics are important, a large amount of vehicular traffic, such as that carried by interstate highways, may decrease the suitability of a bridge as a night roost (Pierson et al. 1996).

Within a particular bridge, the greatest number of night-roosting bats usually is found in the warmest chambers (Perlmeter 1996), which often occur at either end of the bridge and are located over land. Sloping riverbanks or use of fill to stabilize bridge supports often result in end chambers being closer to the ground than center chambers, and occupied chambers occasionally are less than 2 m above ground (P. Ormsbee, unpubl. data; Renison 2003). Central chambers that are located over water are used less frequently by bats and have lower temperatures, presumably because the space over the water is more open than along the shore, resulting in greater exposure to air currents and an increase in convective heat loss (Adam and Hayes 2000; Perlmeter 1996; Renison 2003).

TREES

Many, perhaps most, forest-dwelling bats use living or dead trees for day roosting (Barclay and Kurta, chap. 2 in this volume; Carter and Menzel, chap. 3 in this volume), but documentation of these structures as specific night roosts is less common and often indirect (table 5.2). For example, an Indiana myotis left a transmitter overnight in a tree that previously had not been documented as a day roost (Murray and Kurta 2004). Indiana myotis, eastern red bats (*Lasiurus borealis*), and western red bats (*L. blossevillii*) have been radiotracked at night to forested areas where trees were the most logical roosting site, but the exact location of the resting bat could not be determined (S. Amelon, pers. comm.; Butchkoski and Hassinger 2002; Murray and Kurta 2004; Pierson et al. 2002). Purdy (2002) collected guano on a daily basis from basal hollows of redwood trees (*Sequoia sempervirens*), and during attempts to mist net bats that produced the feces, she generally caught bats approaching from the outside, presumably to use the hollows for night roosting. Individuals from a few species, however, have been radiotracked to night roosts in trees, including little brown myotis and big brown bats (Brigham 1991; Swier 2003). Even species that normally day roost only in rock crevices or shallow caves, such as pallid bats, occasionally use trees at night (Rambaldini and Brigham 2004).

The true importance of snags and hollow trees to forest-dwelling bats

Figure 5.2.
Examples of bridges used as night roosts by various species of Myotis bats in Oregon (a), Michigan (b), and Indiana (c).
Photos by P. Ormsbee, S. Sherburne, and J. Kiser, respectively

as night roosts, distinct from maternity sites, is not known. Because snags and hollow trees are used by forest-dwelling bats for day roosting, we assume that they also use these structures for night roosting, at least more often than has been documented. This may especially be true for areas where human-made roosts, such as buildings and bridges, are absent. Trees are not as massive as buildings or bridges, but cavities in trees can provide sheltered roosting sites that are a few degrees warmer at night than ambient air (Kalkounis and Brigham 1998; Willis 2003). Although trees lack the permanence of human-made structures, snags and hollow trees are integral components of forests, and one of their values as night roosts may be their large quantity and broad distribution across forested ecosystems (O'Donnell and Sedgeley 1999; Willis 2003).

ROCK CREVICES, CAVES, AND MINES

Bats use an array of rock features for night roosting, including overhangs, shallow cracks, crevices, shelter caves, and deep caves (table 5.2, fig. 5.3). Besides providing protection from predators, wind, and rain, shallow cracks and crevices, like most night-roosting structures, potentially provide a nocturnal environment that is warmer than the surrounding air, because surface rock that is exposed to solar radiation warms during the day and cools more slowly than the external air at night (Lausen and Barclay 2003). Caves and mines, in contrast, often maintain relatively constant temperatures throughout the 24-hour day. In some areas, such as Arizona, this temperature can be quite warm (O'Shea and Vaughan 1977), and such underground retreats provide a suitable night-roosting environment, especially if bats can cluster in crevices or solution cavities that entrap warm air. In northern regions or at high altitudes, however, temperatures inside deep caves or mines often are well below nighttime ambient temperatures in midsummer, consequently few bats, especially reproductive females, night roost there (Betts 1997).

Rocky habitats other than caves or mines are probably used by bats for night roosting more often than is reported. They often are difficult to assess for night-roosting activity and can be widely, yet discretely distributed throughout forested landscapes. Rocky locations such as cliffs or lavabeds can be accessed easily by flying bats but not by humans, making it difficult to study use of these features as night roosts.

Figure 5.3.
A rock shelter in eastern Kentucky used as a night roost by various species of bats.
Photo by J. Kiser

Any structure that provides cavernous or crevicelike habitat is warm at night, is somewhat protected from human activity and predation, and occurs in forested habitat has the potential to be used as a night roost by bats. These structures may be used occasionally or consistently and may shelter large numbers of individuals (Kunz 1982; Kunz and Lumsden 2003). For example, colonies containing hundreds of southeastern myotis (*Myotis austroriparius*) night roost in abandoned cisterns in Mississippi (Sherman 2004). Railroad tunnels, culverts, bat or bird boxes, picnic shelters, abandoned wells, and deserted military bunkers are other types of sites that bats use for night roosting (Butchkoski and Hassinger 2002; Keeley and Tuttle 1999; Kunz 1982; Sherman 2004; Swier 2003).

NUMBER OF ROOSTS AND TEMPORAL ASPECTS OF USE
NUMBER OF NIGHT ROOSTS

Members of a day-roosting colony of bats may use several different night roosts in the same season, and individual bats may switch among various roosts even in a single night (Butchkoski and Hassinger 2002; Lewis 1995; Perlmeter 1995; Swier 2003). For example, female little brown myotis in South Dakota night roosted in trees and in a picnic shelter (Swier 2003), and individual Indiana myotis in Michigan used up to three different trees as night roosts in the same night (Murray and Kurta 2004). One factor that may affect the number of night roosts used by bats and the frequency of switching by individuals of gregarious species may be whether a centralized, spacious structure is available for use as a communal night roost. We predict that bats that night roost in bridges or buildings will use alternate sites less often than bats that night roost in trees (Lewis 1995), but more data are needed to test this hypothesis.

FREQUENCY AND DURATION OF BOUTS OF NIGHT-ROOSTING

Few studies have documented the frequency and duration of night roosting by North American bats because of the difficulty in observing individuals (table 5.1). Nevertheless, radiotracking studies indicate that individuals of most species experience multiple bouts of night roosting each night, with total duration of night roosting usually summing to less than two hours per bat per night. For example, Indiana myotis in Pennsylvania and Michigan performed only two to three short bouts (10–30 min each) of night roosting each night (Butchkoski and Hassinger 2002; Murray and Kurta 2004). Similarly, Yuma myotis in western Washington were active throughout most of the night, roosting only for brief periods (20–30 sec) when foraging (Falxa 2004), and little brown myotis in Quebec night roosted for 1.7 and 1.3 hours during pregnancy and lactation, respectively (Henry et al. 2002). Temperate, insectivorous species of bats on other continents also spend most of the night in flight and use multiple, brief periods of night roosting (Audet 1990; Catto et al. 1996; Krull et al. 1991; O'Donnell 2002).

Variation in length of night roosting, however, occurs within and between species of bats. Spotted bats (*Euderma maculatum*) in British Columbia, for example, foraged continuously for 5–6 hours and never night roosted before returning to the day roost for a prolonged period before dawn (Wai-Ping and Fenton 1989). In contrast, spotted bats in Arizona night roosted for 1–3 hours (C. Chambers, pers. comm.; Rabe et al.1998). Although pregnant little brown myotis night roosted for about half the night in Ontario (Barclay 1982), this same species spent only 1.3 hours resting per night in Quebec (Henry et al. 2002).

INTERANNUAL FIDELITY

There are a number of examples of North American bats displaying multiyear fidelity to night roosts, especially bridges. Banded pallid bats return to night roosts during and between years (Lewis 1994). In California, long-legged myotis, big brown bats, and Yuma myotis have been recaptured while night roosting under the same bridge 2, 3, and 4 years, respectively, after banding (Pierson et al. 1996). Individual long-eared myotis (*Myotis evotis*), long-legged myotis, and fringed myotis (*M. thysanodes*), originally banded in 1958 and 1959 by Albright (1959), have been recaptured multiple times over 18–24 years at Oregon Caves National Monument in Oregon (S. Cross, pers. comm.). Long-legged myotis and little brown myotis that were banded from 1991 to 2002 at night roosts under bridges in Oregon were recaptured 6–10 years later at the same bridges (P. Ormsbee, unpubl. data; Perlmeter 1999).

VARIATION IN NIGHT-ROOSTING BEHAVIOR
GENDER DIFFERENCES

Males are not constrained by the physical demands of pregnancy, lactation, or pup rearing during summer and apparently display more flexibility when night roosting than reproductive females in terms of the use of torpor, occupancy of cooler sites, and tendency to cluster (Perlmeter 1995). In Oregon, long-legged myotis and little brown myotis night roosted under bridges during summer and, although both sexes were present, males were fewer in number (<15%) and, in general, roosted in cooler portions of bridges than did females. Bats that roosted in clusters were overwhelmingly female (99%), whereas solitary animals were equally likely to be male or female (Perlmeter 1996). Male Indiana myotis in south-central Indiana also tended to roost alone more often than females that occupied the same bridges (Kiser et al. 2002).

DIFFERENCES AMONG FEMALES

The number of bats using a night roost often is highly variable from night to night, but some large-scale patterns of use appear related to the reproductive condition of females. For example, deposition of feces at night roosts of the little brown myotis in Ontario and New Hampshire was lower during lactation than pregnancy, suggesting decreased use of night roosts by females with dependent young (Anthony et al. 1981; Barclay 1982). In addition, radiotracking studies in Quebec showed that the number of night-roosting bouts away from the maternity site by little brown myotis during lactation was less than during pregnancy, even though total time spent flying, or, conversely, cumulative time spent inactive in the day roost and night roosts, did not differ between pregnant and lactating individuals (Henry et al. 2002). Similarity in time of flight for pregnant and lactating females occurs in other temperate zone species of bats (Audet 1990; Murray and Kurta 2004) and suggests a reorganization of the time spent inactive by lactating females rather than an increase in total duration of roosting, likely due to the demand of caring for pups (Henry et al. 2002).

Other differences in night-roosting patterns may be associated with reproductive condition, but the effect of environmental conditions, such as weather or insect abundance, cannot be ruled out. In New Hampshire, the night-roosting activity of bats peaked after pups were weaned (Anthony et al. 1981), but the largest population of night-roosting individuals in Ontario was observed during pregnancy (Barclay 1982). Similarly, counts of night-roosting bats indicated that the greatest number of little brown myotis and long-legged myotis under bridges in Oregon occurred during late pregnancy (Perlmeter 1996). Such observations clearly indicate changes in the pattern of night roosting by the populations that were studied, but without data from individuals, it is impossible to know whether these differences represent changes in duration, frequency, or location of night roosting.

EFFECTS DUE TO TYPE OF ROOSTING STRUCTURE

The type of structure available for night roosting may influence behavior and help explain some interspecific and intraspecific variation in roosting patterns. If permanent structures with spacious roost sites are available, many bats may night roost in them, forming large colonies. However, when small cracks or crevices of less-enduring structures, such as those in snags, are the only types of night roosts available, bats may be more likely to roost in small groups and perhaps frequently move between sites (Lewis 1995; Willis and Brigham 2004).

Although this hypothesis has not been rigorously tested, there is some supporting evidence from studies on different populations within the same species. For example, at a site in Indiana, Indiana myotis night roosted in clusters under bridges (Kiser et al. 2002), but in Michigan, members of this species consistently night roosted solitarily in trees (Murray and Kurta 2004). There was no bridge, abandoned building, or rocky cliff within the home range of the bats in Michigan (Murray and Kurta 2004), and perhaps lack of a suitable, centralized roosting site explains the difference in roosting behavior between the two studies.

EFFECTS OF WEATHER AND CLIMATE

Cold or wet conditions can affect the pattern of night-roosting behavior that is displayed by individuals or populations of bats. Some bats only visit the day roost at night during rain (Murray and Kurta 2004) and, as mentioned earlier, some species only occupy night roosts away from the day roost during inclement weather (Shiel et al. 1999). In Ontario, the amount of night roosting behavior by little brown myotis increased as ambient temperatures fell below 12°C (Barclay 1982). Similarly, Indiana myotis in Pennsylvania apparently night roosted for longer periods and foraged less during periods of rain, fog, and low temperatures (Butchkoski and Hassinger 2002). We assume that these effects of weather on night roosting in bats are directly related to avoidance of getting cold or wet, or indirectly related to the negative effect of inclement weather on insect abundance (Anthony et al. 1981) or the ability of bats to echolocate.

Factors that vary broadly with latitude might affect night-roosting patterns across large geographic areas. Shorter nights in northern latitudes may force individuals to maximize the amount of time spent foraging and reduce or forgo night roosting to meet daily energetic demands (O'Donnell 2002). Conversely, in areas like the southwestern United States, where climates are mild, nights are long, and foraging seasons are extended, some bats can afford to night roost for longer periods as reported for pallid bats (O'Shea and Vaughan 1977) and spotted bats (Rabe et al. 1998; table 5.1). Further, we hypothesize that bats at more southerly latitudes have more flexibility in choice of night-roosting structures because these bats do not experience the cool nighttime temperatures of northern climates, permitting use of a wider range of sites to meet their energetic requirements while roosting.

DISTANCE TO FORAGING SITES

As commuting distances between day roosts and foraging sites increase, it may be more energetically conservative for bats to night roost in the vicinity of a foraging area instead of returning to the day roost (Kunz 1982; Lacki et al., chap. 4 in this volume). Spotted bats in Arizona commuted 20–38 km from their day roosts to forage in meadows and night roost in adjacent forest stands (C. Chambers, pers. comm.; Rabe et al. 1998), and gray bats (*Myotis grisescens*) foraged along waterways and night roosted under bridges that were up to 40 km from the caves where they day roosted (Johnson et al. 2002). Pregnant females and females with pups that are somewhat independent often commute longer distances than females during lactation, and also occasionally night roost separately from maternity roosts (Adam et al. 1994; Butchkoski and Hassinger 2002; Kunz 1973; Perlmeter 1995). In one study, lactating little brown myotis restricted their nocturnal activity to within 600 m of their maternity roost and used this roost exclusively at night; however, during pregnancy, they traveled over 2 km from the maternity site and night roosted in alternate sites (Henry et al. 2002). Similarly, Brazilian free-tailed bats (*Tadarida brasiliensis*) from small colonies often return to the day roost for night roosting, whereas those that live in large colonies seldom return before dawn; presumably members of large colonies forage farther from home to lessen competition with roost mates (Kunz 1982).

CONSERVATION AND RESEARCH

Most North American species of bats are associated with forests, and understanding the purpose of night-roosting structures and their relationship to foraging habitat and day roosts is necessary to address fully the dynamics and complexities associated with summer-use areas. Research on the use of natural structures for night roosting, such as snags and hollow trees, is essential for creating forest-management practices that conserve appropriate habitat for bat species under natural conditions. Knowledge of inter- and intraspecific differences in nightly activity, daily and seasonal patterns of occupancy of night-roosting structures, thermal conditions within night roosts, and the effect of night roosts on fitness of individual bats or colonies is needed for proper management and to guide attempts at mitigation in response to the inevitable development of natural areas. To obtain the necessary knowledge, field techniques must evolve beyond the collection of data incidental to day-roosting studies or detailed observations of bats roosting in human-made structures.

Neglecting to include night-roosting structures in conservation strategies could have deleterious effects on local populations. Lack of appropriate night roosts could increase exposure to predation or inclement weather, increase energetic demands, disrupt social bonds, or displace bats to areas with increased competition for food and roosts. Forest managers must be cautious so that removal of dilapidated buildings, filling of neglected cisterns, and closure of abandoned mines are not treated as perfunctory tasks, but considered new opportunities to assess night-roosting activity and protect resources that are essential for local populations of bats. If an assessment of night-roosting activity is too costly or impractical, it may be prudent to assume that night roosting is occurring and to protect the structure rather than eliminate a potentially critical resource.

During forest-management activities, resource personnel should consider the night-roosting needs of all species within the local community and remember that trees probably provide night roosts for many kinds of bats. Consequently, decisions regarding retention of large green and hollow trees and creation of snags should consider their potential use as night roosts, as well as day roosts. In addition, managers should remember that fidelity to specific night-roosting areas and structures occurs and that existing dead or hollow trees may not be adequately replaced by artificial structures, such as bat houses, or even human-made snags or tree hollows. The art of creating snags and hollow trees is not well refined, and fashioning such structures may not be an equitable exchange for retention of existing trees that are naturally suitable as roosts. For example, some aspects of hollow trees can be created using a chainsaw, but biologists do not yet know how to produce a hollow tree that reflects real ecological processes, such as heartwood rotting from species-specific fungi (Parks et al. 1997). These natural processes may be fundamental to developing roost-site characteristics essential to bats, even if they are unknown to biologists.

ACKNOWLEDGMENTS

We thank all who provided material for this chapter, in particular, R. Barclay, M. Brigham, C. Chambers, S. Cross, D. Dalton, G. Falxa, T. Kunz, J.

MacGregor, M. Painter, D. Purdy, and D. Sparks. We thank L. Templeman and A. Hart for assistance in preparing this chapter and S. Weber for editorial assistance. We extend special thanks to A. Kurta, T. Kunz, T. Weller, and an anonymous reviewer for helpful edits and advice.

LITERATURE CITED

Adam, M.D., and J.P. Hayes. 2000. Use of bridges as night roosts by bats in the Oregon Coast Range. Journal of Mammalogy 81:402–407.

Adam, M.D., M.J. Lacki, and T.G. Barnes. 1994. Foraging areas and habitat use of the Virginia big-eared bat in Kentucky. Journal of Wildlife Management 58:462–469.

Albright, R. 1959. Bat banding at Oregon Caves. Murrelet 40:26–27.

Anthony, E.L.P., and T.H. Kunz. 1977. Feeding strategies of the little brown bat, *Myotis lucifugus,* in southern New Hampshire. Ecology 58:775–786.

Anthony, E.L.P., M.H. Stack, and T.H. Kunz. 1981. Night roosting and the nocturnal time budget of the little brown bat, *Myotis lucifugus:* effects of reproductive status, prey density, and environmental conditions. Oecologia 51:151–156.

Audet, D. 1990. Foraging behavior and habitat use by a gleaning bat, *Myotis myotis* (Chiroptera: Vespertilionidae). Journal of Mammalogy 71:420–427.

Barclay, R.M.R. 1982. Night roosting behavior of the little brown bat, *Myotis lucifugus.* Journal of Mammalogy 63:464–474.

———. 1989. The effect of reproductive condition on the foraging behavior of female hoary bats, *Lasiurus cinereus.* Behavioral Ecology and Sociobiology 24:31–37.

Bell, G.P., G.A. Bartholomew, and K.A. Nagy. 1986. The roles of energetics, water economy, foraging behavior and geothermal refugia in the distribution of the bat, *Macrotus californicus.* Journal of Comparative Physiology B 156:441–450.

Betts, B.J. 1997. Microclimate in Hell's Canyon mines used by maternity colonies of *Myotis yumanensis.* Journal of Mammalogy 78:1240–1250.

Brack, V., Jr., J.O. Whitaker Jr., C.W. Stihler, and J.D. Kiser. 2006. Summer ecology of eastern small-footed myotis, *Myotis leibii,* in Virginia, Kentucky, and West Virginia. Northeastern Naturalist (in press).

Brigham, R.M. 1991. Flexibility in foraging and roosting behaviour by the big brown bat (*Eptesicus fuscus*). Canadian Journal of Zoology 69:117–121.

Burnett, C.D., and P.V. August. 1981. Time and energy budgets for dayroosting in a maternity colony of *Myotis lucifigus.* Journal of Mammalogy 62:758–766.

Butchkoski, C.M., and J.D. Hassinger. 2002. Ecology of a maternity colony roosting in a building, pp. 130–142, *in* The Indiana bat: biology and management of an endangered species (A. Kurta and J. Kennedy, eds.). Bat Conservation International, Austin, TX.

Catto, C.M.C., A.M. Hutson, P.A. Racey, and P.J. Stephenson. 1996. Foraging behaviour and habitat use of the serotine bat (*Eptesicus serotinus*) in southern England. Journal of Zoology (London) 238:623–633.

Catto, C.M.C., P.A. Racey, and P.J. Stephenson. 1995. Activity patterns of the serotine bat (*Eptesicus serotinus*) at a roost in southern England. Journal of Zoology (London) 235: 635–644.

Chapman, K., K. McGuiness, and R.M. Brigham. 1994. Status of the pallid bat in British Columbia. Province of British Columbia, Ministry of Environment, Lands, and Parks, Wildlife Branch, Wildlife Working Report No. WR-61.

Clark, M.K., A. Black, and M. Kiser. 1998. Roosting and foraging activities of *Corynorhinus rafinesquii* and *Myotis austroriparius* in the Francis Beidler Forest in South Carolina. Bat Research News 39:162–163.

Clem, P.D. 1992. Seasonal population variation and emergence patterns in the evening bat, *Nycticeius humeralis,* at a west-central Indiana colony. Proceedings of the Indiana Academy of Sciences 101:33–44.

Cross, S.P. 1977. A survey of the bats of Oregon Caves National Monument. Un-published report. U.S. National Park Service, Oregon Caves National Monument, Cave Junction, OR.

———. 1998. Assessment of historic mining sites in Burns District BLM for bat use potential. Unpublished report. U.S. Bureau of Land Management, Burns District, Burns, OR.

Cross, S.P., H. Lauchstedt, and M. Blankenship. 1998. Numerical status of Townsend's big-eared bats at Salt Caves in Klamath River Canyon and other sites in southern Oregon, 1997. Unpublished report. U.S. Bureau of Land Management, Klamath Falls District, Klamath Falls, OR.

Cross, S.P., and D.L. Waldien. 1995. Survey of bats and their habitats in the Roseburg District of the BLM in 1994. Unpublished report. U.S. Bureau of Land Management, Roseburg District, Roseburg, OR.

Dalton, C.D. 2001. Foraging habitat and activity of the California leaf-nosed bat, *Macrotus californicus,* located on the eastern section of the Barry M. Goldwater Air Force Range, Arizona. Unpublished report. U.S. Air Force, 56th Range Management Office, Natural Resources Program, Luke Air Force Base, Glendale, AZ.

Davis, W.H., M.D. Hassell, and C.L. Rippy. 1965. *Myotis leibii leibii* in Kentucky. Journal of Mammalogy 46:683–684.

Davis, W.H., and H.B. Hitchcock. 1965. Biology and migration of the bat, *Myotis lucifugus,* in New England. Journal of Mammalogy 46:296–313.

Falxa, G.A. 2004. Rethinking Yuma bat and little brown bat foraging endurance. Program and Abstracts, 77th Annual Meeting of the Northwest Scientific Association, Ellensburg, WA.

Fellers, G.M., and E.D. Pierson. 2002. Habitat use and foraging behavior of Townsend's big-eared bat, (*Corynorhinus townsendii*) in coastal California. Journal of Mammalogy 83:167–177.

Fenton, M.B. 1969. Summer activity of *Myotis lucifugus* (Chiroptera: Vespertilionidae) in Ontario and Quebec. Canadian Journal of Zoology 47:597–602.

Fenton, M.B., and R.M.R. Barclay. 1980. *Myotis lucifugus.* Mammalian Species 14:1–8.

Gould, E. 1955. The feeding efficiency of insectivorous bats. Journal of Mammalogy, 36:399–407.

Hall, J.S., and F.J. Brenner. 1968. Summer netting of bats at a cave in Pennsylvania. Journal of Mammalogy 49:779–781.

Hayes, J.P. 2003. Habitat ecology and conservation of bats in western coniferous forests, pp. 81–119, *in* Mammal community dynamics in coniferous forests of western North America: management and conservation (C. J. Zabel and R. G. Anthony, eds.). Cambridge University Press, Cambridge, MA.

Heithaus, E.R., and T.H. Fleming. 1978. Foraging movements of a frugivorous bat, *Carollia perspicillata* (Phyllostomatidae). Ecologiocal Monographs 48:127–143.

Henry, M., D.W. Thomas, R. Vaudry, and M. Carrier. 2002. Foraging distances and home range of pregnant and lactating little brown bats (*Myotis lucifugus*). Journal of Mammalogy 83:775–784.

Herbers, J.M. 1981. Time resources and laziness in animals. Oceologia 49:252–262.

Hermanson, J.W. 1998. Chiropteran muscle biology: a perspective from molecules to function, pp. 127–139, *in* Bat biology and conservation (T.H. Kunz and P.A. Racey, eds.). Smithsonian Institution Press, Washington, DC.

Hermanson, J.W., and T.J. O'Shea. 1983. *Antrozous pallidus.* Mammalian Species 213:1–8.

Herreid, C.F., II. 1967. Temperature regulation, temperature preferences and tolerance, and metabolism of young and adult free-tailed bats. Physiological Zoology 40:1–22.

Hill, R.W., G.A. Wyse, and M. Anderson. 2004. Animal physiology. Sinauer Associates, Sunderland, MA.

Humphrey, S.R., and J.B. Cope. 1976. Population ecology of the little brown bat, *Myotis lucifugus,* in Indiana and north-central Kentucky. American Society of Mammalogists, Special Publication 4:1–81.

Humphries, M.M., D.W. Thomas, and D.L. Kramer. 2003. The role of energy availability in mammalian hibernation: a cost-benefit approach. Physiological and Biochemical Zoology 76:165–179.

Johnson, J.B., M.A. Menzel, J.W. Edwards, and W.M. Ford. 2002. Gray bat night-roosting under bridges. Journal of the Tennessee Academy of Science 77:91–93.

Kalcounis, M.C., and R.M. Brigham. 1998. Secondary use of aspen cavities by tree-roosting big brown bats. Journal of Wildlife Management 62:603–611.

Keeley, B.W., and M.D. Tuttle. 1999. Bats in American bridges. Bat Conservation International, Resource Publication 4:1–41.

Kerth, G., and K. Reckardt. 2003. Information transfer about roosts in female Bechstein's bat: an experimental field study. Proceedings of the Royal Society of London B 1514:511–515.

Kerth, G., K. Weissmann, and B. König. 2001. Roosting together, foraging apart: information transfer about food is unlikely to explain sociality in female Bechstein's bats (*Myotis bechsteinii*). Behavioral Ecology and Sociobiology 50:283–291.

Kiser, J.D., J.R. MacGregor, H.D. Bryan, and A. Howard. 2002. Use of concrete bridges as nightroosts, pp. 208–215, *in* The Indiana bat: biology and management of an endangered species (A. Kurta and J. Kennedy, eds.). Bat Conservation International, Austin, TX.

Krull, D., A. Schumm, W. Metzner, and G. Nueweiler. 1991. Foraging areas and foraging behavior in the notch-eared bat, *Myotis emarginatus* (Vespertilionidae). Behavioral Ecology and Sociobiology 28:247–253.

Kunz, T.H. 1973. Resource utilization: temporal and spatial components of bat activity in central Iowa. Journal of Mammalogy 54:14–32.

———. 1974. Feeding ecology of a temperate insectivorous bat (*Myotis velifer*). Ecology 55:693–711.

———. 1982. Roosting ecology of bats, pp. 1–56, *in* Ecology of bats (T. H. Kunz, ed.). Plenum Publishing Corporation, New York.

Kunz, T.H., and L.F. Lumsden. 2003. Ecology of cavity and foliage roosting bats, pp. 3–89, *in* Bat ecology (T.H. Kunz and M.B. Fenton, eds.). University of Chicago Press, Chicago, IL.

Kunz, T.H., and D.S. Reynolds. 2003. Bat colonies in buildings, pp. 91–102, *in* Monitoring bat populations in the United States and territories: problems and prospects (T.J. O'Shea and M.A. Bogan, eds.). U.S. Geological Survey Information Technology Report ITR 2003-0003.

Kunz, T.H., J.O. Whitaker Jr., and M.D. Wadonoli. 1995. Dietary energetics of the insectivorous Mexican free-tailed bat (*Tadarida brasiliensis*) during pregnancy and lactation. Oecologia 101:407–415.

Kurta, A. 1985. External insulation available to a non-nesting mammal, the little brown bat (*Myotis lucifigus*). Comparative Biochemistry and Physiology A, Comparative Physiology 8:413–420.

———. 1986. Factors affecting the resting and post-flight body temperature of little brown bats, *Myotis lucifugus*. Physiological Zoology 59:429–438.

Kurta, A., G.P. Bell, K.A. Nagy, and T.H. Kunz. 1989. Water balance of free-ranging little brown bats (*Myotis lucifigus*) during pregnancy and lactation. Canadian Journal of Zoology 67:2468–2472.

Kurta, A., and T.H. Kunz. 1988. Roosting metabolic rate and body temperature of male little brown bats (*Myotis lucifugus*) in summer. Journal of Mammalogy 69:645–651.

Lacki, M.J., M.D. Adam, and L.G. Shoemaker. 1993. Characteristics of feeding roosts of Virginia big-eared bats in Daniel Boone National Forest. Journal of Wildlife Management 57:539–543.

Lausen, C.L., and R.M.R. Barclay. 2003. Thermoregulation and roost selection by reproductive female big brown bats (*Eptesicus fuscus*) roosting in rock crevices. Journal of Zoology (London) 260:235–244.

Lewis, S.E. 1994. Night roosting ecology of pallid bats (*Antrozous pallidus*) in Oregon. American Midland Naturalist 132:219–226.

———. 1995. Roost fidelity of bats: a review. Journal of Mammalogy 76:481–496.

Lingle, S. 2001. Anti-predator strategies and grouping patterns in white-tailed deer and mule deer. Ethology 107:295–314.

Mayrberger, S., and V. Mayrberger. 2002. Flight frequencies and durations and exit/ entry sequences of pregnant and lactating big brown bats at a summer roost. Bat Research News 43:165–166.

McNab, B.K. 1982. Evolutionary alternatives in the physiological ecology of bats, pp. 151–200, *in* Ecology of bats (T.H. Kunz, ed.). Plenum Publishing Corporation, New York.

Murray, S.W., and A. Kurta. 2004. Nocturnal activity of the endangered Indiana bat (*Myotis sodalis*). Journal of Zoology (London) 262:1–10.

Nagorsen, D.W., and R.M. Brigham. 1993. Bats of British Columbia. UBC Press, Vancouver, BC.

Norberg, U.M. 1987. Wing form and flight mode in bats, pp. 43–56, *in* Recent advances in the study of bats (M.B. Fenton, P. Racey, and J.M.V. Rayner, eds.). Cambridge University Press, New York.

O'Donnell, C.F.J. 2002. Influence of sex and reproductive status on nocturnal activity of long-tailed bats (*Chalinolobus tuberculatus*). Journal of Mammalogy 83:794–803.

O'Donnell, C.F.J., and J.A. Sedgeley. 1999. Use of roosts by the long-tailed bat, *Chalinolobus tuberculatus,* in temperate rainforest in New Zealand. Journal of Mammalogy 80:913–923.

O'Farrell, M.J., and G.W. Bradley. 1977. Comparative thermal relationships of flight for some bats in the southwestern United States. Comparative Biochemistry and Physiology 58A:223–277.

O'Farrell, M.J., and E.H. Studier. 1980. *Myotis thysanodes*. Mammalian Species 137:1–5.

Ormsbee, P.C. 1996. Selection of day roosts by female long-legged myotis (*Myotis volans*) in forests of the central Oregon Cascades. M.S. thesis, Oregon State University, Corvallis, OR.

O'Shea, T.J., and T.A. Vaughan. 1977. Nocturnal and seasonal activities of the pallid bat, *Antrozous pallidus*. Journal of Mammalogy 58:269–284.

Parks, C.G., E.L. Bull, and T.R. Torgersen. 1997. Field guide to the identification of snags and logs in the interior Columbia River Basin. USDA Forest Service, PNW-GTR-390. Pacific Northwest Research Station, Portland, OR.

Pearl, D.L., and M.B. Fenton. 1996. Can echolocation calls provide information about group identity in the little brown bat (*Myotis lucifigus*)? Canadian Journal of Zoology 74:2184–2192.

Perlmeter, S.I. 1995. Bats and bridges: patterns of night roost activity in the Willamette National Forest. M.S. thesis, York University, Toronto, ON.

———1996. Bats and bridges: patterns of night roost activity in the Willamette National Forest, pp. 132–150, *in* Bats and forests symposium (R.M.R. Barclay and R.M. Brigham, eds.). Research Branch, British Columbia Ministry of Forests, Victoria, BC.

———. 1999. Is this my bridge? A study of night roost fidelity on the Willamette National Forest. Bat Research News 40:186.

Pierson, E.D., P.W. Collins, W.E. Rainey, P.A., Heady, and C.J. Corben. 2002. Distribution, status and habitat associations of bat species on Vandenberg Air

Force Base, Santa Barbara County, California. Santa Barbara Museum of Natural History Technical Reports, No.1.

Pierson, E.D., W.E. Rainey, and R.M. Miller. 1996. Night roost sampling: a window on the forest bat community in northern California, pp. 151–163, *in* Bats and forests symposium (R.M.R. Barclay and R.M. Brigham, eds.). Research Branch, British Columbia Ministry of Forests, Victoria, BC.

Pierson, E.D., M.C. Wackenhut, J.S. Altenbach, P. Bradley, P. Call, D.L. Genter, C.E. Harris, B.L. Keller, B. Lengus, L. Lewis, B. Luce, K.W. Navo, J.M. Perkins, S. Smith, and L. Welch. 1999. Species conservation assessment and conservation strategy for the Townsend's big-eared bat. Unpublished report. Idaho Conservation Effort, Idaho Dept. of Fish and Game, Boise, ID.

Pulliam, H.R. 1973. On the advantages of flocking. Journal of Theoretical Biology 38:419–422.

Purdy, D.M. 2002. Bats use of old-growth redwood basal hollows: a study of capture methods and species use of redwoods. M.S. thesis, Humbolt State University, Arcata, CA.

Rabe, M.J., M.S. Siders, C.R. Miller, and T.K. Snow. 1998. Long foraging distance for a spotted bat (*Euderma maculatum*) in northern Arizona. Southwestern Naturalist 43:266–269.

Racey, P.A. 1973. Environmental factors affecting the length of gestation in heterothermic bats. Journal of Reproduction and Fertility 19 (suppl.):175–189.

Rambaldini, D.A., and R.M. Brigham. 2004. Habitat use and roost selection by pallid bats (*Antrozous pallidus*) in the Okanagan Valley, British Columbia. Unpublished report. British Columbia Ministry of Land, Water and Air Protection, Penticton, British Columbia, and the Osoyoos Indian Band, Oliver, BC.

Renison, N.A. 2003. Bats and bridges of the Methow Valley: night roost selection and bridge temperatures. M,S, thesis, Western Washington University, Seattle, WA.

Richner, H., and P. Heeb. 1996. Communal life: honest signaling and the recruitment center hypothesis. Behavioral Ecology 7:115–119.

Roverud, R.C., and M.A. Chappell. 1991. Energetic and thermoregulatory aspects of clustering behavior in the neotropical bat *Noctilio albiventris.* Physiological Zoology 64:1527–1541.

Rydell, J. 1989. Feeding activity of the northern bat Eptesicus nilssoni during pregnancy and lactation. Oecologia 80:562–565.

Sherman, A.R. 2004. *Corynorhinus rafinesquii* and *Myotis austroriparius* artificial roost characteristics in southwestern Mississippi. M.S. thesis, Jackson State University, Jackson, MS.

Shiel, C.B., R.E. Shiel, and J.S. Fairley. 1999. Seasonal changes in the foraging behavior of Leisler's bat (*Nyctalus leisleri*) in Ireland as revealed by radio-telemetry. Journal of Zoology (London) 249:347–358.

Siders, M.S., and R. Steffensen. 1998. Spotted and western mastiff bat roost study of the North Kaibab Ranger District (Coconino County, Arizona). Unpublished report. North Kaibab Ranger Station, Fredonia, AZ.

Sparks, D., W. Dale, and J.R. Choate. 2000. Distribution, natural history, conservation status, and biogeography of bats in Kansas, pp. 173–228, *in* Reflections of a naturalist: papers honoring Professor Eugene D. Fleharty. Fort Hays Studies, Special Issue. Indiana State University, Terre Haute, IN.

Stihler, C.W. 1994. Radio telemetry studies of the endangered Virginia big-eared bat (Plecotus townsendii virginianus) at Cave Mountain Cave, Pendleton County, West Virginia. Unpublished report. U. S. Forest Service, Monongahela National Forest, WV.

Swier, V.J. 2003. Distribution, roost site selection and food habits of bats in eastern South Dakota. M.S. thesis, South Dakota State University, Brookings, SD.

Swift, S.M. 1980. Activity patterns of pipistrelle bats (*Pipistrellus pipistrellus*) in north-east Scotland. Journal of Zoology (London) 190:285–295.

Taylor, J.R. 2002. Use and function of a bat night roost within an abandoned mine in central Utah. M.S. thesis, Department of Integrative Biology, Brigham Young University, Provo, UT.

Thomas, D.W., M.B. Fenton, and R.M.R. Barclay. 1979. Social behavior of the little brown bat, *Myotis lucifugus* I. Mating behavior. Behavioral Ecology and Sociobiology 6:129–136.

Thomas, S.P. 1987. The physiology of bat flight, pp. 75–99, *in* Recent advances in the study of bats (M.B. Fenton, P. Racey, and J.M.V. Rayner, eds.). Cambridge University Press, New York.

Tigner, J. 2001. Bat habitat in the Black Hills: protection of abandoned mines. Unpublished report. South Dakota Game, Fish and Parks July Report 35. Rapid City, SD.

Tigner, J., and W.C. Aney. 1994. Report of black hills bat survey. Unpublished report. Nemo/Spearfish Ranger District Black Hills National Forest.

Tuttle, M.D., and D. Stevenson. 1982. Growth and survival of bats, pp. 105–150, *in* Ecology of bats (T.H. Kunz, ed.). Plenum Press, New York.

Wai-Ping, V., and M.B. Fenton. 1989. Ecology of spotted bat (*Euderma maculatum*) roosting and foraging behavior. Journal of Mammalogy 70:617–622.

Waldien, D.L., and J.P. Hayes. 2001. Activity areas of female long-eared myotis in western Oregon. Northwest Science 75:307–314.

Wilde, C.J., C.H. Knight, and P.A. Racey.1999. Influence of torpor on milk protein composition and secretion in lactating bats. Journal of Experimental Zoology 284:35–41.

Wilkinson, G.S. 1992. Information transfer at evening bat colonies. Animal Behaviour 44:501–518.

Willis, C.K.R. 2003. Physiological ecology of roost selection in female, forest-living big brown bats (*Eptesicus fuscus*) and hoary bats (*Lasiurus cinereus*). Ph.D. dissertation, University of Regina, Regina, Saskatchewan.

Willis, C.K.R., and R.M. Brigham. 2004. Roost switching, roost sharing and social cohesion: forest-dwelling big brown bats, *Eptesicus fuscus,* conform to the fission-fusion model. Animal Behaviour 68:495–505.

Zembal, R., and C. Gall. 1980. Swarming, reproduction, and early hibernation of *Myotis lucifugus* and *M. volans* in Alberta, Canada. Journal of Mammalogy 61:347–350.

MIGRATION AND USE OF AUTUMN, WINTER, AND SPRING ROOSTS BY TREE BATS

6

Paul M. Cryan and Jacques P. Veilleux

Compared with the tremendous progress made during recent years in determining the importance of trees to bats in summer, our understanding of the use of forests by bats during other seasons of the year is limited. Even so, certain patterns are apparent from the fragmentary information available on use of roosts by bats in autumn, winter, and spring. In this chapter we discuss the various roosting habits, thermoregulatory strategies, seasonal movements, migration behaviors, and habitat needs of forest-dwelling bats during winter and migration. Species of bats that use trees as roosts can be categorized into two general groups: cave bats, those that use trees in summer and hibernate in subterranean sites during winter, and tree bats, those that mostly use trees as roosts year-round. This review emphasizes the latter group, because information on the winter use of trees by cave bats is extremely limited. Species that hibernate in subterranean sites tend to make local, directionally scattered migrations, whereas tree bats make longer, latitudinal migrations. Evidence indicates that tree bats form larger aggregations during migration and that migration behavior differs between spring and autumn. Roosts used by tree bats vary during migration, although quantitative studies are lacking, as are data on winter behavior and habitat requirements of tree bats. Because winter and migration are periods during which mortality among bats may be high, it is important to determine whether the roosting and foraging needs of tree bats are satisfied during the colder months of the year.

Temperate insectivorous bats that inhabit forests during summer are faced with major challenges as winter approaches. Unable to meet the thermoregulatory costs of staying active without a consistent energy supply, most bats in temperate areas avoid harsh winter conditions by escaping in either time (by hibernating at local sites) or space (by migrating to warmer areas). As ambient temperatures and insect food supplies decrease during autumn, bats must prepare for winter by accumulating fat reserves and sometimes simultaneously mating and moving to wintering sites. Because of the potential risks to populations of bats associated with individuals surviving harsh temperate winters, it is important that we understand not only the needs of bats at wintering sites, but also during times when bats are moving to and from such areas.

In this chapter we discuss how bats that roost in forests during summer cope with the challenges posed by winter in the temperate zone, with a focus on species that depend on trees throughout the year. Our specific objectives are to provide overviews of the following aspects of the ecology of tree bats: (1) roosting patterns and thermoregulatory strategies; (2) patterns of movement and distribution; (3) migration behaviors and roost use during autumn, winter, and spring; and (4) potential threats to these bats. We close with a discussion of needs for further research and management considerations.

ROOSTING PATTERNS AND THERMOREGULATION STRATEGIES

More than half ($n = 24$) of the 46 species of North American bats are known to use trees as roosts during some part of the year (Kunz and Reynolds 2003). However, many of these species inhabit trees only during summer (June–August) before moving to caves, mines, or buildings for the winter. Such species undergo prolonged bouts of torpor during winter within thermally stable sites (i.e., subterranean structures and buildings) and are often referred to as either "hibernating" or "cave" bats. The general thermoregulatory strategy of these bats is to select wintering sites that are consistently cold enough to sustain low metabolic rates and body temperatures, allowing them to survive on accumulated fat during the winter. Essentially, these bats "wait out" the food shortage caused by winter cold. To avoid confusion and overgeneralization in this chapter, clear definitions of torpor and hibernation are necessary. Herein, torpor refers to any controlled lowering of body temperature and metabolic rate below levels maintained by resting euthermic individuals (Barclay et al. 2001). Bouts of torpor can last from hours to months. Hibernation is a form of torpor that is seasonally induced and differs from other forms in that the daily arousal process is inhibited (Ransome 1990). The phenomenon of hibernation has been well documented for species of bats in the temperate zone that winter in subterranean structures and buildings. Excellent reviews are provided in Davis (1970), Ransome (1990), and Speakman and Thomas (2003).

Only a small proportion of the species of bats that inhabit temperate regions are presently known to use trees as roosts during winter. Little is known about the wintering habits of some species of bats that roost in trees during summer, and trees likely provide winter roosts even for individuals of some species that are currently known only to use subterranean hibernacula (Hayes 2003). This is particularly likely in regions where temperatures rarely drop below freezing during winter. For example, species of *Myotis* may use trees during winter along the Pacific Coast of North America (Cowan 1942; Gellman and Zielinski 1996). Observations of such behavior are rare or vague though (Mearns 1898; Merriam 1884). As with the relatively recent discovery of widespread use of trees by bats during summer, future research efforts that monitor bats during the colder months will likely reveal additional species wintering in trees.

Species currently known to rely on trees as roosts throughout the year are commonly referred to as "tree bats" (Griffin 1970). In North America, these species include the western red bat (*Lasiurus blossevillii*), eastern red bat (*L. borealis*), hoary bat (*L. cinereus*), southern yellow bat (*L. ega*), northern yellow bat (*L. intermedius*), Seminole bat (*L. seminolus*), western yellow bat (*L. xanthinus*), and silver-haired bat (*Lasionycteris noctivagans*). The evening bat (*Nycticeius humeralis*) may also depend on trees throughout the year (Boyles et al. 2003; Robbins et al. 2004). In general, lasiurines (genus *Lasiurus*) roost alone or in small family groups (<5), consisting of mother and young, in foliage and other tree canopy structures (e.g., Spanish moss, *Tillandsia usneoides;* Barbour and Davis 1969), whereas silver-haired bats and evening bats form larger (>5) summer aggregations in the bole of large-diameter trees (e.g., under bark and in cavities; Kunz 1982; Watkins 1972). Tree bats occasionally appear in caves and mines during winter, although this behavior, with the exception of silver-haired bats, is rare and typically involves males and juveniles (Myers 1960).

Tree bats use torpor to limit energy expenditure during unfavorable conditions. Use of torpor at low temperatures has been documented in the silver-haired bat (Neuhauser and Brisbin 1969), eastern red bat (Davis and Lidicker 1956; Genoud 1993), hoary bat (Bowers et al. 1968; Brisbin 1966; Cryan and Wolf 2003; Genoud 1993), Seminole bat (Genoud 1990, 1993), and evening bat (Genoud 1990, 1993). Data on distribution indicate that few tree bats in North America leave the continent during winter (Cryan 2003), and it is likely that many use torpor on their wintering grounds. Because of this, differences in thermoregulatory strategies between migratory tree bats and other hibernating bats are by no means clear (Griffin 1970). The specifics of how tree bats differ in patterns of winter torpor from other species are not known. The current assumption is that tree bats winter in areas that are less prone to freezing temperatures and prey shortage, exhibit short-duration bouts of torpor, and continue foraging throughout the winter. However, such an assumption may be unwarranted, because few studies of tree bats in winter have been carried out. Available information does not conclusively demonstrate that tree bats are capable of entering torpor for long periods (e.g., >1 week) without spontaneous arousal (Genoud 1993; M. Milam, pers. comm.), but it is possible that some may choose wintering locations and roost structures that facilitate prolonged (>1 month) bouts of torpor. Hibernation, as defined above, may be a thermoregulatory strategy used by tree bats in some situations, but such behavior has yet to be discovered.

MOVEMENT PATTERNS
MOVEMENTS OF TREE BATS

Much of what we know about bat migration comes from banding studies of more gregarious species (e.g., Brazilian free-tailed bat, *Tadarida brasiliensis;* cave myotis, *Myotis velifer;* little brown myotis, *M. lucifugus*) that winter in caves. Unfortunately, our current understanding of migration in tree bats is limited because of a lack of technology that allows for long-term monitoring of migratory movements. Banding or tagging efforts directed toward less colonial species, like tree bats, typically prove ineffective. However, indirect methods of research such as mapping occurrence records (Cryan 2003; Findley and Jones 1964) or analysis of stable isotopes (Cryan et al. 2004) have helped reveal patterns of bat migration. Using museum data, Cryan (2003) summarized the seasonal movements of the western red bat, eastern red bat, hoary bat, and silver-haired bat in North America. From this work, several patterns with important management implications emerged in the seasonal distributions of these wide-ranging species. First, each moves from northern summering grounds to more southern latitudes during winter. This pattern reflects the presumed thermoregulatory strategy of inhabiting warmer areas and remaining active throughout the winter. Second, the migration route of each species is apparently contained within the continent of North America. Third, although range maps typically depict these species as widespread, they do not occur in all parts of their range during any single season. Furthermore, differences between sexes in the migratory movements of tree bats often exist. Females tend to migrate earlier and farther than males during spring, and differences in distribution between sexes during summer are sometimes large (Cryan 2003). These patterns suggest that there are regional differences in the period of occupancy, nature of use, and demographics of tree bats occurring in forests of North America.

Other lasiurines that use trees during winter in North America inhabit more southern latitudes ($<34°$ N) throughout the year. Data pertaining to the seasonal whereabouts and migratory movements of these species, if they occur, are lacking. Evening bats are known to winter in trees at latitudes as high as $36°$ N (Robbins et al. 2004; J. Boyles, pers. comm.), and evidence indicates that some females migrate into northern parts of the range during summer while males remain within the same area during both winter and summer (Bain 1981; Bain and Humphrey 1986).

MOVEMENTS OF HIBERNATING BATS

Several studies of species that roost in trees during summer but overwinter in caves have demonstrated patterns of long-distance movement. For example, banding studies of little brown myotis (Humphrey and Cope 1976) and Indiana myotis (*Myotis sodalis;* Kurta and Murray 2002) revealed travel distances between winter and summer habitats of 455 and 532 km, respectively. Detailed reviews of movement patterns by season for colonial bats that hibernate in caves can be found in Griffin (1970), Baker (1978), and Fleming and Eby (2003). In general, species of bats that winter in subterranean structures tend to make shorter migrations, which are less influenced by latitude, than tree bats (Baker 1978). Subterranean roosts are cold and thermally stable; they possess roost microclimates relatively independent of latitude compared with aboveground structures. Hence, the autumn migratory movements of species that hibernate during winter in underground sites are typically oriented toward nearby regions with suitable conditions for hibernation rather than areas with warm surface temperatures. Furthermore, migration along gradients of elevation may occur in hibernating species that inhabit mountainous regions. For example, big brown bats (*Eptesicus fuscus*), which spend the warmer months in buildings around Fort Collins, Colorado (elevation, 1,500 m), move into the nearby Rocky Mountains during autumn where they spend the winter in rock crevices at higher elevations (>1,600 m) (D. Neubaum, pers. comm.).

BEHAVIOR AND ROOST USE IN AUTUMN

As detailed elsewhere in this volume, reproductively active female bats require more energy than males and nonreproductive females during summer (Barclay and Kurta, chap. 2 in this volume; Lacki et al., chap. 4 in this volume). Such energy needs often lead to differential roosting requirements between the sexes during the warmest months. For example, reproductive females often seek out warm sites to limit thermoregulatory costs and expedite fetal growth and lactation, whereas males and nonreproductive females may choose cooler roosts that facilitate torpor and energy conservation (Veilleux et al. 2004). This variation in energy need and associated habitat selection among bats can lead to differential distribution of the sexes at various spatial scales. Bat communities at higher elevations in temperate areas are often male biased, with reproductive females occurring more frequently at lower sites (Cryan et al. 2000; Russo 2002). At larger scales, hoary bats and silver-haired bats exhibit distributional differences between sexes during summer on the order of hundreds of kilometers (Cryan 2003; Findley and Jones 1964). Such differences in the summer distribution and habitat needs of male and female bats cannot be overlooked when incorporating habitat considerations for bats into forest management plans. As summer turns to autumn, differences in energy need and behavior of male and female bats likely diminish, suggesting that outside of the maternity season the roosting and foraging needs for male and female bats should be similar, leading to more simplified management considerations.

GROUPING

As with other species of bats that roost alone or in small groups during summer, tree bats tend to form larger aggregations during migratory periods (Fleming and Eby 2003). The best evidence for increased gregariousness among tree bats during migration is the numerous observations of bat flocks during autumn. One of the most dramatic reports of a flock of bats in autumn comes from Mearns (1898), who observed "great flights of red bats during the whole day" in the Hudson Highlands of New York. Howell (1908) observed a diurnal migration of what he presumed to be eastern red bats or silver-haired bats during late September in Washington, D.C. Diurnal flights of hoary bats were observed in both Minnesota (Jackson 1961) and Nevada (Hall 1946). Murphy and Nichols (1913) reported a flock of bats flying southwest over Long Island, New York. Some accounts of flocks leave little doubt that the bats were migrating. For example, two eastern red bats were taken from a flock of approximately 200 bats that circled a ship 104 km off the New England coast during late September (Carter 1950). Both silver-haired bats and eastern red bats were collected from a group of approximately 100 bats that landed on a ship 32 km off the coast of North Carolina during early September (Thomas 1921). Small groups ($n = 2–4$) of silver-haired bats and eastern red bats were seen during autumn mornings as they arrived onshore during migration over Lake Michigan (Byre 1990). The latter records indicate that mixed-species flocks of bats sometimes occur.

Observations of roosting bats also provide evidence of larger aggregations and mixed-species groups during migration. Groups of migrating hoary bats roosting on Southeast Farallon Island, approximately 32 km off the coast of California, sometimes number up to 60 individuals in a single tree (A. Brown, pers. comm.). During late August in the North Bay Area of California, Constantine (1959) found a group of approximately 15 western red bats roosting in an apricot (*Prunus* sp.) tree, but none was found in the area later in winter. Grinnell (1918) noted "many" western red bats roosting together with hoary bats during April in California.

Capture efforts have documented both spring and autumn migratory "waves" of tree bats through the coincident trapping of multiple individuals (Barclay et al. 1988; Findley and Jones 1964; Mumford 1963, 1973; Vaughan 1953). The details of how tree bats in North America form and maintain aggregations during migratory periods are unknown, but evidence of communication exists. Downes (1964) observed eastern red bats using specific roost sites during autumn and noted that different individuals somehow found and used the same roost on subsequent days. Constantine (1966) observed a similar phenomenon, where both eastern red bats and hoary bats used the same foliage roost on different days. In Georgia, Seminole bats and eastern red bats also used the same roost, even though other roosting sites were available (Constantine 1958). Barclay et al. (1988) noted that migrating silver-haired bats somehow found roosts previously used by others but, as with all of these cases, were unable to determine the method of communication. The European tree bat, *Nyctalus noctula*, migrates long distances, apparently alone, but gathers in resting groups that often number up to 1,000 individuals (Strelkov 1969). Further research into repeated use of roosts by migrating bats may help determine whether such roosts are visited more frequently because they are (1) remembered from prior visits; (2) limited in availability and, thus, more likely to house a bat; or (3) discovered by means of communication among bats. The ultimate goal of such research should be to determine whether revisitation is due to a special significance of these roosts or roosting areas, such as increased protection from predators, beneficial temperatures, or proximity to sources of prey.

MATING

Observations of mating tree bats indicate that copulation occurs during migration periods. Copulating eastern red bats were observed during August in Michigan and New York (Murphy and Nichols 1913; Stuewer 1948), as well as during both autumn and spring in Arkansas (Saugey et

al. 1998). Migrating hoary bats were observed copulating in trees during
autumn on Southeast Farallon Island (A. Brown, pers. comm.). We do not know whether breeding during migration takes place haphazardly en route or at certain rendezvous sites that are familiar to migrating bats. Baker (1978) suggested that male bats may set up new territories for mating every day along their migration route and that females visit these territories to breed. If males establish territories along the migratory routes of females, it is reasonable to propose that males may form leks (Alcock 2001). Somewhat contradictory to the evidence for mating of bats during autumn, Mearns (1898) observed segregated flocks of eastern red bats during autumn in New York. However, Mumford (1973) noted that sex-segregated groups of eastern red bats captured during autumn in Indiana were predominantly comprised of juvenile bats.

ROOSTS USED DURING AUTUMN MIGRATION

Our knowledge of the roosting needs of tree bats during autumn migration is based on limited and fragmentary information (see also Carter and Menzel, chap. 3 in this volume). Eastern red bats are known to use a variety of roosts during autumn. In California, western red bats have been observed using fig (*Ficus* spp.) and apricot trees (*Prunus* spp.; Constantine 1959), mallow (*Malva* spp; A. Brown, pers. comm.), *Sparmannia africana* (Orr 1950b), and orange trees (*Citrus* spp.; Grinnell 1918). Eastern red bats have been found roosting in a willow bush (*Salix* spp.) in Saskatchewan (Hall 1938), as well as in blue beech (*Carpinus caroliniana*) and wild black cherry trees (*Prunus* spp.) in New York during autumn (Terres 1956). Presumably, migrating eastern red bats were observed hanging from the exposed roots of trees on a cliff face during October in New York (Murphy and Nichols 1913). Hoary bats were found using cherry (*Prunus* spp.), pine (*Pinus* spp.), and cypress trees (*Cupressus* spp.) in California (Constantine 1959; Tenaza 1966), a squirrel nest in Georgia (Neill 1952), and shrubs and various fruit trees along the Pacific Coast (Brown 1935; Dalquest 1943). A Seminole bat was found roosting in a winged-elm tree (*Ulmus alata*) during late September in central Oklahoma (Caire and Thies 1987). Unlike lasiurines, silver-haired bats are often encountered in human-made structures during autumn migration (Barbour and Davis 1969; Glass 1961; Murphy and Nichols 1913; Schowalter et al. 1978). Although silver-haired bats undoubtedly use a variety of trees throughout their range during autumn, the only tree species to be documented as roosts of this species in autumn include *Tamarix* spp. in California (Grinnell 1937) and green ash (*Fraxinus pennsylvanica*) in Manitoba (Collister 1995).

FEEDING

Although tree bats sometimes possess fat reserves during autumn and winter (Gosling 1977; Layne 1958; Tenaza 1966; Van Gelder 1956), some species apparently feed during autumn migration. Miller (1897) observed both silver-haired bats and eastern red bats foraging during a migration stopover on the Atlantic Coast, and a female hoary bat collected while migrating through Florida was feeding during late October (Zinn and Baker 1979). An interesting question is whether migrating bats feed opportunistically or rely on some consistent food supply that is available along their migration route. Curiously, the three known instances of hoary bats attacking or eating eastern (*Pipistrellus subflavus*) and western pipistrelles (*P. hesperus*) occurred during autumn (Bishop 1947; Orr 1950a; J. Barnes, pers. comm.). As with tree bats, little is known about the habitat needs of cave bats as they move from summer roosts in trees to their hibernacula. The limited data on roosts used by migrating bats might suggest opportunistic roosting. Although this may prove true, such judgments are premature because of potential bias associated with the observations. For example, juvenile tree bats making their first migration may end up in unusual situations where they are more likely to be encountered. Clearly, more studies of roosts used by bats during migration are needed.

BEHAVIOR AND ROOST USE IN WINTER

During the winter, tree bats in North America generally occur at latitudes below 40° N and in coastal regions where freezing temperatures are infrequent. Figures 6.1–6.3 show winter (December–February) occurrence records for the western red bat, eastern red bat, hoary bat, and silver-haired bat within the United States (see Cryan [2003] for additional winter records from Mexico and Canada). Unfortunately, much of what we know about the distribution of tree bats during winter is based on occurrence records, which can be biased. Such records only indicate the presence of bats in an area and not absence. Tree bats use torpor, roost in situations where they are not readily observed, and are rarely sought out by biologists during winter. Because of this, bats undoubtedly occur in many areas depicted in Figures 6.1–6.3 for which there is no winter record. Only systematic surveys targeting wintering tree bats will refine our understanding of their distributions during the colder months.

WINTER ROOSTS

Compared with what we know of the summer habits of tree bats, we know very little about the use of roosts and activity during winter. Documentation of winter roosts is rare and quantitative data are lacking. Lasiurines apparently use a variety of roosts during winter. Orr (1950b) regularly observed western red bats roosting 2.4–4.6 m above the ground in large-leafed shrubs (*Sparmannia africana*) from September through early May in California, and Grinnell (1918) reported western red bats using orange trees in California. Eastern red bats, hoary bats, and Seminole bats are known to roost in Spanish moss during winter in the southeastern United States (Constantine 1958; Sherman 1956), and western yellow bats roost in palm trees (family Arecaceae) during the winter in Arizona (Cockrum 1961). In one of the few studies of the use of winter roosts by tree bats to date, Saugey et al. (1998) tracked a female eastern red bat during December in Arkansas. The bat regularly switched between a variety of roosts including a shortleaf pine (*Pinus echinata*), a

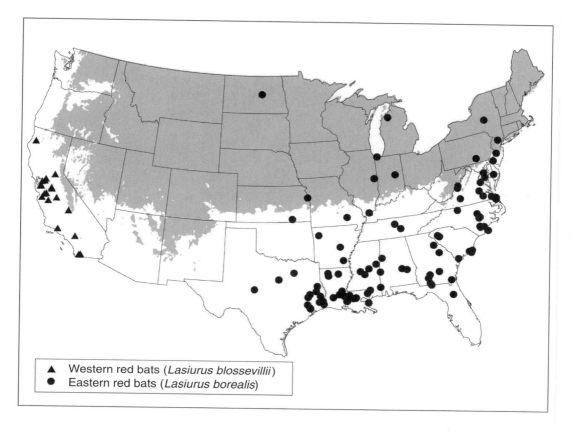

Figure 6.1.
Winter (December–February) distribution of museum records for western and eastern red bats (*Lasiurus blossevillii* and *L. borealis*, respectively) in the United States. Data based on those presented in Cryan (2003). Shaded area shows regions of the United States where winter 24-hr surface temperatures average = 0°C. *Climate Source, Corvallis, OR*

small shrub, and hardwood-pine leaf litter on the forest floor. Moorman et al. (1999) observed what were believed to be eastern red bats emerging from deciduous leaf litter during prescribed burns in South Carolina in winter. Koontz and Davis (1991) found eastern red bats roosting in the branches of American beech (*Fagus grandifolia*) during winter in Kentucky.

Silver-haired bats have been found wintering in a variety of trees throughout their range including a Douglas fir snag (*Pseudotsuga menziesii;* Nagorsen et al. 1993), giant cedars (*Thuja plicata;* Cowan 1933), bald cypress (*Taxodium distichum;* Padgett and Rose 1991), a cottonwood stump (*Populus deltoides;* Tyler and Payne 1982), and unspecified species of trees (Brimley 1897; Murphy and Nichols 1913). Silver-haired bats are sometimes found in winter roosts that are not trees. A variety of records indicate that individuals use buildings during winter (Bartsch 1956; Brigham 1995; Gosling 1977; Izor 1979; Nagorsen et al. 1993), including a report by Clark (1993) who found a communal winter roost of both sexes of silver-haired bats in an abandoned building in North Carolina. Murphy and Nichols (1913) reported regularly encountering silver-haired bats in ships at harbor in New York. There are also records of silver-haired bats from caves and mines; these records typically involve individual bats roosting inside crevices, but small groups (<10) have also been noted (Beer 1956; Krutzsch 1966; Pearson 1962; Smith and Parmalee 1954; Szewczak et al. 1998; Turner 1974). The extent to which

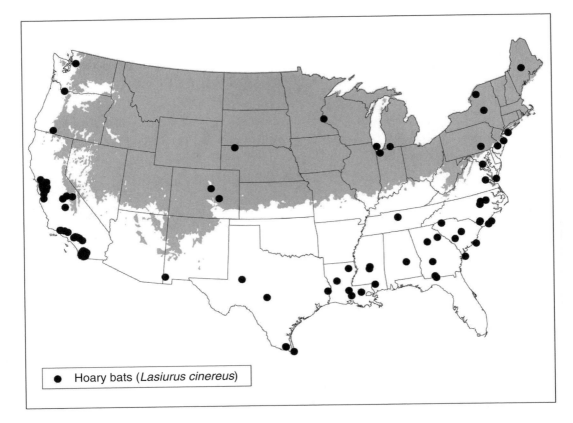

Figure 6.2.
Winter (December–February) distribution of museum records for hoary bats (*Lasiurus cinereus*) in the United States. Data based on those presented in Cryan (2003). Shaded area shows regions of the United States where winter 24-hr surface temperatures average = 0°C. *Climate Source, Corvallis, OR*

silver-haired bats use nontree roosts during winter remains to be determined. Most observations of silver-haired bats using caves and mines during winter are from relatively high latitudes (mean = 39.7° N; $n = 7$) and it may be that this species switches to subterranean sites when trees get too cold. It is also possible, however, that silver-haired bats do not use trees during winter on a regular basis in some parts of their range.

TREES AND COLD

Do bats roost in trees in areas that experience subfreezing temperatures for prolonged periods? The shaded areas of Figures 6.1–6.3 show regions of the United States where 24-hour surface temperatures during winter (December–February) average ≤0°C. Although tree bats are occasionally collected in these areas during winter, their ability to survive under such conditions for prolonged periods is unknown. Regardless, there is little doubt that lasiurines are capable of withstanding subfreezing temperatures for short periods (<1 month). Under laboratory conditions, eastern red bats recovered from exposure to temperatures <9°C even after body tissues potentially froze (Davis 1970; Genoud 1993). Field observations also reveal tree bats surviving subfreezing conditions. Saugey et al. (1998) followed a female eastern red bat in Arkansas during December when nighttime temperatures frequently dropped below freezing and found the bat torpid beneath leaf litter for 18 consecutive days. Rob-

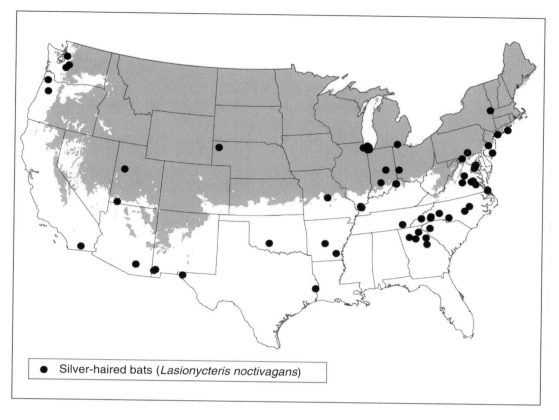

Figure 6.3.
Winter (December–February) distribution of museum records for silver-haired bats (*Lasionycteris noctivagans*) in the United States. Data based on those presented in Cryan (2003). Shaded area shows regions of the United States where winter 24-hr surface temperatures average = 0°C. *Climate Source, Corvallis, OR.*

Silver-haired bats (*Lasionycteris noctivagans*)

bins et al. (2004) monitored several eastern red bats during winter in southern Missouri and found them roosting in tree foliage on warmer days, only to move to sites beneath leaf litter when temperatures dropped below about 8–10°C. Experimental laboratory studies indicate that several species of *Lasiurus* are capable of maintaining stable and relatively energy-efficient torpor at air temperatures between 5–10°C, but are reluctant to remain in torpor for extended periods at temperatures <5°C (Cryan and Wolf 2003; Genoud 1993). Eastern red bats likely choose roosts beneath ground debris because of insulating, and potentially (due to microbial action) thermogenic effects in comparison with aboveground air temperatures (Robbins et al. 2004). Observations of eastern red bats during winter at northern latitudes are rare (Shump and Shump 1982), but do occur (Baird 1960; Banfield 1974; Bowers et al. 1968; Grimm 1977; Keith 1978). Whitaker (1967) presented evidence for overwintering by a male hoary bat found active in Indiana during January. The alimentary canal of this bat contained grass and a shed snakeskin. Perhaps this is evidence of litter-roosting behavior by hoary bats during winter. Although natural fires in forests of the southeastern United States historically occur during summer, controlled burns are now carried out during winter. If roosting beneath leaf litter is a common behavior among wintering lasiurines, current forest management practices may have negative impacts on their populations (Carter et al. 2002; Carter and

Menzel, chap. 3 in this volume). It remains to be determined whether lasiurines also roost in litter on the forest floor during winter in coniferous
forests.

More is known about roost use by migratory bats in Europe than in
North America. The noctule bat hibernates within tree cavities in coastal
regions of western Europe but uses buildings in central areas of the continent (Sluiter et al. 1973; van Heerdt and Sluiter 1965). In southern
Kazakhstan, the noctule bat hibernates in tree cavities at sites generally
below 43° N latitude (Strelkov 1969). Sluiter et al. (1973) monitored temperatures within both small and large sections of trees occupied by colonies of hibernating noctule bats. Their work showed that larger trees
offered bats better protection from external temperature fluctuations
than smaller trees during winter. Hibernating colonies of noctule bats in
trees sometimes exceed 100 individuals (van Heerdt and Sluiter 1965),
which is similar to winter roosting behavior of evening bats in North
America (Robbins et al. 2004). Colonial behavior may be a requisite characteristic of species of bats that overwinter in trees at northern latitudes.
Perhaps silver-haired bats also form hibernating colonies in trees, but
such groups have not yet been observed.

Gellman and Zielinski (1996) found evidence of year-round use of
redwood hollows in coastal California during winter, but their methodology precluded species identification. As mentioned earlier, it is possible that species currently known to only roost in caves during winter also
use trees in some regions. In New York, Mearns (1898) found eastern
small-footed myotis (*Myotis leibii*) hibernating in trees and Merriam
(1884) noted that woodsmen regularly encountered bats while cutting
firewood during winter in northeastern New York.

FEEDING

Available evidence suggests that winter behavior of tree bats consists of
bouts of torpor interspersed with opportunistic foraging. Visual observation, capture data, and diet analysis provide evidence of winter feeding
in several species. Both eastern red bats and silver-haired bats are known
to feed during winter in the Great Dismal Swamp of Virginia and North
Carolina (Padgett and Rose 1991; Whitaker et al. 1997), and both species
have been observed flying during winter in southeastern Arkansas (Baker
and Ward 1967). Observations of eastern red bats show that the species
is sometimes active in states as far north as Missouri (LaVal and LaVal
1979) and Kentucky (Davis 1970) during winter. Acoustic monitoring
offers much promise as a means of studying the winter activity and feeding habits of tree bats.

BEHAVIOR AND ROOST USE IN SPRING

As winter gives way to spring, migratory bats must make their way back
to the summering grounds. Spring migration may be a particularly challenging time for bats, as pregnant females sometimes migrate hundreds
of kilometers from their wintering to summering grounds during a time
when fat reserves and insect prey are not always available (Kurta and

out their spring migration.

ROOSTS USED DURING SPRING

Unlike many observations from autumn, it is sometimes difficult to discern if tree bats found during spring are migrants or residents. Saugey et al. (1998) tracked an adult male eastern red bat during April in Arkansas and found it awake and alert beneath leaf litter for several days. A male tracked during May switched between roosts in white oak trees (*Quercus alba*) during a 3-day period before leaving the study area. A male eastern red bat captured during early March in southern Missouri was tracked to its roost beneath leaf litter, where it remained for 4 days (Boyles et al. 2003). Fassler (1975) found a male eastern red bat torpid in a woodpecker hole in a tree stump during late March in southern Kentucky. Our limited understanding of the timing of spring migration by tree bats precludes determination of whether these individuals were still on their wintering grounds or on the move. In the only study to date on use of roosts by migrating bats, Barclay et al. (1988) examined the roosting habits of silver-haired bats (mostly females) moving through Manitoba during spring. A total of 177 bats was located in 36 roosts in nearly as many trees ($n = 32$). Most bats roosted alone, although 15 groups of two bats and 8 groups of three to six bats were observed. Bats roosted in folds of bark and crevices in the bole of the tree, preferentially choosing large trees of species that were likely to have furrowed bark, splits, and cracks. Some roost sites were used on multiple occasions both within and among years. On several occasions, bats did not emerge from roosts on cold nights, suggesting that they wait for warmer temperatures before they continue migrating. Other documented roosts of silver-haired bats during spring include an individual found torpid beneath ground debris (Sanborn 1953), three bats found torpid beneath the bark of a small (~15 cm diameter) oak snag (*Quercus* sp.; A. Kerwin, in litt.) in western Oregon, a female found in an underground rodent burrow in Indiana (Brack and Carter 1986), as well as several taken from crevices in sandstone ledges and a cave in West Virginia (Frum 1953). The latter bats had enough food in their systems to suggest they had recently fed (Frum 1953).

FEEDING

Tree bats apparently feed during spring migration. Among 128 hoary bats captured while migrating through New Mexico during spring 2002, most produced guano that was comprised predominantly of moth parts (Valdez and Cryan, in litt.). The 3-week period during which these bats were captured coincided with a local eruption of moths and hints at the possibility that movement by tree bats may sometimes be synchronized with periods of prey availability.

SITE AND ROUTE FIDELITY

Available evidence indicates that bats follow fairly regular migration circuits (Baker 1978) and, during spring, return to specific roosts, roosting areas, and foraging sites used during the previous summer. Most of this evidence comes from studies on species other than tree bats. Barclay and Brigham (2001) examined patterns of year-to-year reuse of tree snags by California bats in British Columbia during summer months. They found that seven of eight roost trees used by reproductive females in 1995 were occupied again in 1996 and 1998 through 2000, although it was not determined if specific bats returned. In central Canada, Willis et al. (2003) found that reuse of aspen trees over several summers by big brown bats was common and that marked individuals used the same trees during multiple years. Furthermore, these bats tended to roost in the same small (1–2 km^2) areas year after year. Veilleux and Veilleux (2004) also found evidence of interannual fidelity to tree-roost areas by eastern pipistrelles. Two adult female eastern pipistrelles observed via radiotelemetry during the summers of 1999 and 2000 roosted within the same 2.3- and 0.8-ha areas, respectively, during both years of monitoring. In addition, a female eastern pipistrelle tracked during 2000 roosted 141 m from the site where it was captured as a juvenile during the previous year, hinting at the possibility of female natal philopatry. Regardless, all of these examples suggest autumn relocation to winter quarters followed by a return to the specific roost trees occupied during the previous summer.

Some studies leave little doubt that bats make long-distance movements away from and back to their summer roost sites. Kurta and Murray (2002) reported recovering four banded Indiana myotis that traveled an average of 460 km between summering grounds and winter hibernacula. Two of the banded bats subsequently returned to the area where they roosted during the previous summer (Kurta and Murray 2002; Kurta et al. 2002). In light of these observations, it seems entirely possible that bats regularly follow migration routes between established summering and wintering areas.

Evidence for migration to and from specific sites by tree bats is more limited. In Saskatchewan, Canada, Willis (2003) noted two cases of banded hoary bats returning to specific sites used during previous years; an adult female used the same tree as in the previous summer, and a bat that was banded as a juvenile returned to the same area two years later. In addition, hoary bats and western red bats have regularly appeared during autumn in the few trees on Southeast Farallon Island for the past 35 years of observation (A. Brown, pers. comm.).

POTENTIAL THREATS AND FUTURE DIRECTIONS

Certain factors make migratory bats particularly susceptible to population decline (Fleming and Eby 2003). First, migratory bats often require contiguous, yet seasonally distinct, habitats that sometimes span hundreds of kilometers along their annual migration pathway. Degradation of a single region along such annual circuits has the potential to negatively impact populations that move through that area. For example, if some disturbance along a migration corridor disrupts the ability of bats to locate summering grounds, hibernacula, or mating grounds, individual fitness may be reduced or mortality increased. Second, bat populations may concentrate in small areas during migration, rendering them vulnerable to mass mortality events. There is currently no means by which to monitor the population status of migratory tree bats (Carter et al. 2003), nor do we possess a clear understanding of their habitat needs or mortality risks during migration or in winter.

PERILS OF MIGRATION

Evidence indicates that tree bats sometimes migrate with, or under similar conditions as, birds and are likely to be susceptible to similar mortality factors. For example, dead eastern red bats were found among migratory birds that washed ashore after both spring and autumn storms on Lake Michigan (Mumford 1973; Mumford and Whitaker 1982). There are numerous reports of tree bats found among dead birds that collided with human-made structures. Most of these incidents transpired during autumn and involved multiple species: eastern red bats, hoary bats, and silver-haired bats at a lighthouse on Lake Erie (Saunders 1930); eastern red bats at television towers in Kansas, Tennessee, and North Dakota (Avery and Clement 1972; Ganier 1962; Van Gelder 1956); eastern red bats, hoary bats, northern yellow bats, and Seminole bats at a television tower in Florida (Crawford and Baker 1981); eastern red bats and silver-haired bats at a building in Chicago (Timm 1989); and eastern red bats at the Empire State Building in New York City (Terres 1956). For many of these collision events tens to hundreds of birds were reported, whereas only a few bats were encountered. For example, Crawford and Baker (1981) reported 54 bats killed on 49 nights over a 25-year monitoring period at a television tower in Florida, and Timm (1989) reported 79 bats killed over an 8-year period at a building in Chicago. Mortality rates for bats killed at wind turbines present a different story. In contrast to the low number of deaths during collision events listed above, 458 bats of at least six species were killed during the autumn of 2003 at a wind facility in West Virginia (J. Kerns, pers. comm.). Migratory tree bats comprise most of the bats killed at wind turbines in North America with the majority of collisions occurring in autumn (Gruver 2002; Johnson et al. 2003; J. Kerns, pers. comm.). The reasons for such disproportionate kills during autumn are unknown. Curiously, unusual encounters with migrating tree bats typically happen during autumn rather than spring (Cryan 2003). It is possible that spring migration by tree bats is relatively low-altitude, whereas autumn movement occurs at greater heights. For example, hoary bats fly low (1–5 m off the ground) within riparian areas while migrating through New Mexico during spring, but apparently not during autumn (P. Cryan, in litt.). In contrast, a hoary bat recently collided with an airplane 2,438 m above Oklahoma in October (Peurach 2003). Emerging technologies that allow for the long-term tracking of tree bats during migration may shed light on seasonal differences in movement and help us understand such disproportionate mortality during autumn.

In addition to the perils of collisions during flight, migrating bats may be susceptible to predation both during migration and on the wintering grounds. Stomach contents of predators captured during winter revealed the remains of both silver-haired bats and eastern red bats (Sperry 1933). If trees with adequate roost sites are not available during migration or on the wintering grounds, torpid bats may be vulnerable to higher rates of predation.

RESEARCH AND MANAGEMENT NEEDS

Compared with the tremendous progress made during recent years in determining the importance of trees to bats in summer, our understanding of the use of forests by bats during other seasons is limited. Among 29 radiotracking studies on the use of roosts by bats in North America, few (<5) investigated roosting habits outside of the summer months. It is important to develop an understanding of the habitat needs of migratory bats, because a lack of suitable habitat may lead to population declines. Tree bats undoubtedly depend on forests during autumn, spring, and winter to a greater extent than we currently comprehend, but only targeted studies conducted during all seasons will determine the degree to which this is true (Sedgeley 2001).

More than anything, the scattered information compiled for this review highlights the pressing need for increased efforts to uncover the seasonal habits and whereabouts of migratory tree bats under natural conditions. Much of the information that is currently available is biased toward urban areas (e.g., numerous records of bats roosting in fruit trees; Carter and Menzel, chap. 3 in this volume). Tree bats are undoubtedly present in the forests of North America throughout the year, but go undetected in many regions during the colder months. Combined efforts of both researchers and managers could help overcome this lack of knowledge.

RESEARCH NEEDS

In terms of resource management, one of the most pressing research needs is to determine the winter roosting and foraging needs of tree bats. For example, we know little about the winter roosts used by hoary bats and silver-haired bats so we currently have no way of determining how current forest management practices affect them. The discovery of eastern red bats roosting beneath leaf litter during winter in the southeastern United States (Boyles et al. 2003; Moorman et al. 1999; Robbins et al. 2004; Saugey et al. 1998) illustrates the potential rewards, and immediate management implications, of targeted studies. Winter roosting and foraging behavior of migratory tree bats is a relatively unexplored frontier of bat research.

Another important research need is to determine whether migratory "corridors" exist. Findings of individual bats returning to familiar summering and wintering areas indicate that bats may also travel along familiar routes between sites. Future research should determine whether migratory tree bats concentrate during migration along particular routes and use those routes repeatedly, or simply move in a dispersed fashion across the landscape. Determining such patterns would help answer important questions. Such as, do bats killed at wind turbines represent small proportions of dispersed migratory groups or large proportions of more concentrated groups? This information would be crucial to managers charged with assessing sites for the potential impact of landscape change on migratory bats. Where might corridors exist? Riparian areas may be particularly important to migratory tree bats, as they provide a constant source of drinking water, insect prey, roost sites in trees, and landmarks to follow during migration. Recent evidence from wind turbine sites also indicates that migratory bats may concentrate along ridgelines and other topographic features that interact with surface winds, especially during autumn.

MANAGEMENT NEEDS

Our current ability to develop management recommendations for migratory tree bats is hampered by a lack of information regarding their needs in spring and autumn. However, there are several things that land managers can do to assess the potential importance of lands in their care to migratory bats. First and foremost, it should be determined if and when tree bats occur at a site (Carter and Menzel, chap. 3 in this volume). Initial surveys could be carried out using echolocation detectors to determine the presence and timing of the occurrence of tree bats. If tree bats are detected at a site, follow-up capture surveys during the periods of occurrence would help determine the sex, age, and body condition of individuals present. These relatively simple steps may reveal the presence of species previously unknown from the management area. Such information should aid practical land management decisions (e.g., timing of burns, timber harvest, or surface water availability). Furthermore, when combined with similar information from other management areas, these data will help develop our understanding of the seasonal whereabouts of bat populations.

Once tree bats are known to occur in an area, efforts could be made to disclose how they are using local habitats. Methods of determining habitat use range from simple techniques, such as searching trees for roosting bats in the sense of Barclay et al. (1998) or watching bat activity at sunset, to more sophisticated techniques like radiotracking, microclimate analysis of roosts (Willis and Brigham 2005), or thermal imaging. Because winter and migration are periods during which mortality among bats may be high, determining whether the roosting and foraging needs of bats in a management unit are being met during these times should be given as much priority as during the warmer months of the year.

ACKNOWLEDGMENTS

We thank D. Neubaum, A. Brown, and J. Kerns for sharing unpublished data. PMC would like to thank T. O'Shea for helpful discussions during the development of this review. We would also like to express our appreciation for the pioneering work of the late Dr. Donald R. Griffin, who led the way in so many fields, including the study of bat migration.

Alcock, J. 2001. Animal behavior, 7th ed. Sinauer Associates, Sunderland, MA.

Avery, M., and T. Clement. 1972. Bird mortality at four towers in eastern North Dakota—fall 1972. Prairie Naturalist 4:87–95.

Bain, J.R. 1981. Roosting ecology of three Florida bats, *Nycticeius humeralis, Myotis austroriparius,* and *Tadarida brasiliensis.* M.S. thesis, University of Florida, Gainesville, FL.

Bain, J.R., and S.R. Humphrey. 1986. Social organization and biased sex ratios of the evening bat, *Nycticeius humeralis.* Florida Scientist 49:22–31.

Baird, J. 1960. A record of the hoary bat for Rhode Island. Narragansett Naturalist 13:56.

Baker, R.J., and C.M. Ward. 1967. Distribution of bats in southeastern Arkansas. Journal of Mammalogy 48:130–132.

Baker, R.R. 1978. The evolutionary ecology of animal migration. Holmes & Meier Publishers, New York.

Banfield, A.W.F. 1974. The mammals of Canada. University of Toronto Press, Toronto, Canada.

Barbour, R.W., and W.H. Davis. 1969. Bats of America. University Press of Kentucky, Lexington, KY.

Barclay, R.M.R., and R.M. Brigham. 2001. Year-to-year reuse of tree-roosts by California bats (*Myotis californicus*) in southern British Columbia. American Midland Naturalist 146:80–85.

Barclay, R.M.R., P.A. Faure, and D.R. Farr. 1988. Roosting behavior and roost selection by migrating silver-haired bats (*Lasionycteris noctivagans*). Journal of Mammalogy 69:821–825.

Barclay, R.M.R., C.L. Lausen, and L. Hollis. 2001. What's hot and what's not: defining torpor in free-ranging birds and mammals. Canadian Journal of Zoology 79:1885–1890.

Bartsch, P. 1956. An interesting catch. Journal of Mammalogy 37:111.

Beer, J.R. 1956. A record of a silver-haired bat in a cave. Journal of Mammalogy 37:282.

Bishop, S.C. 1947. Curious behavior of a hoary bat. Journal of Mammalogy 28:293–294.

Bowers, J.R., G.H. Heidt, and R.H. Baker. 1968. A late autumn record for the hoary bat in Michigan. Jack-Pine Warbler 46:33.

Boyles, J.G., J.C. Timpone, and L.W. Robbins. 2003. Late-winter observations of red bats, *Lasiurus borealis,* and evening bats, *Nycticeius humeralis,* in Missouri. Bat Research News 44:59–61.

Brack, V., Jr., and J.C. Carter. 1985. Use of an underground burrow by *Lasionycteris.* Bat Research News 26:28–29.

Brigham, R.M. 1995. A winter record for the silver-haired bat in Saskatchewan. Blue Jay 53:168.

Brimley, C.S. 1897. An incomplete list of mammals from Bertie, Co., N.C. American Naturalist 31:237–239.

Brisbin, I.L. 1966. Energy-utilization in a captive hoary bat. Journal of Mammalogy 47:719–720.

Brown, D.E. 1935. Hoary bat taken at Westport, Washington. The Murrelet 16:72.

Byre, V.J. 1990. A group of young Peregrine Falcons prey on migrating bats. Wilson Bulletin 102:728–730.

Caire, W., and M.L. Thies. 1987. The Seminole bat, *Lasiurus seminolus* (Chiroptera ; Vespertilionidae), from Central Oklahoma. Southwestern Naturalist 32:273–276.

Carter, T.C., W.M. Ford, and M.A. Menzel. 2002. Fire and bats in the Southeast and Mid-Atlantic: more questions than answers? pp. 139–143, *in* The role of fire for nongame wildlife management and community restoration: traditional uses and new directions (W.M. Ford, K.R. Russell, and C.E. Moorman, eds.). USFS General Technical Report NE-288.

Carter, T.C., M.A. Menzel, and D.A. Saugey. 2003. Population trends of solitary foliage-roosting bats, pp. 41–47, *in* Monitoring trends in bat populations of the United States and territories: problems and prospects (T.J. O'Shea and M.A. Bogan, eds.). U.S. Geological Survey Information Technology Report ITR 2003-0003.

Carter, T.D. 1950. On the migration of the red bat, *Lasiurus borealis borealis.* Journal of Mammalogy 31:349–350.

Clark, M.K. 1993. A communal winter roost of silver-haired bats, *Lasionycteris noctivagans* (Chiroptera; Vesperitilionidae). Brimleyana 19:137–139.

Cockrum, E.L. 1961. Southern yellow bat from Arizona. Journal of Mammalogy 42:97.

Collister, D. 1995. Silver-haired bat migration at Matlock, Manitoba. Blue Jay 53:110–112.

Constantine, D.G. 1958. Ecological observations on bats in Georgia. Journal of Mammalogy 39:64–70.

———. 1959. Ecological observations on lasiurine bats in the North Bay area of California. Journal of Mammalogy 40:13–15.

———. 1966. Ecological observations on lasiurine bats in Iowa. Journal of Mammalogy 47:34–41.

Cowan, I. M. 1933. Some notes on the hibernation of *Lasionycteris noctivagans.* Canadian Field-Naturalist 48:74–75.

———. 1942. Notes on the winter occurrence of bats in British Columbia. Murrelet 23:61.

Crawford, R.L., and W.W. Baker. 1981. Bats killed at a north Florida television tower: a 25-year record. Journal of Mammalogy 62:651–652.

Cryan, P.M. 2003. Seasonal distribution of migratory tree bats (*Lasiurus* and *Lasionycteris*) in North America. Journal of Mammalogy 84:579–593.

Cryan, P.M., M.A. Bogan, and J.S. Altenbach. 2000. Effect of elevation on distribution of female bats in the Black Hills, South Dakota. Journal of Mammalogy 81:719–725.

Cryan, P.M., M.A. Bogan, R.O. Rye, G.P. Landis, and C.L. Kester. 2004. Stable hydrogen isotope analysis of bat hair as evidence for seasonal molt and long-distance migration. Journal of Mammalogy 85:995–1001.

Cryan, P.M., and B.O. Wolf. 2003. Sex differences in the thermoregulation and evaporative water loss of a heterothermic bat, Lasiurus cinereus, during its spring migration. Journal of Experimental Biology 206:3381–3390.

Dalquest, W.W. 1943. Seasonal distribution of the hoary bat along the Pacific Coast. Murrelet 24:21–24.

Davis, W.H. 1970. Hibernation: ecology and physiological ecology, pp. 265–300, *in* Biology of bats (W.A. Wimsatt, ed.). Academic Press, New York.

Davis, W.H., and W.Z. Lidicker Jr. 1956. Winter range of the red bat, *Lasiurus borealis.* Journal of Mammalogy 37:280–281.

Downes, W.L., Jr. 1964. Unusual roosting behavior in red bats. Journal of Mammalogy 45:143.

Fassler, D.J. 1975. Red bat hibernating in woodpecker hole. American Midland Naturalist 93:254.

Findley, J.S., and C. Jones. 1964. Seasonal distribution of the hoary bat. Journal of Mammalogy 45:461–470.

Fleming, T.H., and P. Eby. 2003. Ecology of bat migration, pp. 156–208, *in* Bat ecology (T.H. Kunz and M.B. Fenton, eds.). University of Chicago Press, Chicago, IL.

Frum, W.G. 1953. Silver-haired bat, *Lasionycteris noctivagans,* in West Virginia. Journal of Mammalogy 34:499–500.

Ganier, A.F. 1962. Bird casualties at a Nashville T-V tower. The Migrant 33:58–60.

Gellman, S.T., and W.J. Zielinski. 1996. Use by bats of old-growth redwood hollows on the north coast of California. Journal of Mammalogy 77:255–265.

Genoud, M. 1990. Seasonal variations in the basal rate of metabolism of subtropical insectivorous bats (*Nycticeius humeralis* and *Lasiurus seminolus*): a comparison with other mammals. Revue Suisse de Zoologie 97:77–90.

———. 1993. Temperature regulation in subtropical tree bats. Comparative Biochemistry and Physiology 104A:321–331.

Glass, B.P. 1961. Two noteworthy records of bats for Oklahoma. Southwestern Naturalist 6:200–201.

Gosling, N.M. 1977. Winter record of silver-haired bat, *Lasionycteris noctivagans* (Le-Conte), in Michigan. Journal of Mammalogy 58:657.

Griffin, D.R. 1970. Migration and homing of bats, pp. 233–264, *in* Biology of bats (W.A. Wimsatt, ed.). Academic Press, New York.

Grimm, C.T. 1977. Hoary bat in Niantic, Connecticut, in January. Proceedings of the Linnaean Society of New York 73:84–86.

Grinnell, H.W. 1918. A synopsis of the bats of California. University of California Publications in Zoology 17:223–404.

Grinnell, J. 1937. Mammals of Death Valley. Proceedings of the California Academy of Sciences 23:115–169.

Gruver, J.C. 2002. Assessment of bat community structure and roosting habitat preferences for the hoary bat (*Lasiurus cinereus*) near Foote Creek Rim, Wyoming. M.S. thesis, University of Wyoming, Laramie, WY.

Hall, E.R. 1938. Mammals from the Touchwood Hills, Saskatchewan. Canadian Field-Naturalist 52:108–109.

———. 1946. Mammals of Nevada. University of California Press, Berkeley, CA.

Hayes, J.P. 2003. Habitat ecology and conservation of bats in western coniferous forests, pp. 81–119, *in* Mammal community dynamics in coniferous forests of western North America: management and conservation (C. J. Zabel and R. G. Anthony, eds.). Cambridge University Press, Cambridge, MA.

Howell, A.H. 1908. Notes on diurnal migrations of bats. Proceedings of the Biological Society of Washington 21:35–37.

Humphrey, S.R., and J.B. Cope. 1976. Population ecology of the little brown bat, *Myotis lucifugus,* in Indiana and north-central Kentucky. Special Publications of the American Society of Mammalogists 4:1–81.

Izor, R.J. 1979. Winter range of the silver-haired bat. Journal of Mammalogy 60:641–643.

Jackson, H.H.T. 1961. Mammals of Wisconsin. University of Wisconsin Press, Madison, WI.

Johnson, G.D., W.P. Erickson, M.D. Strickland, M.F. Shepherd, and D.A. Shepherd. 2003. Mortality of bats at a large-scale wind power development at Buffalo Ridge, Minnesota. American Midland Naturalist 150:332–342.

Keith, A.R. 1978. Hoary bat at Martha's Vineyard in winter. Cape Naturalist 7:52–53.

Koontz, T., and W.H. Davis. 1991. Winter roosting of the red bat, *Lasiurus borealis.* Bat Research News 32:3–4.

Krutzsch, P.H. 1966. Remarks on silver-haired and Leib's bats in eastern United States. Journal of Mammalogy 47:121.

Kunz, T.H. 1982. *Lasionycteris noctivagans.* Mammalian Species 172:1–5.

Kunz, T.H., and D.S. Reynolds. 2003. Bat colonies in buildings, pp. 91–102, *in* Monitoring trends in bat populations of the United States and territories: problems and prospects (T.J. O'Shea and M.A. Bogan, eds.). U.S. Geological Survey Information Technology Report ITR 2003-0003.

Kurta, A., and S.W. Murray. 2002. Philopatry and migration of banded Indiana bats (*Myotis sodalis*) and effects of radio transmitters. Journal of Mammalogy 83:585–589.

Kurta, A., S.W. Murray, and D. H. Miller. 2002. Roost selection and movements across the summer landscape, pp. 118–129, *in* The Indiana bat: biology and management of an endangered species (A. Kurta and J. Kennedy, eds.). Bat Conservation International, Austin, TX.

LaVal, R.K., and M.L. LaVal. 1979. Notes on reproduction, behavior, and abundance of the red bat, *Lasiurus borealis*. Journal of Mammalogy 60:209–212.

Layne, J.N. 1958. Notes on mammals of southern Illinois. American Midland Naturalist 60:219–254.

Mearns, E.A. 1898. A study of the vertebrate fauna of the Hudson Highlands, with observations on the Mollusca, Crustacea, Lepidoptera, and the flora of the region. Bulletin of the American Museum of Natural History 10:303–352.

Merriam, C.H. 1884. The mammals of the Adirondack Region, northeastern New York. Privately published, New York, NY.

Miller, G.S., Jr. 1897. Migration of bats on Cape Cod, Massachusetts. Science 5:541–543.

Moorman, C.E., K.R. Russell, M.A. Menzel, S.M. Lohr, J.E. Ellenberger, and D.H. Van Lear. 1999. Bats roosting in deciduous leaf litter. Bat Research News 40:74–75.

Mumford, R.E. 1963. A concentration of hoary bats in Arizona. Journal of Mammalogy 44:272.

———. 1973. Natural history of the red bat (*Lasiurus borealis*) in Indiana. Periodicum Biologorum 75:155–158.

Mumford, R.E., and J.O. Whitaker Jr. 1982. Mammals of Indiana. Indiana University Press, Bloomington, IN.

Murphy, R.C., and J.T. Nichols. 1913. Long Island fauna and flora-1: the bats. Science Bulletin of the Museum, Brooklyn Institute of Arts and Sciences 2:1–14.

Myers, R.F. 1960. *Lasiurus* from Missouri caves. Journal of Mammalogy 41:114–117.

Nagorsen, D.W., A.A. Bryant, D. Kerridge, G. Roberts, A. Roberts, and M.J. Sarell. 1993. Winter bat records for British Columbia. Northwestern Naturalist 74:61–66.

Neill, W.T. 1952. Hoary bat in a squirrel's nest. Journal of Mammalogy 33:113.

Neuhauser, H.N., and I.L. Brisbin. 1969. Energy utilization in a captive silver-haired bat. Bat Research News 10:30–31.

Orr, R.T. 1950a. Unusual behavior and occurrence of a hoary bat. Journal of Mammalogy 31:456–457.

———. 1950b. Notes on the seasonal occurrence of red bats in San Francisco. Journal of Mammalogy 31:457–458.

Padgett, T.M., and R.K. Rose. 1991. Bats (Chiroptera: Vespertilionidae) of the Great Dismal Swamp of Virginia and North Carolina. Brimleyana 17:17–25.

Pearson, E.W. 1962. Bats hibernating in silica mines in southern Illinois. Journal of Mammalogy 43:27–33.

Peurach, S.C. 2003. High-altitude collision between an airplane and hoary bat, *Lasiurus cinereus*. Bat Research News 44:2–3.

Ransome, R. 1990. The natural history of hibernating bats. Christopher Helm, London, U.K.

Robbins, L.W., J.G. Boyles, B.M. Mormann, and M.B. Milam. 2004. Fall, winter, and spring roosting behavior of eastern red bats and evening bats in Missouri. Bat Research News 45:69.

Russo, D. 2002. Elevation affects the distribution of the two sexes in Daubenton's bats *Myotis daubentonii* (Chiroptera: Vespertilionidae) from Italy. Mammalia 66:543–551.

Sanborn, C.C. 1953. April record of silver-haired bat in Oregon. The Murrelet 34:32.

Saugey, D.A., R.L. Vaughn, B.G. Crump, and G.A. Heidt. 1998. Notes on the natural history of *Lasiurus borealis* in Arkansas. Journal of the Arkansas Academy of Science 52:92–98.

Saunders, W.E. 1930. Bats in migration. Journal of Mammalogy 11:225.

Schowalter, D.B., W.J. Dorward, and J.R. Gunson. 1978. Seasonal occurrence of silver-haired bats (*Lasionycteris noctivagans*) in Alberta and British Columbia. Canadian Field-Naturalist 92:288–291.

Sedgeley, J. 2001. Winter activity in the tree-roosting lesser short-tailed bat, *Mystacina tuberculata,* in a cold-temperate climate in New Zealand. Acta Chiropterologica 3:179–195.

Sherman, H.B. 1956. Third record of the hoary bat in Florida. Journal of Mammalogy 37:281–282.

Shump, K.A., Jr., and A.U. Shump. 1982. *Lasiurus cinereus.* Mammalian Species 185:1–5.

Sluiter, J.W., A.M. Voute, and P.F. Van Heerdt. 1973. Hibernation of *Nyctalus noctula.* Periodicum Biologorum 75:181–188.

Smith, P.W., and P.W. Parmalee. 1954. Notes on the distribution and habits of some bats from Illinois. Transactions of the Kansas Academy of Science 57:200–205.

Speakman, J.R., and D.W. Thomas. 2003. Physiological ecology and energetics of bats, pp. 430–490, *in* Bat ecology (T.H. Kunz and M.B. Fenton, eds.). University of Chicago Press, Chicago, IL.

Sperry, C.C. 1933. Opossum and skunk eat bats. Journal of Mammalogy 14:152–153.

Strelkov, P.P. 1969. Migratory and stationary bats (Chiroptera) of the European part of the Soviet Union. Acta Zoologica Cracoviensia 14:393–436.

Stuewer, F.W. 1948. A record of red bats mating. Journal of Mammalogy 29:180–181.

Szewczak, J.M., S.M. Szewczak, M.L. Morrison, and L.S. Hall. 1998. Bats of the White and Inyo Mountains of California-Nevada. Great Basin Naturalist 58:66–75.

Tenaza, R.R. 1966. Migration of hoary bats on South Farallon Island, California. Journal of Mammalogy 47:533–535.

Terres, J.K. 1956. Migration records of the red bat, *Lasiurus borealis.* Journal of Mammalogy 37:442.

Thomas, O. 1921. Bats on migration. Journal of Mammalogy 2:167.

Timm, R.M. 1989. Migration and molt patterns of red bats, *Lasiurus borealis* (Chiroptera: Vespertilionidae) in Illinois. Bulletin of the Chicago Academy of Sciences 14:1–7.

Turner, R.W. 1974. Mammals of the Black Hills of South Dakota and Wyoming. University of Kansas Publications of the Museum of Natural History, Miscellaneous Publications 60:1–178.

Tyler, J.D., and L. Payne. 1982. Second Oklahoma record for the silver-haired bat, *Lasionycteris noctivagans.* Southwestern Naturalist 27:245.

Van Gelder, R.G. 1956. Echo-location failure in migratory bats. Transactions of the Kansas Academy of Science 59:220–222.

Van Heerdt, P.F., and J.W. Sluiter. 1965. Notes on the distribution and behavior of the noctule bat (*Nyctalus noctula*) in the Netherlands. Mammalia 29:463–477.

Vaughan, T.A. 1953. Unusual concentration of hoary bats. Journal of Mammalogy 34:256.

Veilleux, J.P., and S.L. Veilleux. 2004. Intra-annual and inter-annual fidelity to summer roost areas by female eastern pipistrelles, *Pipistrellus subflavus.* American Midland Naturalist 152:196–200.

Veilleux, J.P., J.O. Whitaker Jr., and S.L. Veilleux. 2004. Reproductive stage influences roost use by tree roosting female eastern pipistrelles, *Pipistrellus subflavus.* Ecoscience 11:249–256.

Watkins, L.C. 1972. *Nycticeius humeralis.* Mammalian Species 23:1–4.

Whitaker, J.O., Jr. 1967. Hoary bat apparently hibernating in Indiana. Journal of Mammalogy 48:663.

Whitaker, J.O., Jr., R.K. Rose, and T.M. Padgett. 1997. Food of the red bat *Lasiurus borealis* in winter in the great dismal swamp, North Carolina and Virginia. American Midland Naturalist 137:408–411.

Willis, C.K.R. 2003. Physiological ecology of roost selection in female, forest-living big brown bats (*Eptesicus fuscus*) and hoary bats (*Lasiurus cinereus*). Ph.D. dissertation, University of Regina, Regina, Saskatchewan.

Willis, C.K.R., and R.M. Brigham. 2005. Physiological and ecological aspects of roost selection by reproductive female hoary bats (*Lasiurus cinereus*). Journal of Mammalogy 86:85–94.

Willis, C.K.R., K.A. Kolar, A.L. Karst, M.C. Kalcounis-Rüeppel, and R.M. Brigham. 2003. Medium-and long-term reuse of trembling aspen cavities as roosts by big brown bats (*Eptesicus fuscus*). Acta Chiropterologica 5:85–90.

Zinn, T.L., and W.W. Baker. 1979. Seasonal migration of the hoary bat, *Lasiurus cinereus,* through Florida. Journal of Mammalogy 60:634–635.

SILVICULTURAL PRACTICES AND MANAGEMENT OF HABITAT FOR BATS

7

James M. Guldin, William H. Emmingham, S. Andrew Carter, and David A. Saugey

The twenty-first century has seen a shift in the philosophy and practice of forestry. Historic assumptions that prevailed as recently as three decades ago have been challenged in light of new concepts and practices, developed through advances in research and lessons from practical experience. The goals of forest management today encompass a wider array of resources than in the past, and managers are using a wider range of tools and techniques to provide them.

For example, consider the evolution in wildlife management. The old prevailing wisdom held that good forest management, which was focused on timber objectives, was also good wildlife management (Bissonette 1986). This reflected the perceived benefits of using clearcutting to harvest stands, and the resulting response of plant species that provided soft mast and browse for deer (*Odocoileus virginianus*) and wild turkey (*Meleagris gallopavo*). These game species were of special interest to state wildlife management agencies and to outdoor enthusiasts who enjoyed hunting. Today, in general, it is understood that a particular forest management strategy is desirable or undesirable for wildlife, depending on how the habitat for the wildlife species or community of interest is affected (Bissonette 1986; Hunter 1990). Thus, if a harvest creates early successional conditions, it will favor wildlife species that use early successional habitat, but will not favor species that depend on older forests.

The practice of silviculture has evolved as well. Three decades ago, it was socially acceptable and a matter of public policy to maximize timber production through intensive forestry that involved clearcutting followed by the planting of fast-growing trees (Behan 1990; Kessler et al. 1992; Swanson and Franklin 1992). As a result, silviculture came to be associated primarily with intensive forestry for timber production, especially by the public. That view is too narrow by today's standards, where demands of society for forest resources go beyond timber production. Contemporary forest management goals include health, diversity, productivity, and sustainability of ecosystems and the diverse elements they contain; those goals are achieved through silviculture. This view has conceptual advantages beyond those of multiple use, especially if those multiple uses are in conflict (Behan 1990; Franklin 1989; Kessler et al. 1992; O'Hara et al. 1994; Swanson and Franklin 1992).

The value that a given society holds for forests and forestry depends on the availability and abundance of forests, whether forests are or have been exploited, and whether people are concerned about the sustainability of forests and forest resources into the future. The views of the society at large regarding forest management typically mature through four stages (Kimmins 1991, 1992): (1) unregulated exploitation of local forests and clearing of forests for agriculture and grazing; (2) institution of legal and political mechanisms or religious taboos to regulate exploitation; (3) development of an ecological approach to timber management with the goal of sustainable management of the biological resources of the forest; and (4) social forestry, which recognizes the need to manage the forest as a multifunctional resource in response to the diverse demands of modern society.

In North America, steps 1 and 2 can often be seen on private forestlands where owners are unfamiliar with the concepts of forestry and simply cut trees when they can be profitably sold. In many areas of the country, urban sprawl might be classified as step 1 if the definition included clearing for urban development. Industrial forestlands that support intensive management to provide sustainable harvests would be classified at some intermediate stage between steps 2 and 3. Forestry on public lands is generally at step 3, but changing attitudes about public forestry in North America (Kessler et al. 1992) suggest a transition into the fourth step in Kimmins' hierarchy.

It is the fourth step of Kimmins' hierarchy that holds the most promise for managers and biologists to work together. As more demands are placed on forests, the resource attributes needed to meet those demands expand and often conflict. These conflicts can quickly overwhelm managers (Farrell et al. 2000; Kessler et al. 1992). Perhaps the greatest hope for resolving such conflict is to manage forests for patterns and processes that restore and maintain ecosystems (Kessler et al. 1992). Management recommendations could then be related to how they are integral to essential ecosystem function.

In this chapter we seek to familiarize people interested in the ecology and conservation of bats with silvicultural practices used by foresters, and comment on the implications of those practices for forest-dwelling bats. We hope to dispel old notions that equate silviculture exclusively with timber management, to broaden the definition of silviculture to include nontimber habitat management, and to facilitate communication between biologists and foresters about practical modifications of existing silvicultural practices to promote habitat for bats in forests.

SILVICULTURE: AN OVERVIEW

Both the old forestry and the new are implemented using silviculture. Silviculture can be defined as the science and art of manipulating a stand of trees toward a desired future condition. For example, a desired future condition could be old growth forest, foraging habitat for the Indiana myotis (*Myotis sodalis*), pulpwood destined for a mill, or nearly anything someone could describe using quantitative or qualitative descriptions of the stand of forest under management.

There are two main elements to silviculture: the individual treatment practices that make a short-term change in the conditions of a forest stand, and the systems or prescriptions developed by arranging individual treatments over time through the life of the forest stand. The individual practices can often be quantified in great detail and, in general, are based on current scientific literature, experimentation, and implementation. The systems or prescriptions are based less on science and more on creativity, experience, and adaptation, elements that are perhaps better described as art rather than science.

Silvicultural practices are designed to indirectly manipulate supplies of water, nutrients, and incident solar radiation by removing undesired plants that usurp resources from desired plants. A silviculturist sees a forest stand as a collection of individuals of different species, each of which uses resources according to its size and species attributes. Some individuals are to be retained to meet the goals of the forest owner. Other individuals that unduly compete with those to be retained are then removed. Removal can be through felling, girdling, use of herbicides, top-killing with fire, or any other practice that effectively reduces or eliminates the ability of undesirable plants to compete with desired individuals. If the removal of undesired trees can be accomplished by having someone else pay the landowner for them, cut them down, and haul them away to a mill, so much the better for the landowner. But this is a secondary outcome to the primary silvicultural objective of redirecting the flow of resources on a site to the trees that are to be favored.

Increasing growth rates is fundamental to the goals of most forest landowners, whether these goals are economical, social, or ecological. Human longevity is a fraction of the timescale required for forest development to occur; thus, landowners and foresters do not have the luxury of waiting for a desired stand condition to develop on its own. Silviculture enables trees to grow faster, so that desired stand conditions can be achieved more rapidly. For example, suppose a stand with old-growth structural attributes is needed to satisfy a desired future condition. Foresters could wait the requisite centuries for that structure to develop, or silvicultural practices could be used to accelerate development of the desired structural features in a shorter time frame.

Often, a silviculturist must choose among several ways to implement a practice to achieve a certain objective. For example, if the objective is to remove small midstory hardwoods in a mature pine stand, many different treatment alternatives exist such as the use of herbicides, prescribed fire, or manual chainsaw felling. Each alternative will achieve the intended effect, but will result in slightly different conditions following treatment, and those posttreatment differences will be accentuated over time. Moreover, each alternative varies in the costs and human resources needed to implement the treatment.

Silvicultural systems and prescriptions are designed to meet the long-term objectives of the forest owner. In light of the expected changes in stand age and condition, a prescription is little more than a list of the silvicultural practices planned for the stand over time. Prescriptions typi-

Table 7.1. An example of a silvicultural prescription for managing even-aged, naturally regenerated loblolly-shortleaf pine stands under the seed-tree silvicultural system in the upper West Gulf Coastal Plain

Step	Stand age (yr)	Operation and probability of event	Starting condition		Expected results		
			N	BA	N	BA	D
	0	(seedfall from seed trees)	–	–	–	–	–
1	0	prescribed burning	–	–	–	–	–
2	3	removal cut, harvest seed trees	10	2	0	0	0
3	4	precommercial thinning (with rolling chopper)	>8,000	8	4,000	4	3
4	4	precommercial thinning (with brush saw)	4,000	4	1,500	2	3
5	4	pine release (with brush saw)	1,500	2	1,100	1	3
6	14	prescribed burning	–	–	–	–	–
7	14	pulpwood thinning (long pulpwood)	1,100	25	600	14	18
8	15	salvage cut after ice storm (~10% in any given year)	600	15	500	13	18
9	20	prescribed burning	–	–	–	–	–
10	20	pulpwood thinning (long pulpwood)	500	22	300	14	23
11	26	prescribed burning	–	–	–	–	–
12	26	salvage cut after ice storm (~10% in any given year)	300	22	220	16	30
13	31	prescribed burning	–	–	–	–	–
14	31	thinning (sawlogs)	220	22	160	16	36
15	36	prescribed burning	–	–	–	–	–
16	36	thinning (sawlogs)	160	21	125	16	41
17	41	prescribed burning	–	–	–	–	–
18	41	thinning (sawlogs)	125	18	100	16	43
19	45	prescribed burning	–	–	–	–	–
20	45	seed cut, seed-tree reproduction cutting method	100	17	10	2	48

Source: From Zeide and Sharer 2000.
Note: The overall system consists of 20 separate silvicultural treatments imposed over a 45-year period (N, stem density, trees/ha; BA, basal area, m2/ha; D, average stem diameter, cm).

cally include the time frame for which the system is being developed, the treatments that are planned, the expected time when each treatment will be applied, technical details about how each treatment will be applied, and the conditions expected before and after each treatment (table 7.1). In essence, the prescription is a detailed long-term plan for the forest stand being managed.

In general, foresters make a distinction between the practice of silviculture and that of forest management. The usual domain of silviculture is the forest stand, whereas that of forest management is the forest as a whole, containing all stands under a given ownership (Baker et al. 1996; Smith 1986; Wiersum 1995). Under the old paradigm, forest management plans were based primarily on the sustainable yield of timber over the long term (Baker et al. 1996; Behan 1990; Farrell et al. 2000; Kessler et al. 1992; Wiersum 1995). As modern ideas of sustainability expand to embrace all the values that forests provide, the practice of silviculture has grown to reflect an understanding of how landscapes function, and how the treatments conducted within each of the individual forest stands affect that function. This means that a silviculturist planning a treatment in a given stand must take into account the activities occurring in adjacent stands, not only within an ownership but across ownerships as well (Baker et al. 1996; Farrell et al. 2000; Franklin 1989). This is often easier to do in landscapes dominated by public lands than in those that consist of numerous private landowners.

Finally, although some may find it counterintuitive, the modern perspective on silviculture is dramatically enhanced if active markets for timber within a region are present (Kessler et al. 1992). Owning, staffing, and managing a forest are not cost free (Hunter 1990). Because trees have commercial value as lumber and pulp, the costs of conducting silvicultural treatments often can be partly or completely defrayed by selling the harvested trees.

An example of managing for ecosystems and ecosystem attributes on a landscape scale is found in the restoration of the shortleaf pine-bluestem (*Pinus echinata-Andropogon* spp.) ecosystem in the Ouachita Mountains of Arkansas and Oklahoma (Bukenhofer et al. 1994; Bukenhofer and Hedrick 1997; Guldin and Guldin 2003; Hedrick et al. 1998). At the time of European colonization, these forests were much more open than today. Fires no doubt contributed to that openness (Foti and Glenn 1991; Mattoon 1915). Seventy years of fire suppression resulted in denser stands with more trees in the smaller size classes (fig. 7.1), which had detrimental effects on a host of flora and fauna adapted to open understory conditions, including the endangered red-cockaded woodpecker (*Picoides borealis*). To restore the desired historic condition, a silvicultural prescription was developed to thin overstory and midstory pines using commercial timber sales, to remove midstory hardwoods by mechanical treatment, and to reintroduce surface fires on a one- to three-year interval. Funds generated by the timber sales helped defray the cost of the midstory removal and the prescribed fire treatments. As a result, treated stands now support many plant and animal species adapted to open Ouachita woodlands (fig. 7.2), including expanding numbers of the red-cockaded woodpecker. Thus, a desired ecological outcome was achieved using classical silvicultural treatments and existing timber markets.

SILVICULTURAL PRACTICES

Silvicultural treatments can be divided into three categories that are correlated with tree size: reproduction cutting methods (large trees), regeneration treatments (seedlings and saplings), and intermediate treatments (small or immature trees larger than saplings) (Smith 1986). The goal of reproduction cutting is to harvest mature trees to favor the establishment and development of new trees; the species composition, desired spacing, and tolerance to shade of the species being managed determine what kinds of reproduction cutting methods might be effective in the forest

Figure 7.1.
A shortleaf pine (*Pinus echinata*) stand in the western Ouachita Mountains after 70 years of fire exclusion and before restoration treatment.
Photo by J. Guldin

Figure 7.2.
A shortleaf pine (*Pinus echinata*) stand in the western Ouachita Mountains after thinning in the overstory, removal of encroaching midstory hardwoods, and initiation of cyclic prescribed fire.
Photo by J. Guldin

type being managed. The goals of regeneration treatments are to prepare a site for seed or seedlings of the desired species, to either plant seedlings, scatter seed, or encourage natural seedfall of desired species, and to promote proper development of young trees. The focus of intermediate treatments is to reduce competition to levels that favor continued growth of the desired trees.

REPRODUCTION CUTTING METHODS

Reproduction cutting methods are used when a decision is made to harvest all or part of the mature trees in a stand, and to establish new trees to perpetuate a succeeding generation of trees. Even-aged methods are characterized by one or two age classes of the desired species, and operate over a length of time called a "rotation," which lasts from establishment of regeneration to final harvest. Clearcutting, seed-tree, and shelterwood methods are the cutting practices typically used for even-aged systems (Smith et al. 1997). Uneven-aged methods are characterized by three or more age classes of the desired species of trees. Single-tree selection and group-selection methods are used for uneven-aged systems (Baker et al. 1996; Smith et al. 1997). The methods vary by the number and distribution of trees retained on the site and by the ecological conditions for regeneration that are created.

CLEARCUTTING METHOD

In the clearcutting method, all or most of the trees are removed from the stand. Timber sales typically are used to harvest trees of commercial value, and subsequent treatments are used to remove the remaining trees. Depending on ownership objectives, some snags and living trees can be left standing to enhance visual qualities, meet habitat objectives, or provide structural elements that would otherwise be missing from young stands.

In more intensive applications, clearcut stands are generally reforested by plantings. If timber production is an important goal, stands are often planted using genetically improved planting stock of fast-growing species such as Douglas-fir (*Pseudotsuga menziesii*), loblolly pine (*Pinus taeda*), or cottonwood (*Populus deltoides*). In some situations, clearcuts can be reforested by use of direct seeding, although spacing uniformity is often sacrificed. Additionally, because of the cost of genetically improved seeds, direct seeding does not have the same opportunity for genetic improvement as planting does.

Clearcuts also can be reforested through natural regeneration. The most common applications of this are in hardwood stands in eastern North America, where the succeeding stand originates from saplings previously existing in the stand prior to the clearcut and from stump sprouts from harvested trees. In upland oak stands, for example, considerable effort is made to encourage development of regeneration of desired species of suitable size before the clearcut occurs (Sander et al. 1983). In other forest types, such as aspen stands in the Lake States, harvested stands sprout vigorously and little supplemental effort is required to obtain abundant regeneration (Perala and Russell 1983).

It is more problematic to rely on natural regeneration following clearcutting of species that do not sprout, such as pines. One successful approach is to depend on seed fall from adjacent stands. This will work only if the clearcut is sufficiently narrow or otherwise oriented such that all parts of the clearcut are within the effective seeding distance of mature trees in adjacent stands. Harvesting can be relied on to scatter seeds throughout a stand if mature seeds are present within the crowns of trees to be harvested. This approach works well in species that retain mature seeds in their crowns for an extended time, such as jack pine (*Pinus banksiana*), sand pine (*P. clausa*), and lodgepole pine (*P. contorta*). Otherwise, the precise timing required for initiating and completing the clearcut between seed maturation and dispersal is rarely attainable.

From an ecological perspective, the clearcutting method attempts to mimic large-scale disturbances such as crown fires, tornados, and insect outbreaks. It is well suited to tree species that cannot survive shading, such as aspen (*Populus tremuloides, P. grandidentata*), paper birch (*Betula papyrifera*), red alder (*Alnus rubra*), and most pines. It is also well suited for providing habitat for animals that use open conditions. For example, the silver-haired bat (*Lasionycteris noctivagans*) is known to use clearcuts (Patriquin and Barclay 2003).

SEED-TREE METHOD

The seed-tree method is similar to clearcutting, except a small number of mature seed-bearing trees, typically no more than 10–25 trees/ha, are retained to reseed the stand. After the new age class is in place, the seed trees are usually removed. It is not always economically feasible to do so, however, and in these instances seed trees could be used to create snags or be left as relict trees to provide roosting habitat for bats.

Many of the physical characteristics of trees vary among individuals and are highly heritable. Because seed trees are the primary source of seed for the succeeding stand, keeping seed trees that have desirable traits is important. Silviculturists frequently select for desirable economic traits such as straightness of the stem or growth form. Seed-producing ability is also a highly inherited trait, and field personnel are often instructed to evaluate candidate seed trees based on evidence of past fruitfulness. Because traits of trees that are beneficial to bats and those beneficial to timber quality may differ, biologists should seek to identify whether traits that affect bats, such as bark characteristics, are heritable. If so, field crews could be instructed to retain seed trees with those traits when marking a stand.

The seed-tree method works best with tree species that regenerate readily following major disturbance events such as fire or windstorms. These species are shade intolerant, in general, and have light, readily dispersed seeds such as aspen, paper birch, western larch (*Larix occidentalis*), and southern pines (Young and Giese 1990). In the southern United States, the seed-tree method is well suited to loblolly pine, a disturbance-adapted species that is a prolific seed producer (Cain and Shelton 2001). Heavy-seeded species, such as oaks (*Quercus* spp.), hickories (*Carya* spp.), and longleaf pine (*P. palustris*) are poorly adapted to this method because they have irregular seed production, limited seed dispersal capability, and rely on seedlings being present in the understory prior to disturbance. Wildlife species favored by the seed-tree method will be similar to those favored by clearcutting. Regardless, retaining seed trees may provide roosting opportunities for bats that are otherwise not available in clearcuts.

SHELTERWOOD METHOD

The shelterwood method retains some of the mature trees to act both as a seed source and to partially shade the ground. The number of trees retained depends on form and crown shape, but typically varies from 30–60 trees/ha. Because more trees are retained, seeds do not have to travel as far to reforest a site; thus, the shelterwood method can be used with heavy-seeded species such as oaks and longleaf pine. Shading from the residual trees helps ameliorate harsh climatic conditions (e.g., summer frost or high soil-surface temperatures) at ground level.

Smith (1986) describes three specific treatments in the shelterwood method: the preparatory cut, the seed cut, and the removal cut. The preparatory cut is a late-rotation thinning used to promote crown development in future seed trees by removing other trees that compete with them. The seed cut removes all trees except those intended to reseed the stand. The removal cut harvests the seed trees once a new stand is successfully established. This is typically done five to ten years after the seed cut. In some cases, the removal cut may be deferred for half or more of the subsequent rotation, which results in a two-aged stand (Helms 1998; Smith 1986). The shelterwood method mimics relatively small small-scale or moderate-intensity disturbances. Because the number of residual trees retained in the shelterwood method varies depending on the tree species, the method might benefit bats adapted to edges and also those species of bats adapted to open woodlands.

SINGLE-TREE SELECTION METHOD

This uneven-aged method involves periodic harvesting of individual mature trees scattered across the stand to establish new seedlings in small openings created by the harvest. In general, no more than 30% of the stand is harvested at any one time (Young and Giese 1990), and harvesting is typically done every 10 to 20 years. One aspect unique to single-tree selection is that if careful attention is paid to the size distribution of trees, a continuous yield of timber products can be generated from a single stand. This method is particularly well suited to the forest landowner whose ownership is of limited area.

The single-tree selection method imitates conditions caused by the death of an individual tree in an unmanaged mature forest (fig. 7.3). When a tree dies, the sunlight, water, and nutrients that were used by that tree become available for other plants in the immediate vicinity. Shade-

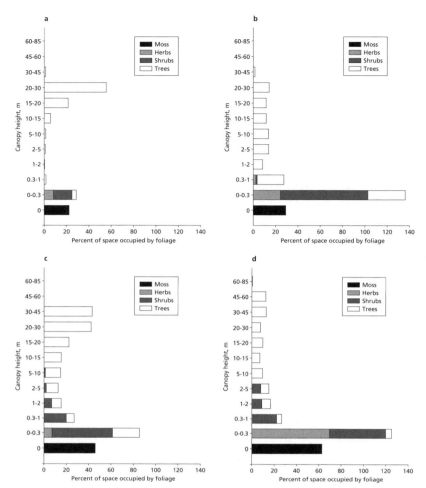

Figure 7.3.
Percent of space occupied by foliage of four different types of vegetation at different heights: (a) a managed even-aged stand; (b) a managed uneven-aged stand; (c) an unmanaged even-aged stand; (d) an old-growth stand. Sample stands are in the Douglas-fir (*Pseudotsuga menziesii*) forest type in the western Cascade Mountains (W. Emmingham, unpubl. data).

tolerant tree seedlings, saplings in the understory and lower canopy of the forest, and trees whose crowns are adjacent to the space that the dead tree occupied in the upper canopy, all respond with faster growth. As this process continues over time, forests of shade-tolerant species are naturally maintained with individual trees in many size classes.

Foresters use the single-tree selection method to improve upon this process, primarily by trying to manage a balance of trees of different sizes and age classes. Size, rather than age, is generally used to describe such stands because age and size are poorly correlated in shade-tolerant species. Typically, individuals will grow until resources are limiting, and then persist with little additional growth until some minor disturbance frees up additional resources to allow the tree to resume growth. In a managed, single-tree selection stand, reproduction cutting is sufficiently frequent to maintain acceptable growth rates over time. Foresters use mathematical models of stem density versus diameter class as a target to regulate harvests in single-tree selection stands, so that the proper numbers of trees of different size classes can be maintained. Several quantitative and ecological approaches exist to determine the appropriate ways to regulate harvests (Baker et al. 1996; Marquis 1978; O'Hara and Gersonde 2004).

Single-tree selection is most appropriate for shade-tolerant trees, such as sugar maple (*Acer saccharum*), yellow birch (*Betula alleghaniensis*), American beech (*Fagus grandifolia*), hemlocks (*Tsuga* spp.), some cedars (*Thuja* spp. and *Chamaecyparis lawsoniana*), and most firs (*Abies* spp.) and spruces (*Picea* spp.). Shade-intolerant tree species generally do not survive in the heavy shade of single-tree selection, and implementing single-tree selection on a stand of shade-intolerant trees generally converts the stand to one comprising shade-tolerant species.

The method has been successfully adapted to intolerant loblolly-shortleaf pine stands in the western Gulf region (Baker et al. 1996). In that region, silvicultural practices feature frequent harvests and retention of relatively low levels of residual basal area, which leads to optimal growth in sawtimber-sized trees. The frequent cutting and low basal area also ensure that midstory and understory trees maintain relatively uniform growth rates over time (Baker et al. 1996; Guldin and Baker 1998).

GROUP-SELECTION METHOD

The group-selection method is similar to the single-tree selection method, except that groups of trees are removed and new age classes are established largely within the group openings. These openings are designed to imitate disturbance events that remove small groups of trees, such as a localized insect infestation, a spot where a surface fire becomes hot enough to kill a few overstory trees, or where a localized gust in a windstorm blows down a small part of the stand. As in single-tree selection, foresters use the group-selection method to balance stem density in all the size classes of the stand, but recognize that the size classes will be aggregated in the group openings created at different times during the life of the stand.

Group openings can range in size from a few trees to a hectare or more. Ecologically, the upper size limit for a circular group opening on level terrain is found when the radius of the opening equals the height of the surrounding trees in the stand (Helms 1998). In openings of this size or smaller, the trees surrounding the opening cast a significant amount of shade; hence, smaller openings favor shade-tolerant species. In openings larger than this, the shade cast by the surrounding trees does not alter regeneration development in the center of the opening; therefore, larger openings are thought to favor species that are intolerant of shade. However, in a large 1-ha group opening (radius, 56.4 m) with surrounding trees 17 m tall or taller, most of the area within the opening still lies within one tree height of the edge of the opening, a zone within which shade-intolerant species would not compete well. Thus, group selection is inefficient relative to even-aged methods for managing shade-intolerant species.

Harvests are regulated in the group-selection method in the same way as in the single-tree selection method (Baker et al. 1996; Marquis 1978; O'Hara and Gersonde 2004), with one notable exception. It is increasingly common for the group-selection method to be implemented with an area-based approach. This is done by harvesting a percentage of the stand equal to the number of years between harvests divided by the average age of a mature tree. For example, in a stand where mature trees are harvested on average at age 50 and harvests are done every 10 years, 20% of the stand would be cut at each entry (10/50 = 20%). However, if the group openings are large, of uniform size within the stand, and placed in a geometric pattern within the stand rather than as dictated by stand conditions, this method is more appropriately identified as variation of the clearcutting method called patch-clearcutting (Smith 1986).

EVEN-AGED VERSUS UNEVEN-AGED METHODS

In general, even-aged cutting methods cause more site disturbance than uneven-aged methods, and the scale of effect is correlated with the intensity of harvest. Thus, the clearcutting method creates the greatest degree of site disturbance, followed in rank order by the seed-tree method, the shelterwood method, and the group-selection method; the least site disturbance occurs using the single-tree selection method. This same gradient generally applies to erosion, fire hazards from slash, and aesthetics. Conversely, uneven-aged systems often require greater care on the part of loggers to prevent damage to the residual trees. To a lesser degree this is true of even-aged management during thinning operations and in implementing seed-tree and shelterwood cuts.

Converting a stand to even-aged structure can be accomplished rapidly by clearcutting and reforesting the site by any of the methods described above. If the desired species occur within the original stand, the seed-tree or shelterwood method can be used to convert the stand. In either case, conversion is accomplished within a few years. However, converting a well-stocked, even-aged stand to uneven-aged structure is a slow and challenging process (Nyland 2003). At least two cutting cycles are needed to obtain the three age classes that minimally define an uneven-aged stand (Smith 1986). If harvests are conducted every 10–20 years, a minimum of several decades may be required and it may take even longer to configure a stand so that it is capable of providing a sustained yield. As a result, uneven-aged methods are often imposed in formerly unmanaged stands, because multiple size classes are often present, and the smaller size classes are in general comprised of shade-tolerant species. Uneven-aged methods can also be applied to rehabilitate understocked stands or stands that were high-graded in the past, provided that some stocking of desired species still remains in the stand (Baker et al. 1996).

Even-aged methods, especially clearcutting, are often favored over uneven-aged methods for timber management because it is easier to regulate harvests and because more wood fiber can be produced especially under intensive plantation practices. However, uneven-aged methods are effective in producing large trees of high volume, value, and quality per tree. Because of the added expenses of site preparation and intermediate treatments typically used with even-aged management, this system is not always more profitable than uneven-aged management (Young and Giese

1990). Nonetheless, the bias that uneven-aged methods are costly and inefficient still persists and may be a hurdle to implementing these methods where they are ecologically appropriate.

REGENERATION TREATMENTS

Regeneration treatments are intended to promote the germination, establishment, and development of seedlings and saplings of the desired species. Two classes of regeneration treatments are generally recognized: artificial regeneration and natural regeneration.

Artificial Regeneration

Artificial regeneration includes planting seedlings, planting cuttings, or sowing seeds. Seedlings and stock for cuttings are typically produced under controlled conditions in a nursery. Planting seedlings is the most common artificial regeneration technique, and the most widespread application is reforesting clearcuts with conifers. Hardwood planting is becoming increasingly popular, however, especially as a way to reforest abandoned or highly erodible agricultural land in the eastern United States.

Planting techniques typically involve raising seedlings for one or more years in a nursery, then transplanting them to the field during the dormant season. For species important to the wood products industry, such as Douglas-fir, loblolly pine, or eastern cottonwood, planting stock has often been selectively bred for desirable attributes such as growth rate or disease resistance (Namkoong et al. 1988). For a host of other species, however, especially heavy-seeded hardwoods such as the oaks, nursery production depends on collection of wild seed, with no control over genetic quality.

The mid-twentieth century was the heyday of direct seeding in North America. It was used to reforest large tracts of abandoned agricultural land that at one time supported forests. It continues to be applied in western North America to reforest areas affected by forest fires. Direct seeding is effective in such circumstances because few trees remain as seed producers and the areas requiring reforestation are vast. Also, these sites are often subject to severe erosion, making it impractical to wait the year or more required to raise seedlings in a nursery. Because no nursery is needed to produce seed for direct seeding, one can reforest large areas at far less cost than by planting. The major disadvantages of direct seeding are that a large number of seeds must be sown to obtain acceptable stocking, opportunities for gain from genetic improvement are less likely to be realized, and there is little control over spacing. Moreover, the need to reforest vast areas of abandoned agricultural land was largely met by the end of the twentieth century.

Natural Regeneration

Natural regeneration occurs when trees grow without people having planted them. This includes trees growing from recently fallen seeds, from seeds stored in the forest floor, and from root and stump sprouts. Natural regeneration occurs in both managed and unmanaged stands. In managed stands this includes regeneration that has occurred both before and after a harvest. With the exception of most clearcutting, all of the reproduction cutting methods rely on natural regeneration. Although people do not actively plant these trees, management affects the species composition, density, and growth potential of natural regeneration. Achieving the natural regeneration goals of a stand requires attention to the timing of treatments, understanding the regeneration biology of the desired species, and the competitive interactions among associated species.

Site Preparation

After a stand is harvested, few sites are well suited for the establishment of regeneration, be it artificial or natural. This is particularly true for the even-aged reproduction methods, and especially for clearcutting. Site preparation is how silviculturists modify the site to make it suitable for regenerating the desired tree species. These modifications include removing slash (branches and crowns of trees left over from harvesting), exposing bare mineral soil and enhancing soil nutrition, and controlling competing vegetation, including any undesirable trees that remain following the harvest. In upland oak stands, treatments can also include cutting stems of desired oak-advanced growth so that a fast-growing seedling sprout is produced. In general, site preparation is implemented through mechanical treatments, prescribed fire, herbicides, or fertilization. Many refinements are associated with each of these, depending on the reproduction cutting method, the species being regenerated, the site conditions, and whether natural or artificial regeneration is being used. The balance of how much and what kinds of site preparation to conduct varies between good and poor seed years in a given species, and is complicated by the timing of reproduction cutting relative to seed dispersal.

Precommercial Thinning

As seedlings of desired species become saplings, they face increasing levels of competition from individuals of both desired and undesired species. If left unchecked, this competition will often compromise tree growth and health and ultimately lead to mortality of desired species. This is particularly true for shade-intolerant species. Precommercial thinning is used to alleviate intraspecific competition, or competition among desired species. In precommercial thinning, stem density is typically reduced through mechanical felling, although herbicides and prescribed fire are also used. This type of thinning is termed "precommercial" because the trees being removed are not commercially valuable and cannot be sold.

Release Treatments

Release treatments are used to control competition between desired and undesired species, and are applied when the desired species are saplings or smaller. These treatments generally aim to control undesirable species that possess faster growth rates than desired species, such as unwanted fast-growing sprouts that are competing with desired slow-growing seedlings. Release treatments can be divided into two categories: liberation cutting and cleaning. Liberation cutting is the removal of undesirable trees that are taller than the desired species. Cleaning is the removal of competing vegetation that is about the same height as the desired trees, but is expected to overtop them. A common scenario in which release is necessary occurs in the maintenance of early successional and midsuccessional stands. In this scenario a stand of early or midsuccessional trees develops a midstory of later successional species. If the overstory is harvested but the understory is not immediately treated, a liberation treatment will become necessary to allow the original overstory species to develop through the taller and more shade-tolerant midstory. Liberation treatment can be avoided if the midstory is also removed when the overstory is cut.

Cleaning becomes important when the midstory species have the ability to stump sprout. They will generally be able to grow faster than seedlings of the desired species because of energy reserves stored in their undisturbed roots. Cleaning frees these desired species from the competing stump sprouts. In the northeastern United States, cleaning is a common treatment used in upland oak and northern hardwood stands, in which sprouts of the ubiquitous red maple (*Acer rubrum*) can, if left untreated, effectively suppress the favored species.

INTERMEDIATE TREATMENTS

Intermediate treatments traditionally involved removing immature trees that were of some commercial value to reinvigorate growth of the remaining trees to hasten a harvest. As the goal of modern forestry has moved from harvesting trees to achieving a desired stand condition, however, the role of intermediate treatments has changed as well. Intermediate treatments might better be defined as treatments that redistribute resources within an established stand to move the stand toward a desired future condition. Intermediate treatments include thinning, improvement cutting, prescribed burning, and fertilizing.

Thinning

Thinning is used to reduce stem density of the desired species in the stand. Typically this is done to release "better" individuals from competition by removing some trees to enhance the growth of the trees that remain (Lundgren 1981). However, thinning can also achieve other goals correlated with reducing stem density, such as promoting understory growth, reducing clutter for bats, or creating downed woody debris. Regardless of the goal, thinning will redistribute resources to the remaining trees and stimulate their growth.

Distinctions among the different kinds of thinning relate to the relative crown positions of the trees being removed. Low thinning, or thinning from below, primarily removes smaller overtopped trees that are likely to be lost to density-dependent mortality. This type of thinning increases the availability of water and soil nutrients to retained trees, but it generally does not increase the amount of light available to the residual trees or understory, unless it is applied in a manner that removes a third or more of the stem density in the stand. Crown thinning, or thinning from above, removes larger trees that reach the upper canopy to favor similar individuals that have better form. This type of thinning increases the availability of all resources and promotes growth in both the understory and overstory.

Most thinning involves a judgment that some attribute of the trees being retained is superior to that found in the trees being cut. Traditionally, this has been associated with timber production such as straight stems, small branches, no damage in the crowns, and so on. But if branchiness, sweep, bark roughness, or crown deformation enhance habitat for bats, thinning can be adapted to favor trees with those traits.

Improvement Cutting

Improvement cutting is similar to thinning, but with the goal of removing undesirable species from a stand. It is analogous to a release treatment; both are imposed to balance competition between desired and undesired species. With improvement cutting the trees being removed are of merchantable size, but whether they can be sold commercially depends on whether there is sufficient volume to make a harvest operationally feasible for a logger. Improvement cutting is often applied in stands that were not subject to release when young. As in release treatments, improvement "cutting" can be carried out through the use of herbicides. Improvement cutting can also be adapted to favor mixed-species stands by leaving trees of different species in the residual stand, and removing trees that compete with the retained trees.

Prescribed Burning

Prescribed burning is frequently associated with its use as a site preparation tool to dispose of slash and litter, to release seeds from serotinous cones, to release nutrients stored in leaf litter and woody material, and to top-kill seedlings and stump sprouts of species not adapted to fire. However, fire is increasingly being used as an intermediate treatment in established stands to maintain attributes of fire-adapted ecosystems and to reduce fuel loads. In fire-adapted ecosystems, fire can be used like an improvement cut to eliminate competing species that are not adapted to fire-prone habitats. At the same time, this will help promote other aspects of fire maintained ecosystems, such as an open, grassy understory (Sparks et al. 1998).

The practice of prescribed burning is limited by safety constraints. Criteria for when fires can be set typically are very strict to ensure that prescribed fires do not turn into wildfires, that air quality is not compromised, and that those working on the fire are not at risk. Fires are often limited further by hunting seasons and public opinion. Because of these constraints, implementing fires on a landscape scale is not always possible.

Although fertilization is frequently confined to regeneration treatments, it is often used in industrial forestry in stands of all ages. The idea behind fertilization is to add a limiting resource to the stand, rather than by releasing it from use by other trees. It is most effective if the nutrients being applied are limiting; effectiveness diminishes quickly if other resources such as soil moisture are more limiting than the nutrients being applied. Fertilization done early in the life of the stand is usually intended to accelerate sapling development and reduce the length of time required to obtain trees of a given size for harvest. Mature stands that are soon to be harvested can also be fertilized economically several years prior to harvest to provide additional growth of trees that are already of large size and value.

SILVICULTURAL SYSTEMS AND PRESCRIPTIONS

A silvicultural system is an integrated combination and sequence of treatments planned over time, designed to carry an existing stand to a desired future stand condition.

SCHEDULING TREATMENTS

An even-aged system has a discrete beginning and end, starting with the establishment of the new stand and culminating in a final harvest when the stand reaches maturity at the appropriate rotation age r (fig. 7.4a). The reproduction cutting method initiates the new stand, regeneration treatments release desired species from competing vegetation, and intermediate treatments regulate stem density to maintain growth of desired species. Eventually the stand reaches rotation age at which time a new system is initiated. Thus, silvicultural systems in even-aged management follow a chronosequential pattern, and each treatment conducted at each point in time is generally applied across the entire managed portion of the stand.

Conversely, uneven-aged silvicultural systems do not have a discrete beginning and end. The basic unit of management is the cutting cycle, defined as the average interval between harvests. Each cutting cycle harvest contains some elements of reproduction cutting, some elements of thinning, and some elements of regeneration treatment, all conducted concurrently within the different size classes appropriate for each (fig. 7.4b). The uneven-aged silvicultural system then becomes defined by the pattern of implementation (single-tree or group selection) and by the description of treatments required in each age cohort of the stand.

Trends in cash flow, investment, and return can also be inferred from these temporal patterns. In even-aged stands, late rotation thinning and the final harvest provide the largest financial returns, but little opportunity for financial return exists in the first half of the rotation. Some practices, such as planting and site preparation, are extremely expensive and must be capitalized over the life of the stand. This creates powerful economic incentive for managers to optimize growth and reduce the length of the rotation. On the other hand, uneven-aged stands often provide pe-

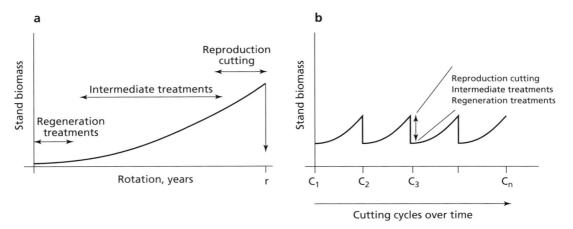

Figure 7.4. Comparison and timing of treatments conducted in even-aged (a) and uneven-aged (b) silvicultural systems over time.

riodic return for landowners and expenses are low. However, the standing volume that must be maintained in an uneven-aged stand does have value, and the compounded value of that standing inventory represents a financial risk if it were to be lost. Landowners must individually decide which cash flow model is appropriate for their respective economic situation.

The temporal patterns (fig. 7.4) also depict trends in canopy cover for each silvicultural system. In the even-aged systems, there is some time early in the rotation when canopy cover, especially cover associated with reproductively mature trees, is limited or absent. Conversely, continuous canopy cover is maintained in uneven-aged systems, especially in the single-tree selection system. If a diversity of canopy conditions across stands is sought, even-aged systems might be appropriate, especially if stands are managed to feature different age classes across the landscape. Conversely, if uniform canopy coverage across a landscape with a minimum of openings in the canopy is sought, uneven-aged systems would be better.

TIMING OF PRESCRIPTIONS

Proper timing of prescriptions is important in both even-aged and uneven-aged systems. In even-aged systems, delays in timely treatment can lead to reduced growth, reduced individual tree vigor, and greater risk of density-dependent mortality. For example, pine bark beetles (*Dendroctonus* spp.; *Ips* spp.) are more likely to cause problems in stands where individual tree vigor is low. Thinning to maintain acceptable vigor and growth rates before stands become overcrowded is an effective tool to promote forest health. In uneven-aged stands, prescriptions must consider the balance between the basal area of small trees, large trees, and the submerchantable seedling and sapling classes. Too many large trees can lead to suppression and mortality of regeneration. If the merchantable size classes become too densely stocked, either by retaining too many residual trees during a cutting cycle harvest or by failing to implement a scheduled cutting cycle harvest in a timely manner, regeneration can become suppressed.

Whether using even-aged or uneven-aged systems, training must be given to field personnel on proper marking techniques. Generally speaking, from intermediate-age classes onward, marking for thinning and reproduction cutting requires attention to the silvicultural axiom "cut the worst trees and leave the best." Describing what is worst or best depends on the ownership objective, and the description might differ if the goal is for maximum timber production versus sustaining habitat of tree-roosting bats. In general, marking in even-aged stands requires attention to uniformity of spacing and to desirable attributes sought in the trees being retained. Conversely, in uneven-aged stands, variability in spacing of overstory trees is necessary to promote openings for regeneration.

Managing even-aged stands often requires fewer visits to the site by field crews than does managing uneven-aged stands. This can be important in organizations where the number of field personnel is declining. Other things being equal, the ideal way to plan a thinning or harvest is to inventory the stand in the field, analyze the inventory data in the office to determine what is to be retained and thus what is to be cut, and then return to the field for the marking. Regardless, it is easier to skip the inventory step in an even-aged stand than an uneven-aged stand. Even-aged stands are more homogeneous in tree size, and under such conditions, the target residual basal area can be easily estimated. In heterogeneous stands, it is more important to use an inventory to develop a target residual stand and to guide operational marking. As a result, management costs are often higher in uneven-aged stands than even-aged stands.

STREAMSIDE MANAGEMENT ZONES

In most forest management applications on public and private lands, special attention is given to the areas immediately adjacent to streams. Because trees draw water from the ground, harvesting trees reduces the water storage capacity of an area. This can lead to increased water flow, which can accelerate erosion. Erosion can adversely affect water quality and aquatic habitats. To protect water quality most public and industrial forests use streamside management zones (SMZs). These are zones surrounding perennial and intermittent streams in which limited or no harvesting of trees is permitted (Wigley and Melchiors 1994).

In addition to water quality, trees overhanging streams in SMZs are important in maintaining colder water temperatures. Because water temperature dictates the amount of dissolved oxygen in water, water temperature is critical to aquatic insects and to some species of fish, such as the salmonids (Beschta et al. 1987). SMZs have also been found to be important for wildlife habitat (Dickson and Wigley 2001). Bats have been observed to use SMZs as travel corridors (Law and Chidel 2002), and may concentrate in SMZs for feeding and drinking (Grindal et al. 1999). Because SMZs have been viewed as critical for many ecological resource values, silvicultural treatments have generally been restricted within these areas. The degree to which SMZs serve as valuable landscape elements for forest-dwelling bats has yet to be clearly determined, and likely will vary among geographic regions.

IMPLICATIONS OF SILVICULTURE FOR BATS

Of all the habitat components required by bats in forests, the two most heavily influenced by forestry are roosting sites and foraging habitat (Hayes and Loeb, chap. 8 in this volume). The management of trees valuable as roosts is not intrinsic to any one silvicultural practice. Rather, most practices can be modified to manage roosting sites. In contrast, different silvicultural practices will promote different foraging habitats, which will favor different species of bats. For example, Patriquin and Barclay (2003) reported that in the boreal forests of Alberta, different species of bats use forests in different conditions; they noted that the silver-haired bat prefers clearcuts and avoids intact forests, the little brown myotis (*Myotis lucifugus*) prefers to forage along the edges of clearcuts, and the northern myotis (*M. septentrionalis*) prefers to forage in intact forests. Such reports support the impression that there is no generic habitat for forest-dwelling bats, but rather, like other wildlife species, different habitats will favor different bat species. Developing adequate forest habitat for bats will require a much better understanding of the habitat that each species prefers than is currently available.

ROOSTING HABITAT

Bats often use trees and snags for roosts (Barclay and Kurta, chap. 2 in this volume; Hayes 2003; Kunz and Lumsden 2003). Some species, such as lasiurine bats, roost in the foliage of live trees, whereas other species roost in a more protected location, such as a cavity, a split trunk, or under sloughing bark (Carter and Menzel, chap. 3 in this volume; Hayes 2003; Kunz and Lumsden 2003; Menzel et al. 2003). Typically, these kinds of roosts are found in relict trees, cull trees (trees made unmerchantable because of decay or deformities), and snags. Historically, silviculturists have not valued such trees.

Relict trees, cull trees, and snags have traditionally been considered impediments to timber production goals. Relict trees are often viewed as trees that should have been harvested earlier and that are compromising the stand by using too many resources. Cull trees, such as those with cavities or a broken crown, are viewed as taking limited resources away from better trees. Snags caused by competition indicate a missed opportunity to profit from the harvest of a tree and lost growth potential for neighboring trees. When snags are created by forces beyond the control of a silviculturist, such as wind, ice, wildfires, insects, and disease, the impulse often is to salvage as much timber as possible. From the vantage of timber production, and even with a more modern set of goals, this is often a logical management decision. For example, salvaging timber after a disease outbreak may be advisable to facilitate reforesting the site and to prevent forest fires and future outbreaks of disease. Changing these attitudes, which run strongly in the profession, may be an impediment for biologists. Nonetheless, from an ecological perspective, the value of relict trees, snags, and cull trees should not be trivialized or marginalized.

We contend that the proper way to deal with relict and cull trees is to quantify their ecological influence and account for that influence in prescription planning. Because water, nutrients, and solar radiation are limited, retaining any tree limits the growth of surrounding trees (Palik and Pregitzer 1994; Thysell and Carey 2000; Traut and Muir 2000). For example, in southern pine stands managed using single-tree selection in southern Arkansas, a typical residual basal area target is 14 m²/ha, with a maximum diameter at breast height (dbh) for residual trees of 50 cm. A tree with a dbh of 80 cm has a basal area of 0.5 m²/ha. Thus, if two of these trees are retained per hectare as relict trees, they represent 7% of the residual basal area of the stand (fig. 7.5). To keep the residual basal area within prescribed limits, more trees would have to be cut within other size classes. Ignoring the prescribed limits would likely lead to reduced growth, mortality in smaller-size classes, and eventual conversion of the stand to more shade-tolerant species.

Because of the influence relict and cull trees have over a stand, it may be desirable to reduce the competitive edge of these retained trees. For example, leaving individuals with poorly formed crowns will minimize the shading of regeneration. Partially girdling, pruning, or disturbing the roots could be used to reduce the competitive ability of retained trees.

Another consideration for relict and cull trees is that they have the potential to reproduce. If these trees are of an undesirable species, or if they have undesirable characteristics that may be heritable, reproduction is often unwanted. Typically this will require a cleaning or thinning to remove the unwanted regeneration, but in some instances it is possible to reduce the likelihood of such trees contributing to the future stand by retaining trees that are poorly adapted to understory conditions. For example, if a stand is being managed for shade-intolerant species, shade-tolerant species could be retained because the regeneration of these trees will likely be overtopped by the faster growing desired species. Alternatively, stands managed for shade-tolerant species could retain shade-intolerant species because their regeneration will likely not survive the shade. It may

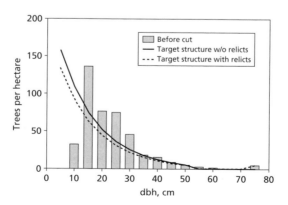

Figure 7.5.
The target, residual stand structure in the 50-cm diameter class and smaller in this hypothetical, uneven-aged stand must be reduced if two relict trees/ha in the 75-cm diameter class are to be retained.

also be possible to reduce the reproductive ability of relict and cull trees by retaining competing vegetation around these trees.

Because snags do not limit resources or reproduce they often may be more desirable to land managers than relict or cull trees. Snags can be unstable, however, and serious injuries and fatalities have occurred from snags falling on forest workers (Myers and Fosbroke 1995). Because of this safety hazard, the Occupational Safety and Health Administration (OSHA) of the U.S. Department of Labor requires that, if work is to be carried out within two tree lengths of a snag, the employer must demonstrate that this will not create a hazard for the employee (Code of Federal Regulations 29, 1910.266(h)(1)(vi)). OSHA further requires all snags deemed dangerous to be felled before any other work is done within two tree lengths of the work zone. Thus, saving a dangerous snag that is 18 m tall requires no harvesting over an area of 0.4 ha surrounding this snag. If snags can be found in clusters, this could reduce the impact of retaining snags.

Additionally, snags can be retained in areas that typically are not harvested, such as SMZs and around historical sites. However, the effectiveness of retaining snags in clusters and in SMZs for forest-dwelling bats remains unknown and should be evaluated.

In any case, because of safety restrictions it is often easier to create snags than to protect them. In nature, snags are created by lightning, wind, ice, disease, and insects, but land managers have little or no control over these forces. Snags can be created by girdling the bole or by topping the tree, without compromising the growth of the stand. Creating snags from cull trees or from species that have little commercial value can further lower this "expense." Also, using cull trees may help reduce the lag time between when a snag is created and when it is of use to bats and other species of wildlife.

It should be possible to develop herbicide technology to promote living snags. A sublethal dose of herbicide injected into a tree would reduce growth and uptake of resources to a negligible level. The tree would remain alive, in a reduced state of vigor, and would essentially become a living snag with marginal effect on the trees that surround it. Assuming this approach provided the characteristics valuable for bats and other wildlife species, such an approach might significantly extend the functional presence and longevity of snag attributes in the stand.

Relict trees, cull trees, and snags are not an innate attribute of any silvicultural treatment. Most treatments can be adapted to include and manage them. Of the reproduction cutting methods, clearcutting is the least well suited to these types of legacy elements, because these trees use resources and suppress regeneration development within their zone of ecological influence. Their retention also hampers forestry operations; snags and relicts are impediments to logging, site preparation, planting, and aerial application of herbicides and fertilizers. Leaving cull trees during thinning operations will reduce the future value of timber when harvested. All of these problems can be overcome at additional expense, which landowners may or may not choose to incur.

Managing relict trees, cull trees, and snags under the seed-tree method is similar to their management in clearcuts. Seed trees can be left as relict trees, or after regeneration has occurred, they can be girdled to create large snags. Often it is only marginally profitable to harvest the seed trees, so this may be a viable alternative. Additional snags can be created at the initial harvest, in particular, if the seed trees are to be retained.

In the shelterwood method, snags can be created with any of the three harvests (preparatory cut, seed cut, or removal cut). Snags created in the first or second harvest will need to be avoided at later harvests, however. Creating snags in clumps may be useful in these situations. Relict trees can be left after the removal cut and cull trees can be left during thinning.

Single-tree selection allows considerable flexibility in retention of relict and cull trees. As described above, however, prescriptions must be adjusted to compensate for the presence of relict trees and care must be taken to ensure cull trees do not become a primary seed source. Because these stands are re-entered on a regular basis, creating and retaining snags can be problematic. It may be that snags created in one cutting cycle would impede operations during the following cutting cycle. However, because most stands managed under single-tree selection are shade tolerant, it may be possible to postpone harvesting around desirable snags without impacting the size distribution and species composition of the stand.

Group selection allows the greatest flexibility in the retention and creation of snags. Groups can be positioned to avoid desirable snags, or to include them within the opening. Openings created with group selection harvests are often used as logging decks, however, and snag retention within them would hinder such operations. The advantage of leaving snags in a group opening is that several cutting cycles will occur before a group opening is harvested again; thus, retained snags would be less likely to hinder subsequent harvests in the cutting cycle than if they were scattered throughout the matrix of the stand between the group openings. Relict trees and cull trees also can be retained, although, just as with snags, allowances must be made for their presence.

FORAGING HABITAT

In flight, bats must contend with physical obstructions or physical clutter (Brigham et al. 1997). These are elements of the habitat that impede flight such as foliage, branches, and tree stems. The degree of physical clutter a bat is able to negotiate depends on body mass and wing morphology (Aldridge and Rautenbach 1987). Additionally, bats must be able to detect physical clutter and prey. Differently structured echolocation calls are useful in different habitats. For example, low-frequency, narrow-bandwidth calls are effective at detecting objects at long distances, but are confounded by even low degrees of clutter. Alternatively, high-frequency, broad-bandwidth calls are effective at detecting objects at only short distances and can contend with higher degrees of clutter (Aldridge and Rautenbach 1987). As a consequence, the degree of clutter a bat can fly and hunt within depends on call structure. Some researchers have grouped bats into "ensembles" based on the degree of clutter found in the habitat they use (Aldridge and Rautenbach 1987; Grindal 1996; Patterson et al. 2003). Aldridge and Rautenbach (1987) found that bats in South Africa segregate into four ensembles: open foragers, woodland edge foragers, intermediate clutter foragers, and clutter foragers. Because silvicultural practices alter clutter, different practices should favor different ensembles.

Clutter in even-aged stands

Throughout the life of an even-aged stand, most of the foliage lies in the main canopy, which gradually increases in height (fig. 7.3a). The area within and below the canopy is typically very dense for approximately the first third of the life of an even-aged stand. During this time, such stands may be too dense for even clutter-adapted bats. As trees reach merchantable size, they often self-prune. This reduction in clutter might create a zone suitable for clutter-adapted species, but this is not always the case, as in trees that do not naturally prune or in stands that develop a midstory. Additionally, few corridors will exist through the canopy. Thinning, improvement cuts, pruning, herbicides, and prescribed fire can be used to further reduce the clutter in the understory and to open corridors through the canopy (fig. 7.6). Depending on the intensity of treatment, these silvicultural practices could be used to increase habitat use by bats (Humes et al. 1999) and create zones beneath the canopy suitable for clutter foragers and intermediate clutter foragers.

Throughout most of the life of even-aged stands, low-clutter habitat suitable for open foragers exists above the canopy. In the seed-tree method slightly less open conditions exist between the first harvest and the removal of the seed trees. Depending on the density of seed trees, these stands might be suitable for open foragers or woodland edge foragers. In the shelterwood method, foraging habitat for intermediate clut-

Figure 7.6. (left) Sharp transition in clutter at the base of the live crown in an even-aged stand of longleaf pine (*Pinus palustris*). Note the extensive area of low clutter beneath the canopy resulting from cyclic prescribed burning. *Photo by J. Guldin*

Figure 7.7. (right) Heterogeneous distribution of clutter within an uneven-aged loblolly-shortleaf pine (*Pinus taeda-Pinus echinata*) stand on the Crossett Experimental Forest, Ashley County, Arkansas. *Photo by J. Guldin*

ter foragers would likely be created during the preparatory cut, while the seed cut would likely favor intermediate clutter foragers and woodland edge foragers.

Clutter in uneven-aged stands

In well-regulated, uneven-aged stands, foliage is present at all strata, from the forest floor to the upper canopy (Baker et al. 1996; Lorimer 1989; Shelton and Murphy 1993; Whitmore 1989). In practice, these different strata form clusters at different heights in the canopy profile (fig. 7.7). Thus, uneven-aged stands generally have zones of low clutter within a matrix of high clutter. These low-clutter zones are larger and extend further into the canopy in the group selection method than in the single-tree selection method. Stands managed under single-tree selection would likely favor clutter-adapted species, while those under group selection might favor clutter- and intermediate-clutter-adapted species. Alternatively, because bats often use roads as travel corridors (Hickey and Neilson 1995; Limpens and Kapteyn 1991; Menzel et al. 2002; Walsh and Brigham 1998), species adapted to more open conditions may be able access stands under uneven-aged management through the road networks that are often maintained to facilitate repeated entries.

Clutter between stands

Factors other than foliage distribution can have a bearing on clutter, especially when areas larger than an individual stand are examined. Because managed and unmanaged stands differ in the distribution of clutter, boundaries between these stands might serve as a filter to exclude ensembles of bats. Similarly, boundaries between stands that differ in silvicultural treatments, management intensity, forest type, or age may also act as filters. This may especially be true of SMZs within individual stands. When managing for bats, some thought should be given to the arrangement of different stands across a landscape. For example, a mature even-aged stand with an open understory, suitable for bats adapted to intermediate levels of clutter, would be inaccessible to such species if it was surrounded by stands suitable to only clutter-adapted species. In such cases, roads may be an important means for gaining access to otherwise inaccessible areas. Thus, we suggest that management of clutter for forest-dwelling bats has both within-stand and between-stand components.

SUMMARY

In the twenty-first century, we expect that the practice of silviculture will broaden to increasingly encompass ecosystem-based goals such as restoration and enhancement of habitat for desired plant and animal species and communities. The array of reproduction cutting methods, regeneration treatments, and intermediate treatments that constitute a silvicultural system can be configured to meet the habitat requirements of bats. The choices among overall reproduction cutting methods, and between even-aged and uneven-aged methods, have implications for bats, especially with regard to roosting and the management of foraging habitat. Special attention needs to be focused on creating and retaining structural and legacy features such as relict trees and snags. Once the type, amount, and distribution of such features are known, they can be incorporated into a variety of silvicultural systems. To satisfy management objectives for species whose habitat requirements transcend individual stands, the forester should plan silvicultural practices in concert across stands and, increasingly, across ownerships.

There are some important hurdles to implementing bat-friendly silviculture. Foremost for bat biologists will be the definition and quantification of those attributes that are of value to bats. Once those needs are understood, biologists and silviculturists can work together to develop prescriptions that meet the needs of bats in forests. The challenge for biologists is to learn as much as possible about roosting, foraging, and other habitat requirements for the bat species of interest. The challenge for silviculturists working with biologists concerned about bats is to incorporate ways to satisfy habitat requirements of bats while meeting other forest management objectives.

LITERATURE CITED

Aldridge, H.D.J.N., and I.L. Rautenbach. 1987. Morphology, echolocation and resource partitioning in insectivorous bats. Journal of Animal Ecology 56:763–778.

Baker, J.B., M.D. Cain, J.M. Guldin, P.A. Murphy, and M.G. Shelton. 1996. Uneven-aged silviculture for the loblolly and shortleaf pine forest cover types. USDA Forest Service, Southern Research Station General Technical Report SO-118: 1–65.

Behan, R.W. 1990. Multiresource forest management: a paradigmatic challenge to professional forestry. Journal of Forestry 88:12–18.

Beschta, R.L., R.E. Bilby, G.W. Brown, L.B. Holtby, and T.D. Hofstra. 1987. Stream temperature and aquatic habitat: fisheries and forestry interactions, pp. 191–232, in Streamside management: forestry and fishery interactions (E.O. Salo and T.W. Cundy, eds.). University of Washington, Institute of Forest Resources Contribution 57, Seattle, WA.

Bissonette, J.A., ed. 1986. Is good forestry good wildlife management? Maine Agricultural Experiment Station, Miscellaneous Publication No. 689, University of Maine, Orono, ME.

Brigham, R.M., S.D. Grindal, M.C. Firman, and J.L. Morissette. 1997. The influence of structural clutter on activity patterns of insectivorous bats. Canadian Journal of Zoology 75:131–136.

Bukenhofer, G.A., and L.D. Hedrick. 1997. Shortleaf pine/bluestem grass ecosystem renewal in the Ouachita Mountains. Transactions of the North American Wildlife and Natural Resources Conference 62:509–513.

Bukenhofer, G.A., J.C. Neal, and W.G. Montague. 1994. Renewal and recovery: shortleaf pine/bluestem grass ecosystem and red-cockaded woodpeckers. Proceedings of the Arkansas Academy of Science 48:243–245.

Cain, M.D., and M.G. Shelton. 2001. Twenty years of natural loblolly and shortleaf pine seed production on the Crossett Experimental Forest in southeastern Arkansas. Southern Journal of Applied Forestry 25:40–45.

Dickson, J.G., and T.B. Wigley. 2001. Managing forests for wildlife, pp. 83–94, in Wildlife of southern forests: habitat and management (J.G. Dickson, ed.). Hancock House, Blaine, WA.

Farrell, E.P., E. Führer, D. Ryan, F. Andersson, R. Hüttl, and P. Piussi. 2000. European forest ecosystems: building the future on the legacy of the past. Forest Ecology and Management 132:5–20.

Foti, T.L., and S.M. Glenn. 1991. The Ouachita Mountain landscape at the time of settlement, pp. 49–65, *in* Proceedings of the conference on Restoration of old growth forests in the Interior Highlands of Arkansas and Oklahoma (L. Hedrick and D. Henderson, eds.). Ouachita National Forest, Winrock International Institute for Agricultural Development, Little Rock, AR.

Franklin, J. 1989. Toward a new forestry. American Forests Nov–Dec:37–44.

Grindal, S.D. 1996. Habitat use by bats in fragmented forests, pp. 260–272, *in* Bats and forests symposium (R.M.R. Barclay and R.M. Brigham, eds). Research Branch, British Columbia Ministry of Forests, Victoria, BC.

Grindal, S.D., J.L. Morissette, and R.M. Brigham. 1999. Concentration of bat activity in riparian habitats over an elevational gradient. Canadian Journal of Zoology 77:972–977.

Guldin, J.M., and J.B. Baker. 1998. Uneven-aged silviculture, southern style. Journal of Forestry 96: 22–26.

Guldin, J.M., and R.W. Guldin. 2003. Forest management and stewardship, pp. 179–220, *in* Introduction to forest ecosystem science and management: third edition (R.A. Young and R.L. Giese, eds.). John Wiley and Sons, New York.

Hayes, J.P. 2003. Habitat ecology and conservation of bats in western coniferous forests, pp. 81–119, *in* Mammal community dynamics in coniferous forests of western North America: management and conservation (C.J. Zabel and R.G. Anthony, eds.) Cambridge University Press, Cambridge, MA.

Hedrick, L.D., R.G. Hooper, D.L. Krusac, and J.M. Dabney. 1998. Silvicultural systems and red-cockaded woodpecker management: another perspective. Wildlife Society Bulletin 26:138–147.

Helms, J.A., ed. 1998. The dictionary of forestry. Society of American Foresters, Bethesda, MD.

Hickey, M.B.C., and A.L. Neilson. 1995. Relative activity and occurrence of bats in southwestern Ontario as determined by monitoring with bat detectors. Canadian Field-Naturalist 109:413–417.

Humes, M.L., J.P. Hayes, and M.W. Collopy. 1999. Bat activity in thinned, unthinned, and old-growth forests in western Oregon. Journal of Wildlife Management 63:553–561.

Hunter, M.L., Jr. 1990. Wildlife, forests, and forestry: principles of managing forests for biological diversity. Prentice Hall, Englewood Cliffs, NJ.

Kessler, W.B., H. Salwasser, C.W. Cartwright, Jr., and J.A. Caplan. 1992. New perspectives for sustainable natural resources management. Ecological Applications 2:221–225.

Kimmins, J.P. 1991. The future of the forested landscapes of Canada. Forestry Chronicles 67:14–18.

———. 1992. Balancing act-Environmental issues in forestry. UBC Press, Vancouver, BC.

Kunz, T.H., and L.F. Lumsden. 2003. Ecology of cavity and foliage roosting bats, pp. 3–89, *in* Bat ecology (T.H. Kunz and M.B. Fenton, eds.). University of Chicago Press, Chicago, IL.

Law, B., and M. Chidel. 2002. Tracks and riparian zones facilitate the use of Australian regrowth forest by insectivorous bats. Journal of Applied Ecology 39:605–617.

Limpens, H.J.G.A., and K. Kapteyn. 1991. Bats, their behavior and linear landscape elements. Myotis 29:63–71.

Lorimer, C.G. 1989. Relative effects of small and large disturbances on temperate hardwood forest structure. Ecology 70:565–567.

Lundgren, A.L. 1981. The effect of initial number of trees per acre and thinning densities on timber yields from red pine plantations in the lake states. USDA Forest Service Research Paper NC-193.

Marquis, D.A. 1978. Application of uneven-aged silviculture and management on public and private lands, pp. 25–61, *in* Uneven-aged silviculture and management in the United States. USDA Forest Service, General Technical Report, WO-24, Washington, DC.

Mattoon, W.R. 1915. Life history of shortleaf pine. Bulletin 244. U.S. Department of Agriculture, Washington, DC.

Menzel, M.A., T.C. Carter, J.M. Menzel, W.M. Ford, and B.R. Chapman. 2002. Effects of group selection silviculture in bottomland hardwoods on the spatial activity patterns of bats. Forest Ecology and Management 162:209–218.

Menzel, M.A., J.M. Menzel, J.C. Kilgo, W.M. Ford, T.C. Carter, and J.W. Edwards. 2003. Bats of the Savannah River site and vicinity. USDA Forest Service, Southern Research Station General Technical Report SRS-68:1–69.

Myers, J.R., and D.E. Fosbroke. 1995. The Occupational Safety and Health Administration logging standard: what it means for forest managers. Journal of Forestry 93:34–37.

Namkoong, G., H.C. Kang, and J.S. Brouard. 1988. Tree breeding: principles and strategies. Monographs on theoretical and applied genetics II. Springer-Verlag, Heidelberg, Germany.

Nyland, R.D. 2003. Even-to uneven-aged: the challenges of conversion. Forest Ecology and Management 172:291–300.

O'Hara, K.L., and R.F. Gersonde. 2004. Stocking control concepts in uneven-aged silviculture. Forestry 77:131–143.

O'Hara, K.L., R.S. Seymour, S.D. Tesch, and J.M. Guldin. 1994. Silviculture and our changing profession: leadership for shifting paradigms. Journal of Forestry 92: 8–13.

Palik, B.J., and K.S. Pregitzer. 1994. White pine seed-tree legacies in an aspen landscape: influences on post-disturbance white pine population structure. Forest Ecology and Management 67:191–201.

Patriquin, K.J., and R.M.R. Barclay. 2003. Foraging by bats in cleared, thinned, and unharvested boreal forest. Journal of Applied Ecology 40:646–657.

Patterson, B.D., M.R. Willig, and R.D. Stevens. 2003. Trophic strategies, niche partitioning, and patterns of ecological organization, pp. 536–579, *in* Bat ecology (T.H. Kunz and M.B. Fenton, eds.). University of Chicago Press, Chicago, IL.

Perala, D.A., and J. Russell. 1983. Aspen, pp. 113–115, *in* Silvicultural systems for the major forest types of the United States (R.M. Burns, tech. comp.). USDA Forest Service, Agriculture Handbook No. 445. Washington, DC.

Sander, I.L., C.E. McGee, K.G. Day, and R.E. Willard. 1983. Oak-hickory, pp. 116–120, *in* Silvicultural systems for the major forest types of the United States (R.M. Burns, tech. comp.). USDA Forest Service, Agriculture Handbook No. 445. Washington, DC.

Shelton, M.G., and P.A. Murphy. 1993. Pine regeneration and understory vegetation 1 year after implementing uneven-aged silviculture in pine-hardwood stands of the silty uplands of Mississippi, pp. 333–341, *in* Proceedings of the 7th Biennial southern silvicultural research conference (J.C. Brissette, ed.). USDA Forest Service, Southern Forest Experiment Station.

Smith, D.M. 1986. The practice of silviculture, 8th ed. John Wiley and Sons, New York.

Smith, D.M., B.C. Larson, M.J. Kelty, and P.M.S. Ashton. 1997. The practice of silviculture: applied ecology, ninth edition. John Wiley and Sons, New York.

Sparks, J.C., R.E. Masters, D.M. Engle, M.W. Palmer, and G.A. Bukenhofer. 1998. Effects of late growing-season and late dormant-season prescribed fire on herbaceous vegetation in restored pine-grassland communities. Journal of Vegetation Science 9:133–142.

Swanson, F.J., and J.F. Franklin. 1992. New forestry principles from ecosystem analysis of Pacific Northwest forests. Ecological Applications 2:262–274.

Thysell, D.R., and A.B. Carey. 2000. Effects of forest management on understory and overstory vegetation: a retrospective study. USDA Forest Service, Pacific Northwest Research Station, General Technical Report PNW-GTR-488.

Traut, B.H., and P.S. Muir. 2000. Relationships of remnant trees to vascular undergrowth communities in the western Cascades: a retrospective approach. Northwest Science 74:212–223.

Walsh, A.L., and R.M. Brigham. 1998. Short-term effects of small-scale habitat disturbance on activity by insectivorous bats. Journal of Wildlife Management 62:996–1003.

Whitmore, T.C. 1989. Canopy gaps and the two major groups of forest trees. Ecology 70:536–538.

Wiersum, K.F. 1995. 200 years of sustainability in forestry: lessons from history. Environmental Management 19:321–329.

Wigley, T.B., and M.A. Melchiors. 1994. Wildlife habitat and communities in streamside management zones: a literature review for the eastern United States, pp. 100–121, *in* Riparian ecosystems in the humid U.S.: functions, values, and management. National Association of Conservation Districts, Washington, DC.

Young, R.A., and G.L. Giese. 1990. Introduction to forest science, second edition. John Wiley and Sons, New York.

Zeide, B., and D. Sharer. 2000. Good forestry at a glance: a guide to managing even-aged loblolly pine stands. Arkansas Forest Resources Center Series 003. University of Arkansas, Division of Agriculture, Arkansas Agricultural Experiment Station, Fayetteville, AR.

THE INFLUENCES OF FOREST MANAGEMENT ON BATS IN NORTH AMERICA

8

John P. Hayes and Susan C. Loeb

In recent years, interest in the ecology of bats and the influences of forest management on bat populations has increased substantially. This interest stems from the interplay of technological advances opening up new areas of research, a greater understanding of the importance of ecological roles played by bats in forest ecosystems, an increased recognition of the potential sensitivity of bats to environmental impacts, and a heightened awareness of the potential consequences of land management activities on biodiversity.

Reflecting conservation concern about the status of bats, many public agencies and conservation groups in the United States and Canada list one or more species of bats as sensitive, of special concern, or endangered. Many of these species use forests to meet some or all of their life-history requirements. Although information on population trends of forest-dwelling bats is generally lacking, their potential sensitivity, coupled with the close association of bats with various elements of forest structure (Hayes 2003), suggest that careful consideration of the influences of forest management on bats is warranted.

Forest management activities can have both a direct and an indirect impact on bats (Guldin et al., chap. 7 in this volume). Activities having a direct impact are those that result in the immediate mortality of individuals. For example, felling a roost structure could result in the mortality of bats inside the roost. Although we know little about the winter ecology of many bats and the extent to which trees and snags are used as winter roost sites by bats (Cryan and Veilleux, chap 6 in this volume), species that use living or dead trees as winter roosts could be particularly vulnerable to the direct impact from the felling of trees and snags during winter logging operations. Management activities such as control of vegetation and insects, or road building, also have a direct impact on bats. For example, eastern red bats (*Lasiurus borealis*) roosting in leaf litter (Mager and Nelson 2001; Saugey et al. 1998) could be killed from prescribed burns set to control understory vegetation. In addition, disruption of boulders in quarries during construction or maintenance of forest roads could result in the death of bats roosting in rock crevices.

Although the mortality of bats stemming directly from forest management activities could sometimes have important local effects on bat populations, especially for rare and highly sensitive species (Belwood

2002), there is little documentation of such occurrences. This lack of documentation could be the result of the infrequency of direct impact, the sense that such occurrences are anecdotal and do not warrant publication, or the cryptic nature of site-specific events that are often undetected and poorly suited to scientific investigation.

Although the scope and magnitude of direct impacts of forest management on bats are poorly known, forest management clearly affects vegetation and forest structure directly. Indeed, manipulation of vegetation and forest structure is both a tool and a goal of many aspects of forest management. The cascading effect of forest management on vegetation modifies habitat characteristics and many ecological processes; these effects can result in either positive or deleterious consequences for bat populations. In this chapter we refer to indirect influences as the effects of forest management on the behavior or demographics of bats that result from changes in ecological conditions or resources. Although direct impacts are likely critical in some situations, we believe indirect influences probably have a greater overall impact on bat populations and communities.

Four ecological factors are important in shaping habitat suitability for bats in North American forests: characteristics and abundance of roost sites, amount of clutter, availability of prey, and availability of water. Forest management can influence all these factors. In this chapter we consider four aspects of forest management that play especially important roles in influencing bat populations: management of timber, riparian areas, roads and bridges, and fire. We focus on the implications of these management activities for bats at the level of the population, because we believe population-level influences are of primary interest when considering management effects. Although we touch on the natural history and ecology of bats as they relate to forest management, details of the ecology of bats are presented elsewhere in this book and are not emphasized here.

FOREST MANAGEMENT, ROOST SITES, AND BAT POPULATIONS

As highlighted by Barclay and Kurta (chap. 2 in this volume), roosting ecology is central to the natural history of bats and management of roosts is key in conservation of bats in forests. Bats use a variety of roost structures in forests including caves, bridges, logs, root wads, foliage, leaf litter, rocks, and cavities and crevices in trees and snags. Although gaps exist in our understanding of the roosting ecology of bats, knowledge of the characteristics of roosts used by bats in forests has increased dramatically during the past decade (Kunz and Lumsden 2003). Large-diameter trees and snags are particularly important roost sites for many species in forest settings (Barclay and Kurta, chap. 2 in this volume). Forest management influences the characteristics and availability of roosts by affecting current and future numbers of roosts and by altering the habitat surrounding roosts.

One of the most immediate and direct ways that forest management activities influence the habitat of bats is their impact on the number of roosts in an area. In the short term, forest managers influence the number of roosts in an area by removal or destruction of existing roosts, decisions to retain existing roosts, and intentional or unintentional creation of new roosts. Because of their importance for many species of bats, we focus on large-diameter trees and snags.

Removal of actual and potential roost sites is common during timber harvest, in particular, under intensive harvest practices. Roost trees are often removed during timber harvest because of their economic value, and snags used for roosting are frequently removed because of concerns over safety and management of fuels. Dead trees left on a site are sometimes hazardous to forest workers and snags adjacent to roads may fall and obstruct roads or strike passing vehicles. Removing roost trees and snags during timber harvest and inhibiting development of new structures through time by managing forests under short rotations can reduce the numbers of potential roost structures in forests. This reduction in the number of roosts is especially evident for the large-diameter snags that are important roosts for many species of bats. For example, snags having diameters greater than 50 cm dbh (diameter at breast height) were more than 50 times more abundant in old-growth stands than in 21- to 40-year-old managed stands in western Oregon (E. Arnett, unpubl. data). Wilhere (2003) showed that under typical prescriptions for timber management in western Washington, large-diameter snags (>63.5 cm dbh) could be up to 100 times less abundant in industrial forestlands than in unmanaged stands. In red pine (*Pinus resinosa*) forests of the Great Lakes Region, mean volume of snags was 2.5 to 5 times higher in unmanaged stands than in managed stands (Duvall and Grigal 1999). In Arkansas, snag densities were reduced by 82% after group selection harvests in mature hardwood stands in the Ozark National Forest (Nelson and Lambert 1996), and in West Virginia hardwood stands that were heavily thinned had 2.6 to 3 times fewer snags than unthinned control sites (Graves et al. 2000).

Forest managers can influence roost abundance for species using crevices and cavities in trees and snags by augmenting snag availability via snag creation, where existing numbers of snags are inadequate to meet conservation goals, or by providing man-made roost structures. Snags can be created in a variety of ways (Lewis 1998), but the influences of different approaches to snag creation on use of roosts by bats or on bat populations have not been examined. Moreover, basic decay characteristics of trees intentionally killed to provide wildlife habitat are poorly known, especially in eastern deciduous forests. Conifer snags created by topping probably provide roosting habitat for bats within the first few years as exfoliation of bark begins to occur at the top of the snag, similar to patterns seen in stumps used as roosts by bats (Waldien et al. 2003). Regardless, we suspect that significant use of created snags by bats does not occur for several years following their creation because of the time lag associated with the deterioration in the integrity of the bark.

Although some studies have examined the use of artificial roosts, such as bat boxes in urban settings (Brittingham and Williams 2000), little has been published documenting the efficacy of artificial roost structures in North American forests. Unpublished observations of several biologists and some published accounts (Chambers et al. 2002) indicate that bats will use artificial roosts in forests, and man-made structures, such as bridges, sometimes receive very high levels of use in some forest settings (Adam and Hayes 2000; Lance et al. 2001). However, because the thermal requirements and characteristics of roosts used by bats apparently differ with species, reproductive condition, season, and roost function, man-made roosts probably rarely meet the full suite of roosting needs for forest-dwelling bats. Thus, although artificial roost structures can be used to supplement roosting opportunities for bats in some situations (Arnett and Hayes 2000; Burke 1999), and the use of artificial roosts may be particularly important in highly disturbed sites during restoration, we generally advocate using artificial roost structures only as temporary supplements for natural roost structures where natural roost structures are rare. In most situations, we strongly recommend that management and conservation strategies emphasize providing natural roost structures through time.

Individual bats in forests typically use multiple roosts in the course of relatively short periods and frequent roost switching has been documented in every study of bats using trees or snags as roost sites in forests (Barclay and Kurta, chap. 2 in this volume; Carter and Menzel, chap. 3 in this volume). Despite this understanding, the number of roosts typically used by individual bats during the course of a year is not known. The reasons underlying the lack of strong fidelity to individual tree roosts are not clear, but it is likely that bats derive some benefit from roost switching in forests (Lewis 1995) and that providing multiple roosts promotes the well-being of bats in forests. Yuma myotis (*Myotis yumanensis*) roosting in trees in an urban area demonstrate higher fidelity to roosts than do bats roosting in trees in nonurban environments, possibly because of the limited availability of roosts in urban settings (Evelyn et al. 2004). This

Figure 8.1.
Relationships between the abundance of bats and number of roosts are poorly understood and could take a variety of forms: (a) abundance of bats increases steeply with the number of roosts; (b) abundance of bats forms a convex curve with the number of roosts; (c) abundance of bats increases linearly with the number of roosts; and (d) abundance of bats forms a concave curve with the number of roosts. The relationship likely varies among species and forest type.

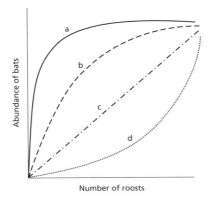

hypothesis and the potential ramifications of roost abundance on the behavioral ecology and fitness of bats warrant further investigation.

Although geographic and species-specific variation exists, snags and trees with certain characteristics, such as a large diameter, a moderate level of decay, the presence of cavities or exfoliating bark, and great height, are important roost sites for many species of bats in forests throughout North America (Barclay and Kurta, chap. 2 in this volume; Carter and Menzel, chap. 3 in this volume). Thus, it is reasonable to infer that landscapes devoid of these structures would be poor or unsuitable habitat for many species of bats. Despite this understanding, a lack of knowledge of numbers of roosts required by individual bats, the extent to which roosts are limiting to bats, and the relationship between number of snags and abundance and viability of bat populations makes it difficult to provide science-based advice on management strategies for bats, beyond the most general recommendations such as "maintain high numbers of roost structures." Some forest plans, such as the Revised Land and Resource Management Plan for the Daniel Boone National Forest in Kentucky, call for retention of all dead and dying, potential, primary roost trees, including all hardwood snags, live shagbark (*Carya ovata*) and shellbark (*C. laciniosa*) hickories, hollow trees, and trees with rot, splits, or cracks (Krusac and Mighton 2002). A similar strategy is being taken in many other national forests within the range of the Indiana myotis (*Myotis sodalis*). Such an approach should be effective in providing for many of the roosting needs of bats, and it should also have significant positive implications for other species that are closely associated with dead trees and sloughing bark. Although this approach is appropriate on lands where conservation is a high priority and areas where considerations related to endangered species, such as the Indiana myotis, are paramount, this approach may be less acceptable to managers in forests where production of wood fiber is a central concern, especially until the linkage between population abundance and roost-site availability is more fully documented.

Most of what we know about the roosting ecology of bats in forested ecosystems comes from the results of radiotelemetry studies. The inference of management implications from these studies is based on the assumption that a positive relationship exists between the abundance of bats and the abundance of structures heavily used or preferred for roosting by bats. Although this is a reasonable hypothesis and a logical expectation, however, to date there are no empirical data available to rigorously evaluate this hypothesis. Even if we assume a positive relationship between bat abundance and number of roosts, this relationship could take a variety of forms (fig. 8.1). The implications of different roost management strategies are related to the nature of the relationship between number of roosts and population abundance. For example, if the abundance of bats increases steeply with the number of roosts and then reaches an asymptote relatively quickly (fig. 8.2a), modest levels of roost retention should result in substantial conservation benefits for bats, but increases beyond a modest level of roost retention will result in small conservation benefits. In contrast, if abundance of bats only increases gradually with the numbers of roosts at low roost densities and then increases more dramatically after some threshold is achieved, modest retention of roosts would only support small populations of bats, and substantial benefits would be derived from increased levels of roost retention (fig. 8.2b). Uncertainty surrounding the nature of the relationship between the number of snags or large trees and bat abundance is compounded by the fact that the relationship likely varies among species of bats, types of roost structures used (i.e., foliage, cavity, or bark), and the composition of trees in the forest.

Developing a quantitative understanding of the relationships between the abundance of bats and the number of roosts is fundamental to understanding relationships between bats and forest management. The limited information available suggests that a positive relationship between the number of large-diameter snags and the abundance of bats is plausible, at least in some forest types (E. Arnett, unpubl. data), but the shape of the relationship remains unclear. We propose that a reasonable working hypothesis is a roughly linear relationship between the number of snags and abundance of bats over the range of snag densities that are typically considered during normal forest management operations, with the relationship reaching an asymptote at very high snag densities. Nonetheless, we recognize that this hypothesis is highly speculative and strongly encourage research to explore this relationship empirically.

At extremely high snag densities, such as those found following stand-replacement fires (Everett et al. 1999), the relationship between number of snags and abundance of bats is likely to be relatively flat over the density of snags present, and some reduction in snags, such as through salvage harvest, is unlikely to have a significant impact on bats. However, the roosting ecology of bats in burned forests is very poorly understood. Just as the suitability of burned forests varies dramatically with time since fire for cavity-nesting birds (Saab et al. 2004), rapid changes in condition of dead trees in the years after the fire probably influence the suitability of snags for bats. Consequently, although burned sites often have large volumes of standing dead wood, only a small amount of this wood may be

Figure 8.2.
Conservation benefits of snag retention vary depending on the nature of the relationship between abundance of bats and number of roosts. Under scenario (a), modest levels of snag retention, such as those retained under management with low levels of snag retention (R_L) should result in habitat for a relatively large population of bats (B_L), and increasing the number of snags retained to high levels (R_H) results in only modest increases in abundance of bats (B_H). In contrast, if the abundance of bats only increases gradually with the number of roosts at low roost levels and then increases more dramatically as in scenario (b), retention of low numbers of roosts (R_L) would only support small populations of bats (B_L), and significantly higher populations (B_H) would result from increased levels of snag retention (R_H).

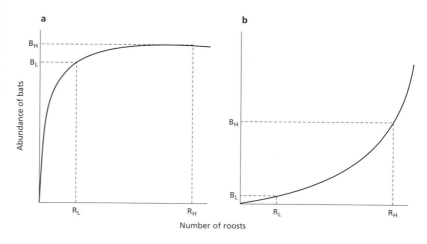

suitable for roosting and the quality of roosts in burned forests likely changes quickly in response to loss of bark from snags over time. Identification of the characteristics of fire-killed trees that provide high-quality roosting habitat over time would help direct postfire management activities.

In contrast to high-intensity burns and stand-replacement fires, in general, low-intensity wildfire and prescribed fires create relatively few snags (Horton and Mannan 1988) and many of these are small in diameter (Tiedemann et al. 2000). Small-diameter snags are relatively poor roost sites for most species of bats (Barclay and Kurta, chap. 2 in this volume), although small-diameter snags can provide roosting habitat for some species in some situations (Menzel et al. 2002). In some cases, prescribed fire and low-intensity wildfire can create basal hollows or fissures in large trees that are important for roosting (Gellman and Zielinski 1996; Taylor and Savva 1988). Although only a few studies have examined the effects of prescribed fire on snags, many snags are often lost during these burns. For example, roughly half of the snags in southeastern Arizona ponderosa pine (*Pinus ponderosa*) forests (Horton and Mannan 1988) and one fifth of snags monitored in north-central Arizona (Randall-Parker and Miller 2002) were consumed in stands subjected to prescribed fires. the vulnerability of snags to burning varies with their decay state, and those with 10–50% bark remaining (characteristic of many bat roosts) can be particularly vulnerable (Horton and Mannan 1988).

Because tree roosts are typically ephemeral, especially snags, development and recruitment of new roosts through time have a strong influence on long-term suitability of sites for bats. By managing stand density, forest management influences the rate of tree mortality, the size of trees that experience mortality, and the rate of growth of trees (Guldin et al., chap. 7 in this volume). In addition, selective harvest, green-tree retention strategies where individual living trees are retained on sites following harvest, and uneven-aged management approaches where a diversity of sizes and ages of trees are perpetually present in an area, can provide continual roost resources in areas managed for timber when implemented in combination with retention and management of dead wood and legacy structures.

CHANGES IN CHARACTERISTICS OF ROOSTS AND AREAS SURROUNDING ROOSTS

In addition to influencing roost abundance, forest management can also affect the environmental context in which a roost occurs. One way that forest management may change roost quality is through alteration of the conditions around a roost, and consequently, the thermal characteristics of roost sites. As bats are small homeotherms with relatively high surface-to-volume ratios and metabolic rates (Speakman and Thomas 2003), maintaining thermoneutrality under fluctuating environmental conditions is key to their survival. Bats have evolved a variety of life-history strategies to address these challenges: use of torpor and hibernation, migration, communal behavior, reproductive strategies, and habitat selection (Neuweiler 2000). Because of the large amount of time that bats spend roosting, thermal dynamics during roosting play a critical role in the ecology of bats (Barclay and Kurta, chap. 2 in this volume; Ormsbee et al., chap. 5 in this volume). The thermal characteristics of roosts can influence the energetic and thermoregulatory costs (Hamilton and Barclay 1994; McNab 1982; Tuttle 1976; Williams and Findley 1979), fetal and postnatal development (Hoying and Kunz 1998; Racey 1973; Racey and Swift 1981), and reproductive success (Lewis 1993) of bats. The thermal characteristics of roosts have been hypothesized to be key factors in the selection of roosts by forest-dwelling bats (Barclay and Kurta, chap. 2 in this volume; Hayes 2003; Kunz and Lumsden 2003).

There are few empirical data on the thermal characteristics of roosts in trees used by bats (Kerth et al. 2000). In one of the most rigorous assessments to date, Sedgeley (2001) compared microclimatic characteristics of roosts used by the New Zealand long-tailed bat (*Chalinolobus tuberculatus*) with ambient conditions and nearby potential, but unused, sites. Temperatures in used and potential roosts were relatively stable despite large fluctuations in ambient conditions. Furthermore, temperatures and humidity in used roosts were consistently higher than in potential but unused sites and used roosts had a relative humidity that was almost constantly 100%. The extent to which the differences in microclimate in used and unused sites can be attributed to intrinsic differences in the sites and how much of the difference is a function of the presence of bats in the roost is not clear. In any case, the microclimatic conditions of roosts used by long-tailed bats apparently provide several benefits to the species.

Although there are few data on roost microclimate conditions of North American bats, several studies have suggested that microclimate plays an important role in roost site selection. For example, thermal characteristics of aspen (*Populus tremuloides*) and conifer cavities differ, and this may influence roost site selection by big brown bats (*Eptesicus fuscus*) in Saskatchewan (Kalcounis and Brigham 1998). Primary roosts of Indiana myotis are usually in open areas and receive considerable solar radiation, whereas alternate roost sites are often located in more closed forests and have more cover from surrounding trees (Britzke et al. 2003; Callahan et al. 1997; Humphrey et al. 1977). Use of alternate roost sites increases during periods of extreme temperature and precipitation, suggesting that the thermal conditions of roosts in open versus closed forests provide a range of microclimates that allow bats to adjust to ambient environmental conditions (Callahan et al. 1997; Humphrey et al. 1977). In Kentucky, eastern red bats (*Lasiurus borealis*) select foliage roosts that have lower roost temperatures than in nearby points in the same roost tree (Hutchinson and Lacki 2001). The lower temperatures of roosts may allow roosting bats to avoid heat stress caused by high summer temperatures or to enter daily torpor to conserve energy.

Thermal characteristics of trees are influenced by a variety of factors, including the size of the tree, the density, conductivity, and specific heat of the wood, the color and characteristics of the bark, and the exposure of the tree to solar radiation and wind (Derby and Gates 1966; Karels and Boonstra 2003; Nicolai 1986; Potter and Andresen 2002). These factors, in turn, interact with the size and location of the crevice or cavity used for roosting and with other factors to shape the thermal characteristics of potential roost sites for bats that roost in crevices or cavities in trees. Ambient temperatures in forests provide the broader thermal environment where potential roosts occur, and these temperatures are influenced by distance from forest openings (Chen et al. 1993, 1995) and vegetative community composition and canopy cover (Xu et al. 2002). Forest openings can act as "heat islands" in forest settings (Xu et al. 2002), although small gaps can have minimal influence on ambient temperatures in some forest types (Clinton 2003).

Microclimatic conditions and thermal characteristics of roosts have been hypothesized to explain the use of roosts protruding above the surrounding forest canopy and roosts located in gaps or areas with reduced canopy cover (Betts 1998; Callahan et al. 1997; Gellman and Zielinski 1996; Ormsbee and McComb 1998; Waldien et al. 2000; Weller and Zabel 2001). Suitable characteristics of roost sites differ among species and sex (Broders and Forbes 2004), however, and optimal thermal conditions at roosts likely vary with species, sex, reproductive status, weather, age of bats, and time of year. Overreliance on one habitat type or topographic setting for retaining roosts is unlikely to provide the conditions necessary to meet the habitat needs for bats across seasons (Hayes 2003). For example, because regulations governing forest practices often result in conservation of streamside buffers or special streamside management zones (SMZs), it is often economically advantageous to allocate leave trees and snags to riparian areas. Because thermal characteristics of riparian areas often differ from those of upslope forests, however, the exclusive retention of snags and wildlife trees in riparian areas is not likely to be in the best interest of bat conservation. Variability in thermal requirements suggests that providing a diversity of roost types situated in a variety of topographic and ecological settings should be most beneficial to conserving bat populations in forested landscapes.

INFLUENCES OF FOREST MANAGEMENT ON OTHER TYPES OF ROOSTS

Although trees and snags are particularly important as roost sites for bats, forest management can influence the availability and characteristics of other roost types. In general, the influence of forest management on the quality of nontree roosts and the resulting effects on bats and bat populations are poorly known. One example of use of nontree roosts in forests is the use of leaf litter as roosts by eastern red bats during winter (Mager and Nelson 2001; Saugey et al. 1998). Prescribed fire in eastern deciduous forests during winter may result in the mortality of these bats, but the magnitude of the influence of burning on litter-roosting bats is unknown (Carter et al. 2002; Carter and Menzel, chap. 3 in this volume). Bridges provide important night roosts for many species of bats (Ormsbee et al., chap. 5 in this volume) and day roosts for a few (Lance et al.

2001; Trousdale and Beckett 2004). Management decisions on the types and characteristics of bridges to construct, bridge placement, and removal of existing bridges could influence bat populations. Forest management in the areas surrounding bridges may also influence use of bridges as roosts by bats. For example, bridges used by Rafinesque's big-eared bats (*Corynorhinus rafinesquii*) as day roosts in Louisiana had significantly more mature deciduous forest within a 0.25-km radius than bridges where no bat was found (Lance et al. 2001). The relationships between use of bridges for roosting and conditions of the surrounding forest are not well understood.

Bats have been documented roosting in crevices in rock quarries created to support the construction and maintenance of forest roads. Mining rock from quarry sites can sometimes trap bats in roosts or crush individual bats. Although we are aware of anecdotal observations where this occurred, the extent to which quarries are used by bats is uncertain.

Finally, caves are critical roost sites for some species of bats and the microclimatic conditions of caves are known to impact the use of caves as roosts by bats (Briggler and Prather 2003; Clark et al. 1996; Hurst and Lacki 1999). It is conceivable that forest management activities adjacent to caves could alter microclimatic conditions of the roost in some situations. Limited research suggests that vegetative structure and habitat surrounding caves may have an influence on use of caves as roosts by some species or in some situations (Raesly and Gates 1987) but not others (Raesly and Gates 1987; Wethington et al. 1997). Increased research is needed to fully understand the interactions between forest management and the quality and use of the nontree roosts of bats.

FOREST MANAGEMENT, CLUTTER, AND BAT POPULATIONS

The amount of clutter or the number of obstacles a bat must detect and avoid in a given area (Fenton 1990) can strongly influence the use of habitat by bats. Differences in maneuverability and morphology influence bats' tolerance of clutter and their ability to exploit habitats differing in the amount of clutter (Aldridge and Rautenbach 1987; Crome and Richards 1988; Kalcounis et al. 1999; Norberg and Rayner 1987). In general, maneuverable species or bats with small bodies and low wing loading are able to use habitats with higher levels of clutter than less maneuverable species or bats with large bodies and high wing loading can use.

Most bats in North America avoid extensive use of highly cluttered habitat. For example, Townsend's big-eared bats (*Corynorhinus townsendii*) rarely forage in dense forests (Dobkin et al. 1995), and Yuma myotis (Brigham et al. 1992) and juvenile little brown myotis (*Myotis lucifugus;* Adams 1997) forage most frequently in areas with low levels of clutter. In an experimental test of clutter and prey availability, small bats were able to take advantage of increased prey availability in a cluttered environment, whereas larger bats were not (Sleep and Brigham 2003). Avoidance of cluttered habitats for foraging was demonstrated experimentally by Brigham et al. (1997), who manipulated the amount of clutter along forest-clearcut edges by constructing artificial "clutter zones" along forest-clearcut edges. Although commuting activity of *Myotis* bats in cluttered and uncluttered habitats did not differ significantly, foraging activity of *Myotis* bats was significantly lower in artificial-clutter zones despite similar insect abundance in clutter zones and control sites.

Forest management activities that influence tree density directly alter the amount of clutter in an area. As result, forest management can directly influence habitat suitability for bats through changes in the amount of vegetative clutter. Several studies have examined the influences of forest management practices on the activity of bats (table 8.1), and many of the findings can be interpreted as a response of bats to clutter. For example, Humes et al. (1999) found that use by bats of 50- to 100-year-old Douglas-fir (*Pseudotsuga menziesii*) forests varied with stand structure and the management history of stands; stands thinned 9–24 years earlier (*n* = 184 trees per hectare) had significantly more bat activity than unthinned stands (*n* = 418 trees per hectare) of the same age. Although the effects of prescribed burning on habitat use by bats has not been studied, some of the structural implications of prescribed fire (Peterson and Reich 2001) are similar to those of thinning, and thus, prescribed burning in young and mid-aged forests may have effects on vegetative structure similar to thinning. However, interactions with prey availability and other factors interplay with clutter to shape habitat quality for bats. For example, Patriquin and Barclay (2003) found that thinning in white fir stands had minimal influence on use of sites by *Myotis* within one year of thinning. Thinning of red pine stands in Michigan also had little effect on bat activity (Tibbels and Kurta 2003), possibly because red pine plantations may be too cluttered and depauperate in insect populations even after thinning to provide suitable habitat for bats.

Bats frequently use edge habitat for commuting and foraging, presumably as a consequence of low tolerance to clutter in combination with prey availability (Clark et al. 1993; Furlonger et al. 1987; Grindal and Brigham 1999; Hogberg et al. 2002; Krusic et al. 1996; Walsh and Harris 1996; Wethington et al. 1996; Zimmerman and Glanz 2000). For example, the amount bat activity in forests of British Columbia was higher along the edges of clearcuts than either within the clearcut or within the uncut forest (Grindal and Brigham 1999). However, the relationship between forest edge and use by bats can be strongly influenced by the morphology of bats. For example, Patriquin and Barclay (2003) found that silver-haired bats (*Lasionycteris noctivagans*), a relatively large species in North America, were more active in clearcuts than in intact patches, whereas little brown myotis foraged most extensively along the forest edge and northern myotis (*Myotis septentrionalis*) foraged most frequently in intact forest.

Small forest gaps resulting from small-scale natural disturbances, variable density thinnings, and uneven-aged management approaches such as group selection (Guldin et al., chap. 7 in this volume) can increase use relative to adjacent undisturbed forest in some situations. For example, Grindal and Brigham (1998) examined patch cuts ranging from 0.5 to 1.5 ha in size in British Columbia and found that bat activity increased

Table 8.1. Published studies examining the influences of silvicultural treatment on bat activity in the United States and Canada

Silvicultural treatment	Study area	Forest type	Habitat comparisons	Source
Thinning and partial harvest	western Oregon	Douglas-fir/western hemlock	old-growth, thinned young, and unthinned young stands	Humes et al. 1999
Partial harvest	central Ontario	Conifer-dominated mixedwood	selectively harvested, old-growth mixedwood, mature white pine-dominated mixedwood, and boreal mixedwood stands	Jung et al. 1999
Thinning and open patches	Michigan	Red pine	unthinned and thinned (5–11 years prior to sampling) stands and wildlife openings	Tibbels and Kurta 2003
Green-tree retention and residual patches	western Washington and Oregon	Douglas-fir/western hemlock	15% (in two leave patterns), 40% (in two leave patterns), 75%, and 100% (control) retention levels, measured pre- and posttreatment	Erickson, J., unpub. data
Small clearcuts and leave-tree patches	northern Alberta	boreal mixedwood	points along the forest edge, in the center, and along the edges of retained patches of trees in 8- to 10-ha clearcuts with two small patches of leave trees	Hogberg et al. 2002
Small patch cuts	southern interior British Columbia	western redcedar/western hemlock	edge of 0.5-, 1.0-, and 1.5-ha patch cuts (pre- and post-treatment), roads, and paired forested sites	Grindal and Brigham 1998
Clearcuts	southern interior British Columbia	(forest type not described)	points in forest (≥50 m from edge), forest-harvest unit edge, and harvested areas (≥50 m from edge) in 12- to 116-ha stands harvested 2–24 years earlier and located in three elevational zones	Grindal and Brigham 1999
Clearcuts and green-tree retention	central British Columbia	subboreal spruce, lodgepole pine, aspen	residual patches and at clearcut edges	Swystun et al. 2001
Recent clearcuts and older stands	western Washington	Douglas-fir/western hemlock	clearcut (2–3 years), young pre-commercially thinned (12–20 years), young unthinned (30–40 years), and mature (50–70 years)	Erickson and West 1996
Small patch cuts and clearcuts	New Hampshire	northern hardwood and spruce-fir	Regeneration (0.1- to 0.8-ha patch cuts and clearcuts up to 12.1 ha combined), sapling/pole, mature and "overmature" stands	Krusic et al. 1996
Two sizes of patch cuts	South Carolina	pine-mixed hardwood	0.03-ha patch cuts, 0.5-ha patch cuts, skidder trails, and unharvested sites	Menzel et al. 2002
Clearcuts, stands thinned to different densities, and different forest types	northern Alberta	boreal mixedwood	edges and centers of 0% (clearcut), 20%, 50%, and 100% (control) retention stands in conifer-dominated, hardwood-dominated, and mixedwood stands	Patriquin and Barclay 2003
Clearcuts and diameter-limit harvests	West Virginia	northern hardwood	Foraging (radiotelemetry) locations in intact, diameter limit harvest, clearcuts, and open sites or roads	Owen et al. 2003

Note: Studies focusing primarily on stand age are not included in this table, but studies that considered stand age in addition to silvicultural treatment and studies evaluating the use of recent clearcuts are included.

in these stands relative to uncut controls. Insect abundance in patch cuts and unlogged forests were similar, suggesting that the differences in activity levels could be attributed to differences in the amount of clutter.

Forest operations often require building and maintenance of roads and trails. Roads often provide flyways for bats, presumably because of reduced clutter over roads. For example, foraging activity of bats in bottomland hardwoods of South Carolina was greater along skidder trails than in intact mature forest (Menzel et al. 2002). The difference in use between habitats was greatest for the larger species such as big brown bats, evening bats (*Nycticeius humeralis*), and lasiurines, whereas smaller eastern pipistrelles (*Pipistrellus subflavus*) tended to use the forest more than the trails. Similarly, in New Hampshire bat activity was higher along forest trails than in forest interiors (Krusic et al. 1996), and bats in Maine made extensive use of gravel roads (Zimmerman and Glanz 2000).

FOREST MANAGEMENT, PREY AVAILABILITY, AND BAT POPULATIONS

As nocturnal insectivores, bats eat large amounts of insect prey (Aubrey et al. 2003), with individuals consuming between 40 and 100% of their body mass in insects per night (Kunz et al. 1995; Kurta et al. 1989, 1990). Challenges in documenting the distribution, abundance, and availability of insects at large spatial scales and relating these to measures of bat activity have precluded rigorous evaluation of the influences of insect availability on habitat use or population abundance of bats (Lacki et al., chap. 4 in this volume). Some studies have examined influences of forest structure and forest management on insect populations (Burford et al. 1999; Humphrey et al. 1999; Lewis and Whitfield 1999), but the high diversity of insect taxa, the concomitant variation in response to forest structure, and regional variability have precluded determination of general patterns that are useful for predicting bat response. Despite this, scientific evidence and anecdotal observations support the hypotheses that bats respond to prey availability, that prey availability is influenced by forest management, and that influences of forest management on prey populations affect bat populations.

Prescribed fire, wildfire, fire suppression, and fire management all influence insect populations and, thus, may affect bat populations. Fire can affect insects directly through mortality or indirectly by changes in soil properties or vegetation characteristics (McCullough et al. 1998). Lepidoptera, important prey for many North American bats, appear to be more vulnerable to fire than are many groups of invertebrates (Siemann et al. 1997). However, the influences of fire on insects depend on the timing of the fire with respect to the life history of insects, the intensity of the fire, its rate of spread, and the area affected by the fire. Although several studies have examined the effects of fire on insect diversity and abundance (Siemann et al. 1997; McCullough et al. 1998), short- and long-term responses within groups can differ considerably (Siemann et al. 1997), and an understanding of the influences of fire on insect populations remains poorly developed. As a result, the impact of fire and fire management on prey availability for bats, and on ecology of bats in general, is poorly understood (Carter et al. 2002).

Similarly, although it is likely that use of insecticides and herbicides influence prey availability for bats, the influence of the chemicals applied, the ecological context, and bat-prey relationships have not been well studied. Insecticides can have a direct impact on prey availability by reducing insect abundance. Herbicides often have a indirect influence on insect populations by changing the abundance and composition of plant communities on which insect communities rely (Guynn et al. 2004). No data are available on the effects of herbicide treatments on insects commonly consumed by bats and, depending on the herbicide used and the implementation of the treatments, herbicide treatments may either have negative or positive effects on bats and their prey.

FOREST MANAGEMENT, AQUATIC HABITAT, AND BAT POPULATIONS

Aquatic habitat plays a critical role in the habitat ecology of bats, both as sources of water and insect prey. Aquatic habitat is particularly important for bats in arid environments, where water can be a limiting resource driving the presence of bats. Bats have relatively high rates of evaporative water loss and consequently require intake of water to maintain their water balance (Kurta et al. 1989, 1990; McLean and Speakman 1999; Webb 1995). As opposed to nectivorous and frugivorous bats, which obtain much of their water from food, insectivorous bats must obtain much of their water intake from free-standing water (Neuweiler 2000). Drinking water may also be important for bats as a source of calcium (Adams et al. 2003).

Riparian areas and areas over water are also important foraging areas for bats. Commuting and foraging activity is typically higher in riparian areas than in upland sites (Furlonger et al. 1987; Grindal et al. 1999; Krusic et al. 1996; Seidman and Zabel 2001; Zimmerman and Glanz 2000) and some species spend significant proportions of their nightly activity foraging and commuting in riparian areas (Barclay 1991; Brigham et al. 1992; Fellers and Pierson 2002; LaVal et al. 1977; Waldien and Hayes 2001). High insect densities in riparian areas are probably a key factor responsible for high levels of bat activity in these habitats (Barclay 1991; Grindal 1996; Thomas 1988).

Many factors affect the suitability of specific riparian and aquatic sites for bat activity. For example, bats often prefer smooth water surfaces instead of riffles or rapidly moving water (Rydell et al. 1999; von Frenckell and Barclay 1987; Warren et al. 2000). They appear to avoid riffles and fast-moving water because of the acoustic clutter associated with rocks or rough water, as well as the high-frequency noise associated with running water (Mackey and Barclay 1989). Bats, such as little brown myotis, which forage within a few meters of the water, are affected both by structural clutter on the stream surface and the noise of running water (Mackey and Barclay 1989). Even big-brown bats, which forage higher (>5 m) above the surface, are affected by the high-frequency noise associated with fast-running water (Mackey and Barclay 1989).

In some regions, considerable attention has been given to management and restoration of instream structure because of the relationship between instream structure and fish populations. Because bat activity has been shown to vary with stream characteristics (Mackey and Barclay 1989; Rydell et al. 1999; von Frenckell and Barclay 1987; Warren et al. 2000), it is likely that forest management activities designed to increase instream habitat complexity for fish populations will influence habitat suitability for bats. For example, increasing pools by adding large wood to streams (Hilderbrand et al. 1997) may create improved foraging and drinking habitat for bats. However, the population-level consequences of these management activities have not been evaluated and many instream manipulations may merely shift the pattern of use by bats up and down stream without significantly influencing the abundance of bats.

As noted earlier, bats frequently use bridges as night roosts, suggesting that the availability of night roosts near riparian areas may be important (Adam and Hayes 2000; Arnett and Hayes 2000; Johnson et al. 2002; Perlmeter 1996; Pierson et al. 1996), although the relationship between location of night roosts and spatial patterns of activity have not been evaluated. The availability of night roosts may also affect the suitability of riparian areas as activity areas for bats (Kunz and Lumsden 2003).

Forest management can directly influence the amount of forest cover in riparian areas, the structure and composition of streamside forests, and the presence and characteristics of roads and bridges in riparian areas. These factors in turn influence the availability of roost sites, the amount of clutter over the stream, the availability and abundance of terrestrial and aquatic invertebrates fed upon by bats, and water quality. Despite the importance of aquatic and riparian habitat for bats and the potential implications of riparian management on bat populations, few studies have examined the effects of riparian management on habitat use by bats. Hayes and Adam (1996) found that the overall amount of activity by bats along streams in western Oregon was lower in clearcut areas than in uncut areas. *Myotis* species showed the strongest negative response to logging, whereas larger species, such as silver-haired bats, were more common in logged areas. Logging also affected insect abundance and size distribution. Large insects (>5 mm), in particular, Lepidopterans, were more abundant in wooded habitat, whereas small insects were more abundant in logged areas. Similarly, in England, Daubenton's bat (*Myotis daubentonii*) and the common pipistelle (*Pipistrellus pipistrellus*), preferred streams with trees on both sides of the river compared with streams lacking trees or with trees on only one side (Warren et al. 2000); these species did not appear to differ in the use of rivers with and without trees in Scotland (Rydell et al. 1994).

Concentrated bat activity around created ponds in upland habitat has been documented (Huie 2002), but it is unclear whether this increased activity reflects a greater abundance of bats in the area, or simply changes in habitat use by bats. The extent to which population changes result from creation of water sources is probably a function of the availability of alternative water sources in the landscape, with population influences being most significant in more arid landscapes.

STAND AGE, FOREST COMPOSITION, AND USE OF UPSLOPE FORESTS BY BATS

There has been considerable interest in the influences of stand age on biodiversity and habitat suitability for a wide range of species. Perhaps this interest has been strongest in the Pacific Northwest, driven largely by concerns over loss of old-growth forests. Thomas's (1988) seminal work pioneering use of echolocation monitoring with bats examined stand age and associations of bats with old-growth forests, priming a generation of echolocation-monitoring studies addressing issues concerning bats and stand age (Crampton and Barclay 1995, 1998; Erickson 1993; Erickson and West 1996; Grindal and Brigham 1999; Humes et al. 1999; Jung et al. 1999; Krusic et al. 1996) (table 8.2). Despite the focus on stand age in several studies, however, it is highly unlikely that bats respond directly to stand age. Rather, bats are almost certainly responding to the structural characteristics of forest stands, the presence and characteristics of habitat elements within those stands, and the broader spatial and environmental context in which those stands occur. In general, the findings of studies examining the use of stands of different ages can be interpreted in terms of distribution and availability of roosts, the response of bats to clutter, and influences of forest structure on prey populations. Because many structural characteristics of forests are often closely associated with stand age, studies focusing on stand age led to an increased understanding of the response of bats to the structural characteristics of forests. However, when considering implications of forest management on bats, the distinction between stand age and stand structure is an important one, as structure and developmental trajectories of stands are labile and can be strongly influenced by silvicultural practices, stand history, and conservation strategies such as retention of legacy structures (Guldin et al., chap. 7 in this volume).

Bats often heavily use open habitat, meadows, and recently cut areas or stands less than 10 years old as activity areas (Brigham et al. 1992; Erickson 1993; Erickson and West 1996; Grindal and Brigham 1998; Krusic et al. 1996; Menzel et al. 2001; Patriquin and Barclay 2003), although high use of open areas does not appear to occur in all forest types or situations (Jung et al. 1999; Lunde and Harestad 1986). High use of open habitat probably reflects reduced clutter or increased abundance of insects. Relatively small openings can provide habitat for bats, as bats often heavily use patch cuts as small as 0.5 ha in forests (Grindal and Brigham 1998; Tibbels and Kurta 2003). Some species, such as the northern myotis, appear to avoid large clearcuts and openings, although they make extensive use of smaller gaps (Owen et al. 2003).

Several studies have documented higher use of old forests than of

Table 8.2. Published studies examining the associations between stand age and levels of bat activity in the United States and Canada

Stand ages considered	Study area	Forest type	Sources
Young (<75 years), mature (100–165 years), and old growth (>200 years)	western Oregon and Washington	Douglas-fir/western hemlock	Thomas 1988; Thomas and West 1991
Clearcut (2–3 years), young precommercially thinned (12–20 years), young unthinned (30–40 years), and mature (50–70 years)	western Washington	Douglas-fir/western hemlock	Erickson and West 1996
Young (0–9 years), sapling/pole (10–59 years hardwood, 10–39 years conifer), mature (60–119 years hardwood, 40–89 years conifer), and "overmature" (>119 years)	New Hampshire	northern hardwood and spruce-fir	Krusic et al. 1996
Young (20–30 years), mature (50–65 years) and old (>120 years)	central Alberta	aspen mixedwood	Crampton and Barclay 1995, 1998
Four age classes: 81–100, 100–120, 121–140, and 141–250 years	southern interior British Columbia	(forest type not described)	Grindal and Brigham 1999
Young (50–100) and old growth (≥200 years)	western Oregon	Douglas-fir/western hemlock	Humes et al. 1999
Young (3–7 years), mature (90–120 years), and old growth (≥120 years)	central Ontario	conifer-dominated mixedwood	Jung et al. 1999

Note: Studies comparing clearcuts or recently cut stands to adjacent uncut forests are not included.

young forests (Crampton and Barclay 1998; Grindal and Brigham 1999; Humes et al. 1999; Jung et al. 1999; Krusic et al. 1996; Perkins and Cross 1988; Thomas 1988; Thomas and West 1991). High use of old-growth stands by bats is often attributed to high availability of roosts, especially large-diameter snags, in these stands (Crampton and Barclay 1996, 1998; Humes et al. 1999; Kalcounis et al. 1999; Perkins and Cross 1988; Thomas 1988; Thomas and West 1991). High levels of activity in old-growth stands (Erickson and West 1996; Hayes and Gruver 2000; Thomas 1988) and other forested sites (Grindal and Brigham 1999; Hayes 1997) shortly after dusk and before dawn has been interpreted as the result of bats commuting to and from stands used for roosting (Erickson and West 1996; Grindal and Brigham 1999; Thomas 1988). In addition, variation in use and partitioning of habitat by bats among vertical strata in forests (Bradshaw 1996; Hayes and Gruver 2000; Jung et al. 1999; Kalcounis et al. 1999; Krusic et al. 1996) suggest that stands with greater vertical complexity may provide increased opportunities for foraging by bats. Stands with relatively simple vertical structure may not provide a diversity of foraging niches for bats, and the structural complexity of old-growth forests may partially account for high levels of activity in old forests (Hayes and Gruver 2000).

Use by bats of upland, intermediate-aged coniferous forests between 10 and 100 years old tends to be relatively low (Crampton and Barclay

1998; Erickson and West 1996; Jung et al. 1999; Krusic et al. 1996; Parker et al. 1996; Thomas 1988; Thomas and West 1991; Tibbels and Kurta 2003). Low use of these stands presumably is the result of high clutter, compounded by low availability of roost sites in these habitats.

Forest composition can interact with structural characteristics of stands to determine habitat suitability for bats. Although not extensively studied, differences in forest type and forest composition appear to have significant influence on activity levels of bats in some areas (Kalcounis et al. 1999; Krusic et al. 1996; Patriquin and Barclay 2003). Causal factors responsible for the differential use by bats of stands varying in vegetative composition are not clear, but may be related to differences in prey and roost availability (Kalcounis et al. 1999; Krusic et al. 1996; Patriquin and Barclay 2003).

LANDSCAPE-LEVEL INFLUENCES OF FOREST MANAGEMENT ON BATS

Although forest operations typically occur at the stand scale, the cumulative effects of decisions at the stand scale determine the composition, configuration, and characteristics of entire landscapes. Given the scale at which bats roost, forage, and travel over the course of a day or season, it is reasonable to propose that changes in landscape characteristics resulting from forest management influence bat populations (Duchamp et al., chap. 9 in this volume). Unfortunately, our understanding of relationships at this scale remains poorly developed. Although the negative influences of fragmentation on bats have been documented in other regions (e.g., Law et al. 1999 and Pavey 1998 in Australia; Estrada and Coates-Estrada 2002 and Schulze et al. 2000 in Mexico and Central America), information about responses of bats to characteristics at the landscape scale in forests in the United States and Canada is particularly weak and the lack of work in this area is striking. The single published paper of which we are aware that focuses on influences of landscape attributes on bats in North American temperate forests (Erickson and West 2003) examined the relationship between bat activity and characteristics measured at the stand and landscape scales in Oregon and Washington. Although a number of characteristics of forest stands measured in this study helped explain the amount of activity at different sites, no significant relationship between bat activity and any measured landscape variable was evident. The final word on the relationship between bats and landscape structure in North American forests, however, is not in sight for several reasons. First, as Erickson and West (2003) point out, the lack of significant associations at the landscape scale could be a function of the stands they selected for study. Erickson and West only examined stands of mature conifers located on public lands. The breadth of landscape conditions found surrounding these stands represents a relatively small part of the spectrum of potential landscape conditions. Consideration of a greater breadth of landscape conditions that include a variety of landscape compositions, including a diverse range of forest ages and stand structure, will be necessary to fully elucidate potential influences of landscape structure on bat populations. Moreover, Erickson and West (2003) defined their landscapes as the 100-ha area around their sample points. This represents a relatively small area in relation to the movement patterns of bats, and the potential influences of larger spatial scales were not tested. Finally, evaluation of relationships between landscape structure and bats needs to be conducted in a variety of forest types and geographical areas differing in bat community composition before patterns and generalities are possible.

CONCLUSIONS

Despite recent conservation concerns, a flurry of studies, and significant progress, our knowledge of the influences of forest management practices on bat populations remains at a rudimentary level for a number of logistical and scientific reasons. As noted in several chapters of this book and elsewhere, bats are difficult to study because of their mobility coupled with their small body mass. As a result, collecting data on the response of bats to forest management practices is challenging. Moreover, estimates of abundance, demographic trends, or reproductive success of bats in forest settings are difficult or impossible to obtain by using current methodologies (Hayes 2003). Because of this, meaningful evaluation of the response of bats to forest management is problematic. Most studies conducted to determine the impact of forest management on bats have either examined the differences in the amount of bat activity in areas varying in forest condition or forest management history (Crampton and Barclay 1998; Grindal and Brigham 1999; Hogberg et al. 2002; Humes et al. 1999; Jung et al. 1999; Kalcounis et al. 1999; Krusic et al. 1996; Menzel et al. 2002; Patriquin and Barclay 2003; Thomas 1988; Zimmerman and Glanz 2000), or have made inferences based on patterns of roost use (Cryan et al. 2001; Lacki and Schwierjohann 2001; Rabe et al. 1998; Waldien et al. 2000). Although these studies are extremely valuable and provide insight on likely management effects, they generally do not provide information that can be used to assess the influences of management on populations directly.

The logic underlying inference of management implications from studies of roost site characteristics and patterns of activity or occurrence was stated clearly and succinctly by Crampton and Barclay (1998): "Given the abundance of bats in old stands and their association with roost trees, we suggest that harvesting will have a negative impact. In particular, if roost trees are limiting in this system, and if this drives habitat selection, it is likely that a decrease in bat abundance will occur as harvesting progresses." We believe the logic underlying this statement is reasonable and sound. To fully assess the underlying assumptions, bat ecologists must develop rigorous estimates of the abundance of bats in different forest conditions, better identify factors responsible for limiting bat populations, and then demonstrate the relationships between the abundance of bats and forest management practices.

Bats are influenced by habitat characteristics at multiple spatial scales, and forest management activities can potentially influence habitat suitability at each of these scales. Unfortunately, because of logistical challenges, the responses of bats often are measured at spatial scales that are small relative to the movement patterns of bats and the scale at which demographic processes are expressed. For example, most studies evaluating the influence of forest management on the amount of bat activity use areas the size of a forest stand (in general, 40 ha) or smaller as the experimental unit for analysis. Although the home ranges of bats in North American forests have not been quantified, many species of bats in forested ecosystems regularly travel several kilometers each night (Butchkoski and Hassinger 2002; Fellers and Pierson 2002; Murray and Kurta 2004). As a result, although studies of activity levels effectively document site-specific influences on patterns of activity, actual consequences at the population level are unclear. For example, in coniferous forests in western Oregon, Humes et al. (1999) showed that the amount of activity of bats in unthinned forests was less than in thinned forests. Does this indicate that thinning forests will increase the abundance of bats? Although this is a likely inference and may be the most parsimonious conclusion, this inference depends heavily on underlying assumptions. Aside from the fact that studies such as Humes et al. (1999) are correlative and do not demonstrate causality, and that echolocation-monitoring studies assess activity levels and not abundance (Hayes 2000), the distribution of thinned stands in an area may merely influence behavioral patterns of bats, rather than directly altering abundance or viability of their populations. We draw attention to this by using the example of Humes et al. (1999) because this work is fundamentally sound, but difficulties in inference from such work is indicative of the challenges faced in attempting to evaluate the influence of forest management on bats.

So where does this leave us? Is the proverbial cup of understanding half-full or half-empty? Clearly, there are a number of gaps in our knowledge. From a scientific perspective we are left without being able to make strong inference and considerable room exists for misinterpretation of the management implications of studies. However, managers almost never have the luxury of having 100% confidence in their approaches and must move forward based on the weight of evidence modified by an ad hoc risk assessment that tempers actions in light of the consequences of error. Thus, applied ecologists must conduct a careful balancing act, helping shape management decisions based on what we know about the ecology of bats, but at the same time cautiously and critically evaluating risk to minimize the chance that hypotheses and weakly supported theories will evolve into dogma.

Understanding quantitative relationships of abundance and viability of bat populations with habitat characteristics at multiple spatial scales is central to understanding the responses of bats to forest management and to shaping management recommendations. We strongly encourage efforts to develop this understanding in a more rigorous way. A critical first step in this is the development of a methodology enabling assessment of population abundance of bats in forests. In the absence of a refined quantitative understanding, we suggest that four key habitat attri-

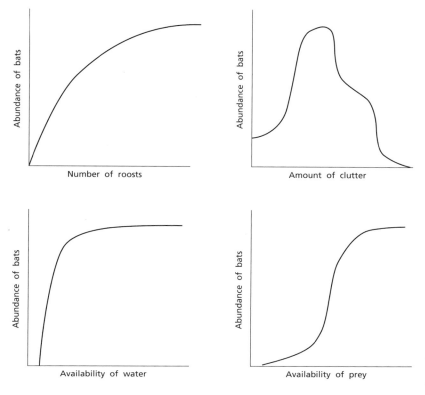

Figure 8.3.
A conceptual model for bats illustrating hypothesized relationships among the abundance of bats and key physical and ecological features that are influenced by forest management. For number of roosts, availability of water, and availability of prey there are response thresholds, beyond which increases in resources do not result in increased numbers of bats. For the amount of clutter, at very low levels of clutter (open conditions, such as in clearcuts) low numbers of bats occur; numbers achieve highest values at some intermediate level of clutter, such as along edges, and then decrease with increasing clutter. The actual shapes of the curves and magnitude of units on the axes vary with location. Under this model, limiting factors for bats can vary with the spatial scale under consideration and region, and the integration of these factors will determine population abundance. In landscapes where all of these factors are optimal for bats, other factors, such as climate, predation, or competition, will limit bat populations.

butes play critical roles in defining niches for bats in forested ecosystems: quality and availability of roosts, water, and prey and the amount of clutter. Although a more rigorous quantitative understanding of these factors is being developed, qualitative understanding and hypothesized relationships supported by empirical evidence provide a useful framework for forest management. As such, our graphical model of habitat relationships (fig. 8.3) may be useful as a heuristic device to help shape management decisions. Future work to incorporate these relationships into formal computer models may be useful to help predict bat responses to forest conditions. Finally, this conceptual model suggests a suite of hypotheses concerning the roles of habitat characteristics in regulation of bat numbers that can be tested in further study.

LITERATURE CITED

Adam, M.D., and J.P. Hayes. 2000. Use of bridges as night roosts by bats in the Oregon Coast Range. Journal of Mammalogy 81:402–407.

Adams, R.A. 1997. Onset of volancy and foraging patterns in juvenile little brown bats, *Myotis lucifugus.* Journal of Mammalogy 78:239–246.

Adams, R.A., S.C. Pedersen, K.M. Thibault, J. Jadin, and B. Petru. 2003. Calcium as a limiting resource to insectivorous bats: can water holes provide a supplemental mineral source? Journal of Zoology 260:189–194.

Aldridge, H.D.J.N., and I.L. Rautenbach. 1987. Morphology, echolocation, and resource partitioning in insectivorous bats. Journal of Animal Ecology 56:763–778.

Arnett, E.B., and J.P. Hayes. 2000. Bat use of roosting boxes installed under flat-bottom bridges in western Oregon. Wildlife Society Bulletin 28:890–894.

Aubrey, K.B., J.P. Hayes, B.L. Biswell, and B.G. Marcot. 2003. Ecological role of arboreal mammals in western coniferous forests, pp. 405–443, *in* Mammal community dynamics in coniferous forests of western North America: management and conservation. (C. J. Zabel and R. G. Anthony, eds.). Cambridge University Press, Cambridge, MA.

Barclay, R.M.R. 1991. Population structure of temperate zone insectivorous bats in relation to foraging behaviour and energy demand. Journal of Animal Ecology 60:165–178.

Belwood, J.J. 2002. Indiana bats in suburbia: observations and concerns for the future, pp. 193–198, *in* The Indiana bat: biology and management of an endangered species (A. Kurta and J. Kennedy, eds). Bat Conservation International, Austin, TX.

Betts, B.J. 1998. Roosts used by maternity colonies of silver-haired bats in northeastern Oregon. Journal of Mammalogy 79:643–650.

Bradshaw, P.A. 1996. The physical nature of vertical forest habitat and its importance in shaping bat species assemblages, pp. 199–212, *in* Bats and forests symposium (R.M.R. Barclay and R.M. Brigham, eds.). Research Branch, British Columbia Ministry of Forests, Victoria, BC.

Briggler, J.T., and J.W. Prather. 2003. Seasonal use and selection of caves by the eastern pipistrelle bat (*Pipistrellus subflavus*). American Midland Naturalist 149:406–412.

Brigham, R.M., H.D.J.N. Aldridge, and R.L. Mackey. 1992. Variation in habitat use and prey selection by yuma bats, *Myotis yumanensis.* Journal of Mammalogy 73:640–645.

Brigham, R.M., S.D. Grindal, M.C. Firman, and J.L. Morissette. 1997. The influence of structural clutter on activity patterns of insectivorous bats. Canadian Journal of Zoology 75:131–136.

Brittingham, M.C., and L.M. Williams. 2000. Bat boxes as alternative roosts for displaced bat maternity colonies. Wildlife Society Bulletin 28:197–207.

Britzke, E.R., M.J. Harvey, and S.C. Loeb. 2003. Indiana bat, *Myotis sodalis,* maternity roosts in the southern United States. Southeastern Naturalist 2:235–242.

Broders, H.G., and G.J. Forbes. 2004. Interspecific and intersexual variation in roost-site election of northern long-eared and little brown bats in the Greater Fundy National Park Ecosystem. Journal of Wildlife Management 68:602–610.

Burford, L.S., M.J. Lacki, and C.V. Covell Jr. 1999. Occurrence of moths among habitats in a mixed mesophytic forest: implications for management of forest bats. Forest Science 45:323–332.

Burke, H.S., Jr. 1999. Maternity colony formation in *Myotis septentrionalis* using artificial roosts: the rocket box, a habitat enhancement for woodland bats? Bat Research News 40:77–78.

Butchkoski, C.M., and J.D. Hassinger. 2002. Ecology of a maternity colony roosting in a building, pp. 130–142, *in* The Indiana bat: biology and management of an endangered species (A. Kurta and J. Kennedy, eds.). Bat Conservation International, Austin, TX.

Callahan, E.V., R.D. Drobney, and R.L. Clawson. 1997. Selection of summer roosting sites by Indiana bats (*Myotis sodalis*) in Missouri. Journal of Mammalogy 78:818–825.

Carter, T.C., W.M. Ford, and M.A. Menzel. 2002. Fire and bats in the Southeast and mid-Atlantic: more questions than answers? pp. 139–143, *in* The role of fire in nongame wildlife management and community restoration: traditional uses and new directions (W.M. Ford, K.R. Russell, and C.E. Moorman, eds.). Proceedings of a special workshop. GTR-NE-288. USDA Forest Service, Northeastern Research Station, Newtown Square, PA.

Chambers, C.L., V. Alm, M.S. Siders, and M.J. Rabe. 2002. Use of artificial roost by forest-dwelling bats in northern Arizona. Wildlife Society Bulletin 30:1085–1091.

Chen, J., J.F. Franklin, and T.A. Spies. 1993. Contrasting microclimates among clearcut, edge, and interior of old-growth Douglas-fir forest. Agricultural and Forest Meteorology 63:219–237.

———. 1995. Growing-season microclimatic gradients from clearcut edges into old-growth Douglas-fir forests. Ecological Applications 5:74–86.

Clark, B.K., B.S. Clark, D.M. Leslie, and M.S. Gregory. 1996. Characteristics of caves used by the endangered Ozark big-eared bat. Wildlife Society Bulletin 24:8–14.

Clark, B.S., D.M. Leslie Jr., and T.S. Carter. 1993. Foraging activity of adult female Ozark big-eared bats (*Plecotus townsendii ingens*) in summer. Journal of Mammalogy 74:422–427.

Clinton, B.D. 2003. Light temperature, and soil moisture responses to elevation, evergreen understory, and small canopy gaps in the southern Appalachians. Forest Ecology and Management 186:243–255.

Crampton, L.H., and R.M.R. Barclay. 1995. Relationships between bats and stand age and structure in aspen mixedwood forests in Alberta, pp. 211–225, *in* Relationships between stand age, stand structure, and biodiversity in aspen mixedwood forests in Alberta (J.B. Stelfox, ed.). Alberta Environmental Centre and Canadian Forest Service, Vegreville, Alberta.

———. 1998. Selection of roosting and foraging habitat by bats in different aged aspen mixedwood stands. Conservation Biology 12:1347–1358.

Crome, F.J.H., and G.C. Richards. 1988. Bats and gaps: microchiropteran community structure in a Queensland rain forest. Ecology 69:1960–1969.

Cryan, P.M., M.A. Bogan, and G.M. Yanega. 2001. Roosting habits of four bat species in the Black Hills of South Dakota. Acta Chiropterologica 3:43–52.

Derby, R.W., and D.M. Gates. 1966. The temperature of tree trunks-calculated and observed. American Journal of Botany 53:580–587.

Dobkin, D.S., R.D. Gettinger, and M.G. Gerdes. 1995. Springtime movements, roost use, and foraging activity of Townsend's big-eared bat (*Plecotus townsendii*) in central Oregon. Great Basin Naturalist 55:315–321.

Duvall, M.D., and D.F. Grigal. 1999. Effects of timber harvesting on coarse woody debris in red pine forests across the Great Lakes states, USA. Canadian Journal of Forest Research 29:1926–1934.

Erickson, J.L. 1993. Bat activity in managed forests in the western Cascade Range. MS thesis, University of Washington, Seattle, WA.

Erickson, J.L., and S.D. West. 1996. Managed forests in the western Cascades: the effects of seral stage on bat habitat use patterns, pp. 215–227, *in* Bats and forests symposium (R.M. R. Barclay and R.M. Brigham, eds.). Research Branch, British Columbia Ministry of Forests, Victoria, BC.

———. 2003. Associations of bats with local structure and landscape features of forested stands in western Oregon and Washington. Biological Conservation 109:95–102.

Estrada, A., and R. Coates-Estrada. 2002. Bats in continuous forest, forest fragments and in an agricultural mosaic habitat-island at Los Tuxtlas, Mexico. Biological Conservation 103:237–245.

Evelyn, M.J., D.A. Stiles, and R.A. Young. 2004. Conservation of bats in suburban landscapes: roost selection by *Myotis yumanensis* in a residential area in California. Biological Conservation 115:463–473.

Everett, R., J. Lehmkuhl, R. Schellhaas, P. Ohlson, D. Keenum, H. Riesterer, and D. Spurbeck. 1999. Snag dynamics in a chronosequence of 26 wildfires on the east slope of the Cascade Range in Washington State, USA. International Journal of Wildland Fire 9:223–234.

Fellers, G.M., and E.D. Pierson. 2002. Habitat use and foraging behavior of Townsend's big-eared bat (*Corynorhinus townsendii*) in coastal California. Journal of Mammalogy 83:167–177.

Fenton, M.B. 1990. The foraging behaviour and ecology of animal-eating bats. Canadian Journal of Zoology 68:411–422.

Furlonger, C.L., H.J. Dewar, and M.B. Fenton. 1987. Habitat use by foraging insectivorous bats. Canadian Journal of Zoology 65:284–288.

Gellman, S.T., and W.J. Zielinski. 1996. Use by bats of old-growth redwood hollows on the north coast of California. Journal of Mammalogy 77:255–265.

Graves, A.T., M.A. Fajvan, and G.W. Miller. 2000. The effects of thinning intensity on snag and cavity tree abundance in an Appalachian hardwood stand. Canadian Journal of Forest Research 30:1214–1220.

Grindal, S.D. 1996. Habitat use by bats in fragmented forests, pp. 260–272, *in* Bats and forests symposium (R.M.R. Barclay and R.M. Brigham, eds.). Research Branch, British Columbia Ministry of Forests, Victoria, BC.

Grindal, S.D., and R.M. Brigham. 1998. Short term effects of small-scale habitat disturbance on activity by insectivorous bats. Journal of Wildlife Management 62:996–1003.

———. 1999. Impacts of forest harvesting on habitat use by foraging insectivorous bats at different spatial scales. Ecoscience 6:25–34.

Grindal, S.D., J.L. Morissette, and R.M. Brigham. 1999. Concentration of bat activity in riparian habitats over an elevational gradient. Canadian Journal of Zoology 77:972–977.

Guynn, D.C., Jr., S.T. Guynn, T.B. Wigley, and D.A. Miller. 2004. Herbicides and forest biodiversity-what do we know and where do we go from here? Wildlife Society Bulletin 32:1085–1092.

Hamilton, I.M., and R.M.R. Barclay. 1994. Patterns of daily torpor and day-roost selection by male and female big brown bats (*Eptesicus fuscus*). Canadian Journal of Zoology 72:744–749.

Hayes, J.P. 1997. Temporal variation in activity of bats and the design of echolocation-monitoring studies. Journal of Mammalogy 78:514–524.

———. 2000. Assumptions and practical considerations in the design and interpretation of echolocation-monitoring studies. Acta Chiropterologica 2:225–236.

———. 2003. Habitat ecology and conservation of bats in western coniferous forests, pp. 81–119, *in* Mammal community dynamics in coniferous forests of western North America: management and conservation (C.J. Zabel and R. G. Anthony, eds.). Cambridge University Press, Cambridge, MA.

Hayes, J.P., and M.D. Adam. 1996. The influence of logging riparian areas on habitat utilization by bats in western Oregon, pp. 228–237, *in* Bats and forests symposium (R.M.R. Barclay and R.M. Brigham, eds.). Research Branch, British Columbia Ministry of Forests, Victoria, BC.

Hayes, J.P., and J.C. Gruver. 2000. Vertical stratification of bat activity in an old-growth forest in western Washington. Northwest Science 74:102–108.

Hilderbrand, R.H., A.D. Lemly, C.A. Dolloff, and K.L. Harpster. 1997. Effects of large woody debris placement on stream channels and benthic macroinvertebrates. Canadian Journal of Fisheries and Aquatic Science 54:931–939.

Hogberg, L.K., K.J. Patriquin, and R.M.R. Barclay. 2002. Use by bats of patches of residual trees in logged areas of the boreal forest. American Midland Naturalist 148:282–288.

Horton, S.P., and R.W. Mannan. 1988. Effects of prescribed fire on snags and cavity-nesting birds in southeastern Arizona pine forests. Wildlife Society Bulletin 16:37–44.

Hoying, K.M., and T.H. Kunz. 1998. Variation in size at birth and postnatal growth in the insectivorous bat *Pipistrellus subflavus* (Chiroptera: Vespertilionidae). Journal of Zoology (London) 245:15–27.

Huie, K.M. 2002. Use of constructed woodland ponds by bats in the Daniel Boone National Forest. MS thesis, Eastern Kentucky University, Richmond, KY.

Humes, M.L., J.P. Hayes, and M.W. Collopy. 1999. Bat activity in thinned, unthinned, and old-growth forests in western Oregon. Journal of Wildlife Management 63:553–561.

Humphrey, J.W., C. Hawes, A.J. Peace, R. Ferris-Kaan, and M.R. Jukes. 1999. Relationships between insect diversity and habitat characteristics in plantation Forests. Forest Ecology and Management 113:11–21.

Humphrey, S.R., A.R. Richter, and J.B. Cope. 1977. Summer habitat and ecology of the endangered Indiana bat *Myotis sodalis*. Journal of Mammalogy 58:334–346.

Hurst, T.E., and M.J. Lacki. 1999. Roost selection, population size and habitat use by a colony of Rafinesque's big-eared bats (*Corynorhinus rafinesquii*). American Midland Naturalist 142:363–371.

Hutchinson, J.T., and M.J. Lacki. 2001. Possible microclimate benefits of roost site selection in the red bat, *Lasiurus borealis,* in mixed mesophytic forests of Kentucky. Canadian Field-Naturalist 115:205–209.

Johnson, J.B., M.A. Menzel, J.W. Edwards, and W.M. Ford. 2002. Gray bat night-roosting under bridges. Journal of the Tennessee Academy of Science 77:91–93.

Jung, T.S., I.D. Thompson, R.D. Titman, and A.P. Applejohn. 1999. Habitat selection by forest bats in relation to mixed-wood stand types and structure in central Ontario. Journal of Wildlife Management 63:1306–1319.

Kalcounis, M.C., and R.M. Brigham. 1998. Secondary use of aspen cavities by tree-roosting big brown bats. Journal of Wildlife Management 62:603–611.

Kalcounis, M.C., K.A. Hobson, R.M. Brigham, and K.R. Hecker. 1999. Bat activity in the boreal forest: importance of stand type and vertical strata. Journal of Mammalogy 80:673–682.

Karels, T.J., and R. Boonstra. 2003. Reducing solar heat gain during winter: the role of white bark in northern deciduous trees. Arctic 56:168–174.

Kerth, G., K. Weissmann, and B. König. 2000. Day-roost selection in female Bechstein's bats (*Myotis bechsteinii*): a field experiment to determine the influence of roost temperature. Oecologia 126:1–9.

Krusac, D.L., and S.R. Mighton. 2002. Conservation of the Indiana bat in national forests: where we have been and where we should be going, pp. 55–65, *in* The Indiana bat: biology and management of an endangered species (A. Kurta and J. Kennedy, eds.). Bat Conservation International, Austin, TX.

Krusic, R.A., M. Yamasaki, C.D. Neefus, and P.J. Pekins. 1996. Bat habitat use in White Mountain National Forest. Journal of Wildlife Management 60:625–631.

Kunz, T.H., and L.F. Lumsden. 2003. Ecology of cavity and foliage roosting bats, pp. 3–89, *in* Bat ecology (T.H. Kunz and M.B. Fenton, eds.). The University of Chicago Press, Chicago, IL.

Kunz, T.H., J.O. Whitaker Jr., and M.D. Wadonoli. 1995. Dietary energetics of the insectivorous Mexican free-tailed bat (*Tadarida brasiliensis*) during pregnancy and lactation. Oecologia 101:407–415.

Kurta, A., G.P. Bell, K.A. Nagy, and T.H. Kunz. 1989. Water balance of free-ranging little brown bats (*Myotis lucifugus*) during pregnancy and lactation. Journal of Mammalogy 67:2468–2472.

Kurta, A., T.H. Kunz, and K.A. Nagy. 1990. Energetics and water flux of free-ranging big brown bats (*Eptesicus fuscus*) during pregnancy and lactation. Journal of Mammalogy 71:59–65.

Lacki, M.J., and J.H. Schwierjohann. 2001. Day-roost characteristics of northern bats in mixed mesophytic forest. Journal of Wildlife Management 65:482–488.

Lance, R.F., B.T. Hardcastle, A. Talley, and P.L. Leberg. 2001. Day-roost selection by Rafinesque's big-eared bats (*Corynorhinus rafinesquii*) in Louisiana forests. Journal of Mammalogy 82:166–172.

LaVal, R.K., R.L. Clawson, M.L. LaVal, and W. Caire. 1977. Foraging behavior and nocturnal activity patterns of Missouri bats, with emphasis on the endangered species *Myotis grisescens* and *Myotis sodalis*. Journal of Mammalogy 58:592–599.

Law, B.S., J. Anderson, and M. Chidel. 1999. Bat communities in a fragmented forest landscape on the south-west slopes of New South Wales, Australia. Biological Conservation 88:333–345.

Lewis, C.N., and J.B. Whitfield. 1999. Braconid wasp (Hymenoptera: Braconidae) diversity in forest plots under different silvicultural methods. Environmental Entomology 28:986–997.

Lewis, J.C. 1998. Creating snags and wildlife trees in commercial forests. Western Journal of Applied Forestry 13:97–101.

Lewis, S.E. 1993. Effect of climatic variation on reproduction by pallid bats (*Antrozous pallidus*). Canadian Journal of Zoology 71:1429–1433.

———. 1995. Roost fidelity of bats: a review. Journal of Mammalogy 76:481–496.

Lunde, R.E., and A.S. Harestad. 1986. Activity of little brown bats in coastal forests. Northwest Science 60:206–209.

Mackey, R.L., and R.M.R. Barclay. 1989. The influence of physical clutter and noise on the activity of bats over water. Canadian Journal of Zoology 67:1167–1170.

Mager, K.J., and T.A. Nelson. 2001. Roost-site selection by eastern red bats (*Lasiurus borealis*). American Midland Naturalist 145:120–126.

McCullough, D.G., R.A. Werner, and D. Neumann. 1998. Fire and insects in northern boreal ecosystems of North America. Annual Review of Entomology 43:107–127.

McLean, J.A., and J.R. Speakman. 1999. Energy budgets of lactating and non-reproductive brown long-eared bats (*Plecotus auritus*) suggest females use compensation in lactation. Functional Ecology 13:360–372.

McNab, B.K. 1982. Evolutionary alternatives in the physiological ecology of bats, pp. 151–200, *in* Ecology of bats (T.H. Kunz, ed.). Plenum Press, New York.

Menzel, M.A., T.C. Carter, J.M. Menzel, W.M. Ford, and B.R. Chapman. 2002. Effects of group selection silviculture in bottomland hardwoods on the spatial activity patterns of bats. Forest Ecology and Management 162:209–218.

Menzel, M.A., J.M. Menzel, W.M. Ford, J.W. Edwards, T.C. Carter, J.B. Churchill, and J.C. Kilgo. 2001. Home range and habitat use of male Rafinesque's big-eared bats (*Corynorhinus rafinesquii*). American Midland Naturalist 145:402–408.

Murray, S.W., and A. Kurta. 2004. Nocturnal activity of the endangered Indiana bat (*Myotis sodalis*). Journal of Zoology (London) 262:197–206.

Nelson, T.A., and M.L. Lambert. 1996. Characteristics and use of cavity trees and snags in hardwood stands. Proceedings of the Southeastern Association of Game and Fish Commissions 50:331–339.

Neuweiler, G. 2000. The biology of bats. Oxford University Press, Oxford, England.

Nicolai, V. 1986. The bark of trees: thermal properties, microclimate and fauna. Oecologia 69:148–160.

Norberg, U.M., and J.M.V. Rayner. 1987. Ecological morphology and flight in bats (Mammalia: Chiroptera): wing adaptations, flight performance, foraging strategy and echolocation. Philosophical Transactions of the Royal Society of London, B, Biological Sciences 316:335–427.

Ormsbee, P.C., and W.C. McComb. 1998. Selection of day roosts by female long-legged myotis in the central Oregon Cascade Range. Journal of Wildlife Management 62:596–603.

Owen, S.F., M.A. Menzel, W.M. Ford, B.R. Chapman, K.V. Miller, J.W. Edwards, and P.B. Wood. 2003. Home-range size and habitat used by the northern myotis (*Myotis septentrionalis*). American Midland Naturalist 150:352–359.

Parker, D.I., J.A. Cook, and S.W. Lewis. 1996. Effects of timber harvest on bat activity in southeastern Alaska's temperate rainforests, pp. 277–292, *in* Bats and forests symposium (R.M.R. Barclay and R.M. Brigham, eds.). Research Branch, British Columbia Ministry of Forests, Victoria, BC.

Patriquin, K.J., and R.M.R. Barclay. 2003. Foraging by bats in cleared, thinned and unharvested boreal forest. Journal of Applied Ecology 40:646–657.

Pavey, C.R. 1998. Habitat use by the eastern horseshoe bat, *Rhinolophus megaphyllus,* in a fragmented woodland mosaic. Wildlife Research 25:489–498.

Perkins, J.M., and S.P. Cross. 1988. Differential use of some coniferous forest habitats by hoary and silver-haired bats in Oregon. Murrelet 69:21–24.

Perlmeter, S.I. 1996. Bats and bridges: patterns of night roost activity in the Willamette National Forest, pp. 132–150, *in* Bats and forests symposium (R.M.R. Barclay and R. M. Brigham, eds.). Research Branch, British Columbia Ministry of Forests, Victoria, BC.

Peterson, D.W., and P.B. Reich. 2001. Prescribed fire in oak savanna: fire frequency effects on stand structure and dynamics. Ecological Applications 11:914–927.

Pierson, E.D., W.E. Rainey, and R.M. Miller. 1996. Night roost sampling: a window on the forest bat community in northern California, pp. 151–163, *in* Bats and forests symposium (R.M.R. Barclay and R.M. Brigham, eds.). Research Branch, British Columbia Ministry of Forests, Victoria, BC.

Potter, B.E., and J.A. Andresen. 2002. A finite-difference model of temperatures and heat flow within a tree stem. Canadian Journal of Forest Research 32:548–555.

Rabe, M.J., T.E. Morrell, H. Green, J.C. DeVos Jr., and C.R. Miller. 1998. Characteristics of Ponderosa pine snag roosts used by reproductive bats in northern Arizona. Journal of Wildlife Management 62:612–621.

Racey, P.A. 1973. Environmental factors affecting the length of gestation in heterothermic bats. Journal of Reproduction and Fertility 19 (suppl.):175–189.

Racey, P.A., and S.M. Swift. 1981. Variations in gestation length in a colony of pipistrelle bats (*Pipistrellus pipistrellus*) from year to year. Journal of Reproduction and Fertility 61:123–129.

Raesly, R.L., and J.E. Gates. 1987. Winter habitat selection by north temperate cave bats. American Midland Naturalist 118:15–31.

Randall-Parker, T., and R. Miller. 2002. Effects of prescribed fire in ponderosa pine on key wildlife habitat components: preliminary results and a method for monitoring, pp. 823–834, *in* Proceedings of the symposium on the ecology and management of dead wood in western forests (W.F. Laudenslayer Jr., P.J. Shea, B.E. Valentine, C.P. Weatherspoon, and T.E. Lisle, eds.). PSW-GTR-181. USDA Forest Service, Pacific Southwest Research Station, Albany, CA.

Rydell, J., A. Bushby, C.C. Cosgrove, and P.A. Racey. 1994. Habitat use by bats along rivers in north east Scotland. Folia Zoologica 43:417–424.

Rydell, J., L.A. Miller, and M.E. Jensen. 1999. Echolocation constraints of Daubenton's bat foraging over water. Functional Ecology 13:247–255.

Saab, V.A., J. Dudley, and W.L. Thompson. 2004. Factors influencing occupancy of nest cavities in recently burned forests. Condor 106:20–36.

Saugey, D.A., B.G. Crump, and G.A. Heidt. 1998. Notes on the natural history of *Lasiurus borealis* in Arkansas. Journal of the Arkansas Academy of Science 52:92–98.

Schulze, M.D., N.E. Seavy, and D.F. Whitacre. 2000. A comparison of the phyllostomid bat assemblages in undisturbed neotropical forest and in forest fragments of a slash-and-burn farming mosaic in Peten, Guatemala. Biotropica 32:174–184.

Sedgeley, J.A. 2001. Quality of cavity microclimate as a factor influencing selection of maternity roosts by a tree-dwelling bat, *Chalinolobus tuberculatus,* in New Zealand. Journal of Applied Ecology 38:425–438.

Seidman, V.M., and C.J. Zabel. 2001. Bat activity along intermittent streams in northwestern California. Journal of Mammalogy 82:738–747.

Siemann, E., J. Haarstand, and D. Tilman. 1997. Short-term and long-term effects of burning on oak savanna arthropods. American Midland Naturalist 137:349–361.

Sleep, D.J.H., and R.M. Brigham. 2003. An experimental test of clutter tolerance in bats. Journal of Mammalogy 84:216–224.

Speakman, J.R., and D.W. Thomas. 2003. Physiological ecology and energetics of bats, pp. 430–490, *in* Bat ecology (T.H. Kunz and M.B. Fenton, eds.). University of Chicago Press, Chicago, IL.

Swystun, M.B., J.M. Psyllakis, and R.M. Brigham. 2001. The influence of residual tree patch isolation on habitat use by bats in central British Columbia. Acta Chiropterologica 3:197–201.

Taylor, R.J., and N.M. Savva. 1988. Use of roost sites by four species of bats in state forest in south-eastern Tasmania. Australian Wildlife Research 15:637–645.

Thomas, D.W. 1988. The distribution of bats in different ages of Douglas-fir forests. Journal of Wildlife Management 52:619–628.

Thomas, D.W., and S.D. West. 1991. Forest age associations of bats in the southern Washington Cascade and Oregon Coast Ranges, pp. 295–303, *in* Wildlife and vegetation of unmanaged Douglas-fir forests (L.F. Ruggiero, K.B. Aubrey, A.B. Carey, and M.H. Huff, eds.). Forest Service, USDA General Technical Report PNW-285, Portland, OR.

Tibbels, A.E., and A. Kurta. 2003. Bat activity is low in thinned and unthinned stands of red pine. Canadian Journal of Forest Research 33:2436–2442.

Tiedemann, A.R., J.O. Klemmedson, and E.L. Bull. 2000. Solution of forest health problems with prescribed fire: are forest productivity and wildlife at risk? Forest Ecology and Management 127:1–18.

Trousdale, A.W., and D.C. Beckett. 2004. Seasonal use of bridges by Rafinesque's big-eared bat, *Corynorhinus rafinesquii,* in southern Mississippi. Southeastern Naturalist 3:103–112.

Tuttle, M.D. 1976. Population ecology of the gray bat (*Myotis grisescens*): factors influencing growth and survival of newly volant young. Ecology 57:587–595.

von Frenckell, B., and R.M.R. Barclay. 1987. Bat activity over calm and turbulent water. Canadian Journal of Zoology 65:219–222.

Waldien, D.L., and J.P. Hayes. 2001. Activity areas of female long-eared myotis in coniferous forests in western Oregon. Northwest Science 75:307–314.

Waldien, D.L., J.P. Hayes, and E.B. Arnett. 2000. Day-roosts of female long-eared myotis in western Oregon. Journal of Wildlife Management 64:785–796.

Waldien, D.L., J.P. Hayes, and B.E. Wright. 2003. Use of conifer stumps in clearcuts by bats and other vertebrates. Northwest Science 77:64–71.

Walsh, A.L., and S. Harris. 1996. Foraging habitat preferences of vespertilionid bats in Britain. Journal of Applied Ecology 33:508–518.

Warren, R.D., D.A. Waters, J.D. Altringham, and D.J. Bullock. 2000. The distribution of Daubenton's bats (*Myotis daubentonii*) and pipistrelle bats (*Pipistrellus pipistrellus*) (Vespertiliionidae) in relation to small-scale variation in riverine habitat. Biological Conservation 92:85–91.

Webb, P.I. 1995. The comparative ecophysiology of water balance microchiropteran bats, pp. 203–218, *in* Ecology, evolution and behaviour of bats (P.A. Racey and S.M. Swift, eds.). Clarendon Press, Oxford, U.K.

Weller, T.J., and C.J. Zabel. 2001. Characteristics of fringed myotis day roosts in northern California. Journal of Wildlife Management 65:489–497.

Wethington, T.A., D.M. Leslie Jr., M.S. Gregory, and M.K. Wethington. 1996. Prehibernation habitat use and foraging activity by endangered Ozark big-eared bats (*Plecotus townsendii ingens*). American Midland Naturalist 135:218–230.

———. 1997. Vegetative structure and land use relative to cave selection by endangered Ozark big-eared bats (*Corynorhinus townsendii ingens*). Southwestern Naturalist 42:177–181.

Wilhere, G.F. 2003. Simulations of snag dynamics in an industrial Douglas-fir forest. Forest Ecology and Management 174:521–539.

Williams, D.F., and J.S. Findley. 1979. Sexual size dimorphism in vespertilionid bats. American Midland Naturalist 102:113–126.

Xu, M., J. Chen, and Y. Qi. 2002. Growing-season temperature and soil moisture along a 10 km transect across a forested landscape. Climate Research 22:57–72.

Zimmerman, G.S., and W.E. Glanz. 2000. Habitat use by bats in eastern Maine. Journal of Wildlife Management 64:1032–1040.

ECOLOGICAL CONSIDERATIONS FOR LANDSCAPE-LEVEL MANAGEMENT OF BATS 9

Joseph E. Duchamp, Edward B. Arnett, Michael A. Larson, and Robert K. Swihart

Bats exhibit a high degree of temporal and spatial mobility across a variety of habitats. This characteristic dictates using a landscape approach for their management. During nightly foraging flights, bats may travel through many distinct habitats. Within a single season, a colony of bats may switch roosts frequently and use roosts located in separate forest stands (Barclay and Kurta, chap. 2 in this volume). Many species of bats migrate long distances between summer roosts and winter hibernation sites (Cryan and Veilleux, chap. 6 in this volume). In general, organisms that require multiple habitats are more sensitive to habitat loss and fragmentation (Swihart et al. 2003), and some species of bats do seem sensitive to habitat fragmentation (Ekman and de Jong 1996; Estrada and Coates-Estrada 2002). To protect and conserve populations of bats effectively, it is important to acknowledge that bats interact with their environment over broad spatial scales comprised of heterogeneous mixtures of habitat. Each of these habitats offers distinct resources that will only meet the needs of resident populations of bats when considered collectively.

During the past 20 years, the field of landscape ecology has emerged, in part, to address the challenges of incorporating spatial patterns into the research and management of ecosystems and their resident organisms (Haila 2002). Turner et al. (2001) define a landscape in a general sense as "an area that is spatially heterogeneous in at least one factor of interest." By this definition, to understand the ecology of bats at a landscape scale it is important to recognize critical factors such as population resources and stressors, along with the size of the area within which the spatial distribution of these factors influence bats including movement, habitat use, survival, and reproduction.

Most of the existing literature on the habitat ecology of bats has focused on understanding resource use at the stand scale, within a single-landscape context (Miller et al. 2003), or a specific habitat characteristic (e.g., roosts). Public and private land management plans are increasingly expected to address resource density and quality, along with the spatial configuration of multiple resources necessary to maintain viable populations of bats (Miller et al. 2003). Unfortunately, the depth of knowledge necessary for effective planning of habitat for bats is lacking (Miller et al. 2003). Manipulative experiments at the landscape scale would provide the most useful information for landscape-level planning, but, in general, are prohibited by the required land area and limited funding (Hargrove and Pickering 1992; McGarigal and Cushman 2002). An alternative option is a mensurative approach, where conclusions are based on measures that are replicated across multiple, spatially distinct landscapes. Under this method, inherent differences among replicate landscapes act as a surrogate for controlled manipulations. Such measures could serve as a basis for predicting the response of bat populations to different landscape management approaches.

Broad-scale efforts necessary for both research and effective management of bat populations at a landscape scale will require cooperation among multiple researchers and managers (Wigley et al., chap. 11 in this volume). Our goal in this chapter is to encourage this process by reviewing the general concepts inherent in landscape ecology and the existing literature on how bats interact with multiple habitat elements across a landscape. We also discuss strategies and landscape-suitability models that might be useful for managing bats at a landscape scale.

FUNDAMENTALS OF LANDSCAPE ECOLOGY
DEFINING THE LANDSCAPE

Heterogeneous landscapes are complex and must be simplified to examine broad-scale interactions in a way that facilitates the recognition of meaningful patterns. Elements within a landscape can be represented either categorically (e.g., forest and grassland) or by using point data where spatially distinct locations are associated with a quantitative measure of a resource of interest (Gustafson 1998). Categorical representations are usually based on data that are relatively easy to acquire, such as existing vegetation maps. They assume that resources within a category are homogeneous and that abrupt transitions occur between categories (Gustafson 1998). In general, point data are more detailed than categorical data and require more labor-intensive methods of data collection (Mitchell and Powell 2003). These data can be used to represent heterogeneity within a category or gradients of transition between habitats (Gustafson 1998; Mitchell and Powell 2003). Despite their distinct approaches, categorical and quantitative representations are complimentary methods for describing heterogeneity in a landscape (Gustafson 1998).

In this chapter we focus primarily on the more intuitive categorical representations, and comment on where more detailed point-based data may be useful. The most general categories describing landscape elements are patch, matrix, and corridor. Rephrasing definitions of Turner et al. (2001), a patch is a landcover that significantly differs from surrounding landcover types, the matrix is the background landcover encompassing most of the landscape, and a corridor is a "relatively narrow strip of landcover that differs from adjacent areas on both sides." These are general definitions useful in quantifying landscape patterns, but they also carry more detailed associations from an organism perspective.

Patches often are described as suitable or unsuitable with regard to meeting the needs of a particular species (Turner et al. 2001). For more sessile organisms, a suitable patch is defined as habitat containing all resources necessary for the survival of an organism. However, bats, like many vagile species, may require resources present in multiple types of habitat patches (Dunning et al. 1992). A more useful method for defining a patch may be resource based, so that a patch is suitable if it contains a particular, exploitable resource (Morrison 2001). A resource-based patch could be further defined using point data, describing the quality or density of resources within the patch. The use of this definition allows for multiple habitat types containing different critical resources to fulfill the needs of an organism. When considering roosting resources, a suitable patch for a bat that roosts primarily in caves would depend on the arrangement of the karst system in an area, with foraging by this species likely to depend on vegetative structure and surface hydrology. Recognizing the distinction between roosting and foraging requirements, and their necessary spatial juxtaposition, is critical to understanding the response of bat populations to land management (Dunning et al. 1992).

As the dominant habitat across the landscape, the matrix habitat separates both suitable and unsuitable patches. To meaningfully separate suitable habitat patches, the matrix also must be unsuitable and either lack resources that make patches suitable or present some risk that impedes access to resources or movement between suitable patches of habitat. An example of a matrix for many species of bats is open, agricultural habitat. Fields of crops are prevalent in many forested ecosystems, lack roosting habitat, and are avoided by many species of bats during foraging (de Jong 1995; Walsh and Harris 1996b; Walsh and Mayle 1991). Note that not all matrix types are equivalent in their isolation effects (Ricketts 2001). For example, in Sweden, Ekman and de Jong (1996) found that the negative influence of the fragmentation of forested habitat on the occurrence of bat species was greater for patches in the agricultural matrix than in forested islands in a matrix of open water formed by a large lake.

Recognizing what habitat functions as a matrix in a landscape is not always straightforward. For example, many forested landscapes are dominated by an inhospitable matrix of young forest surrounding suitable patches of older forest. This human-perceived matrix may in fact offer resources to many organisms (Bunnell 1997), including bats. Some species of bats are known to roost in stumps (Vonhof and Barclay 1997; Waldien et al. 2000), downed logs (Arnett and Hayes 2003), or snags and trees retained in young forests (e.g., Arnett and Hayes 2003; Ormsbee and McComb 1998; Waldien 1998). Additionally, several species of bats forage over young patches of forest or along edges (Clark et al. 1987; Elmore et al. 2005; Erickson and West 1996; Furlonger et al. 1987; Zimmerman and Glanz 2000). Thus, in some forested ecosystems that are not fragmented with urban or agricultural habitats, the theoretical concepts and assumptions of an inhospitable matrix may not apply for some species of bats.

Given an environment of separate suitable patches within an inhospitable matrix, the question of how an organism or population interacts with this fragmented landscape becomes important (Donovan and

Strong 2003). Fragmentation of suitable habitat involves two distinct processes: reduction of the total area of suitable patches of habitat, and increased isolation of the remaining habitat patches. Species-area relationships are well established empirically and demonstrate that smaller patches generally meet the needs of fewer species than larger patches (Connor and McCoy 1979). The isolation of suitable patches can restrict their availability to some species, reducing the chance of recolonization if a species is extirpated (Beier and Noss 1998; Burkey 1989; Morand 2000). Evidence for a negative influence of patch isolation on bat populations is sparse but has been found in some highly fragmented systems. Ekman and de Jong (1996) found that two species of vespertilionids were negatively affected by increased isolation of suitable patches of habitat in a fragmented landscape in Sweden. Additionally, Morand (2000) demonstrated that the species richness of bats declined with increased isolation among islands (patches) in the matrix formed by the Caribbean Sea.

The term corridor has many definitions in ecology (Hess and Fischer 2001). In the context of local landscape fragmentation, a functional definition of a corridor is a habitat that is unusable by the organism for long-term occupancy, but capable of facilitating movement through an inhospitable matrix and, thus, mitigating the effects of isolation (Beier and Noss 1998). To fit this definition, a corridor must be sufficiently narrow to render it unsuitable as a resource patch. Examples of such corridors in an agricultural matrix would be fencerows and tree lines that offer navigational landmarks, protection from predators, and increased densities of prey (Verboom and Spoelstra 1999), and bats use them when traversing fragmented landscapes (Estrada and Coates-Estrada 2001; Verboom and Huitema 1997). Some habitats traditionally viewed as corridors (e.g., narrow riparian buffers between forest patches) may provide both roosting (Veilleux et al. 2003) and foraging sites (Verboom and Spoelstra 1999) for bats and may be better categorized as a patch than a corridor.

Fragmentation of patches also increases the amount of edge at the interface between two landcover types (Murcia 1995). Consider a forest landcover that in the absence of fragmentation is homogeneous in some characteristic. When patches are formed within this landcover type, areas near the edge of the patch can assume substantially different qualities when compared with the rest of the patch (Forman 1995). The remaining homogeneous section of the patch is known as the core area and does not experience edge effects. Edge effects are best documented with regard to changes in vegetation structure, such as increased tree mortality and higher stem density, and abiotic factors, such as increased ambient temperature and light gradients, as one moves from the interior to the edge of a patch of forest (Murcia 1995). Increased tree mortality may benefit bats by increasing roosting opportunities and creating openings in the forest canopy. Trees near forest edges also experience increased windthrow (Guldin et al., chap. 7 in this volume; Hayes 2003; Hayes and Loeb, chap. 8 in this volume; Murcia 1995), which may result in shorter retention of roosting structures over time. Increased temperature near the edge of a patch could also help elevate roost temperature and explain preferences of bats for roosts located closer to the edge of a forest stand (Boonman 2000; Waldien et al. 2000).

Changes related to a forest edge also occur in the opposite direction as one heads away from the forest into an open patch. Lewis (1965, 1970) showed that windbreaks such as tree lines and fencerows alter wind patterns and that density of aerial insects was highest in areas sheltered from wind velocity. Forest edges serve as foraging locations for bats (Furlonger et al. 1987), and sections of edge habitat sheltered from the wind may foster increased densities of insect prey. Although Murcia (1995) noted there is no consensus for defining the intensity or mechanism for edge-organism interactions, the significant relationships that have been found indicate that many species of bats respond to changes in edge habitat.

MEASURING SPATIAL PATTERN

One of the premises of landscape ecology is that the pattern of a landscape influences ecosystem function and, thus, the organisms contained therein. To measure patterns within a landscape, one can quantify the composition and arrangement of patches as represented on a habitat map. This map is usually in raster form where the map consists of square grid cells, each of which has a value indicating a particular landcover category. A group of adjacent cells of the same habitat category is considered a patch. Adjacency can be determined using either the 4- or 8-neighbor rule (Turner et al. 2001). The 4-neighbor rule considers cells sharing a side as adjacent, whereas the 8-neighbor rule also includes cells that share a corner. Given the high mobility of bats, the 8-neighbor rule is probably more appropriate for analysis.

Measurement of landscape patterns depends on the spatial extent of a landscape and on the minimum size of the grid cells, known as the resolution or grain (O'Neil et al. 1996). Spatial extent is the total area encompassed by a landscape (Wiens 1989). Spatial resolution describes the detail of the map and is the smallest unit for which distinctions can be made among landcover types (Wiens 1989). The appropriate spatial extent and resolution depend on the size of the habitat patches of interest (O'Neil et al. 1996). To avoid bias in landscape metrics, spatial extent should encompass an area at least twice as large as the area of the largest habitat patch (O'Neil et al. 1996) and should include an area that is relevant to the organism (Wiens 1989). Conversely, the grain should be at most half the size of the smallest habitat patch that is relevant to the organism or research question under study (O'Neil et al. 1996). This precaution ensures that significant habitat patches are not overlooked or categorized as another landcover type.

Although flight allows bats to travel great distances relatively quickly, small or narrow habitat features such as tree lines or hedgerows may be important for determining how a bat interacts with a landscape (Estrada and Coates-Estrada 2001; Verboom and Spoelstra 1999). These types of resources can be as narrow as a few meters in width. To include these features on a categorical map requires a grain that is smaller than what is available on most habitat maps or existing satellite imagery techniques (Mitchell and Powell 2003). The incorporation of such fine-scale habitat elements may require hand digitization from high-resolution aerial photography, or an estimation of the density of such resources based on field sampling (Mitchell and Powell 2003).

Once an appropriate habitat map is created, several programs including FragStats (McGarigal and Marks 1995) can be used to quantify spatial patterns. Most of these metrics are highly correlated but can be classified into broad, relatively independent categories (Riitters et al. 1995). Two general classes of landscape metrics are composition and spatial configuration (Turner et al. 2001). Compositional measures are not spatially explicit and instead measure diversity and relative amounts of landcover categories within a defined landscape. These can be as straightforward as the proportion of a particular habitat in a landscape, or slightly more complex measures of evenness and dominance (O'Neil et al. 1988).

Measures of spatial configuration are spatially explicit and deal with the arrangement and juxtaposition of landcover types. At the scale of an individual habitat patch, configuration measures include size, perimeter, shape complexity, and isolation. Across a landscape, measures of patch size and perimeter can be summarized along with the number of patches of a landcover type. Shape complexity is measured by a combination of area and perimeter, or by fractal measures that are independent of scale. Broader measures address connectivity among homogeneous patch types and aggregation of landcover types across a landscape.

SAMPLING AND EXPERIMENTAL DESIGN AT A LANDSCAPE SCALE

Few studies have addressed landscape-scale relationships of bats in forests (Estrada and Coates-Estrada 2002; Gehrt and Chelsvig 2004; Gorresen and Willig 2004; Jaberg and Guisan 2001; Walsh and Harris 1996a). Understanding the cause-and-effect relationships among landscape composition, configuration, and bat populations is critical to landscape-scale management plans targeting conservation of bats. Replicated, manipulative experiments with a randomized selection of samples are the ideal method for drawing inferences about causal relationships (Hurlbert 1984; McGarigal and Cushman 2002), but bats are not well suited to direct manipulative experiments at a landscape scale. The fact that many species of bats regularly cover distances of a few to several kilometers in a night (Pierson 1998) would require control of an inordinate amount of land to achieve independent replicate samples. Additionally, any such study would need to be of sufficient duration to witness the response of a population to the landscape manipulations (Burel 1993; McGarigal and Cushman 2002). Because most species of bats have long life spans and low reproductive rates (Barclay and Harder 2003), such large-scale, long-term studies likely will be impractical in the vast majority of situations, but they should remain a priority for future research (Hayes 2003; Miller et al. 2003).

Another option is a mensurative experimental approach (Hurlbert

1984; McGarigal and Cushman 2002). The mensurative approach at a landscape scale relies on preexisting variation in landscape composition at different locations. Sampling occurs within multiple landscapes that differ primarily in the predefined variable(s) of interest. Confounding variation should either be avoided or incorporated by using a randomized block design (Hurlbert 1984). The inferences resulting from such studies are correlative and, therefore, not as strong as results from manipulative experiments. If properly designed and implemented, however, mensurative landscape studies can serve as a basis for generating hypotheses to be tested with more manageable, small-scale experiments.

Defining the spatial extent of a sample landscape that is relevant should be based on the area used by an organism during the ecological process of interest (Addicott et al. 1987; Bissonette 2003; Porter and Church 1987). Radiotransmitters allow for estimation of the amount of area used by bats outside of migration. Given the short life span of radiotransmitters used to study most species of bats (Barclay and Kurta, chap.2 in this volume) monitoring an individual bat throughout its entire summer home range is improbable. However, home ranges or maximum distances traveled for multiple individuals within a species, sex, or age, should provide a spatial context for that group that can be applied to the ecological process of interest. For example, Duff (2004) defined a landscape as a 1,500-m radius circular plot centered on mist net sites, because this area fell approximately in the middle of the home-range sizes of bats known to occupy the study areas. Arnett and Hayes (2003) used radiotelemetry locations of day roosts to define the landscape area potentially available to bats; they determined that a 4.8-km radius circular plot centered on capture sites encompassed >90% of all day roosts for each of three species of bats under study in western Oregon.

Avoiding pseudoreplication is essential for achieving robust results from the experimental process (Hurlbert 1984). Although pseudoreplication has been thoroughly reviewed (Hurlbert 1984), we believe it merits mention here because of its pervasiveness in habitat studies of bats (Miller et al. 2003) and in research on habitat fragmentation in general (McGarigal and Cushman 2002). For landscape-level mensurative experiments, avoiding pseudoreplication includes accounting for spatial autocorrelation (McGarigal and Cushman 2002) and recognizing the implications of the spatial limits of a study (Hurlbert 1984). Because the independence of treatments in a landscape-level study is established spatially, a threshold must be recognized where separate locations are no longer significantly related. If two locations are near to one another relative to a variable of interest, a high degree of correlation can be present among samples merely because of their spatial position. This spatial autocorrelation decreases the independence of samples and the effective sample size. Although a consensus has not been reached on how to deal with spatial autocorrelation, a variety of statistical approaches have been discussed in the literature, including diagnostic tests (Dale and Fortin 2002), auto-Gaussian and autologistic regression (Augustine et al. 1996;

Lichstein et al. 2002), randomization software for independent subsets (Holland et al. 2004), and hierarchical linear models (Raudenbush and Bryk 2002).

When designing and interpreting any study, it is important to remember that results only reflect the nature of the sampled system. The selection of sampling methods must be appropriate for the question or hypothesis to be addressed. As an example, if a study is concerned with the relationship of bat colony size to variation in patch area and isolation, and several patches are sampled within a landscape, the results of such a study are only applicable to the patches within that landscape. To draw more general conclusions applicable to other landscapes, patches from additional landscapes would also need to be sampled. We caution researchers to clearly state the assumptions and limitations of their work, as applying results to an inappropriate scale will inevitably lead to erroneous conclusions (Miller et al. 2003).

SELECTING LANDSCAPE ELEMENTS: HOW DO BATS VIEW THEIR LANDSCAPES?

Studies of landscape ecology typically hinge on a map of relevant landcover types and relate habitat composition and spatial configuration of the map to behavioral and demographic measures of a population (Turner et al. 2001). How the map is defined in composition, spatial resolution, and spatial extent is critical to the results of such studies (Edwards et al. 2003; Mayer and Cameron 2003; Mitchell and Powell 2003). For all these factors, recent findings have emphasized the need to base these decisions on the perspective of the organism, thereby limiting the bias of our own anthropogenic view (Knight and Morris 1996; Wiens 1989). Holland et al. (2004) offer a program (FOCUS) to aid in identifying the appropriate scale(s) of response of a species to a landscape. Using FOCUS, one can compare population measures with landscape metrics at multiple spatial scales. The scale(s) at which a species shows the strongest response is known as the characteristic scale (Holland et al. 2004). When combined with relevant knowledge of the ecology of a species, sampling designs targeting multiple spatial scales may allow us to account for species-specific scale responses to habitat elements within the surrounding landscape (fig. 9.1).

Adopting an organism perspective first requires recognizing the temporal and spatial limits of an ecological process (Bissonette 2003), or what Addicott et al. (1987) described as an ecological neighborhood. For bats, a logical starting point begins with seasonal shifts in behavior. Many North American bats exhibit seasonal movements between summer and wintering grounds (Cryan and Veilleux, chap. 6 this volume; Fleming and Eby 2003). How a bat uses a landscape along with necessary characteristics of a landscape are likely distinct between these two stages, as well as during migration. Each of these seasonal, ecological neighborhoods can be further broken down into more detailed sections examining seasonal or finer-scale (e.g., daily) patterns of activity. Recognizing the critical resources within these neighborhoods should allow the designation of ap-

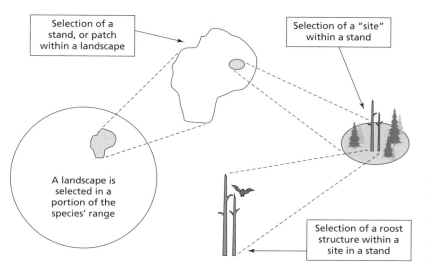

Selection of a stand, or patch within a landscape

Selection of a "site" within a stand

A landscape is selected in a portion of the species' range

Selection of a roost structure within a site in a stand

Figure 9.1.
A conceptual diagram of roost selection by bats at multiple spatial scales in a forested landscape. *Modified from Arnett and Hayes 2003*

propriate landscape maps. The nature of critical resources will differ among species, sex, and life stages, and may depend on behavioral responses of bats to other coexisting organisms (Lima and Zollner 1996). In the rest of this section, we review the current literature to gain insights for addressing such issues. We focus on the summer habitat of bats, because this component has been the most thoroughly studied habitat element of temperate-zone bats.

Roost structures are recognized as an important resource for bats and may be limiting for bat populations in some areas (Barclay and Kurta, chap. 2 in this volume; Humphrey 1975; Pierson 1998). If they are truly limiting, increasing the density of roosts should allow for higher population levels until roosts are abundant enough to no longer be limiting (Hayes and Loeb, chap. 8 in this volume). Bats using tree roosts appear to select characteristics that can be sex and species specific for both the tree and its immediate surroundings (Hayes 2003; Kunz and Lumsden 2003; Lacki and Baker 2003). The specificity of roost-tree requirements for some species of bats may necessitate maps that incorporate point data in addition to simple categorical designations to adequately define resource distribution. Even for a single individual or colony of bats, roost resources require both spatial and temporal considerations. For many species of bats, colonies or individuals are known to switch roosts frequently and to use several trees in an area (Barclay and Kurta, chap. 2 in this volume; Carter and Menzel, chap. 3 in this volume; Hayes 2003; Kunz and Lumsden 2003; Lewis 1995; Vonhof and Barclay 1996). Multiple roost structures probably offer necessary variation in location and microclimate, allowing bats to adjust to changes in their environment (Callahan et al. 1997; Lewis 1995). Trees are also a more ephemeral resource when compared with caves and anthropogenic structures (Kurta et al. 1993). Some species of bats show fidelity to roost sites by reusing roost trees for several years (Barclay and Brigham 2001; Chung-Mac-Coubrey 2003). The fitness cost incurred by a bat population that loses a particular roost tree is unknown. It is reasonable to assume, however, that maintaining relatively stable populations of bats requires that an adequate density of potential roost trees, offering a range of environmental conditions, be maintained through time and with some degree of spatial consistency.

An appropriate roost site must also be within a reasonable proximity to suitable foraging habitat and water sources. Suitable foraging habitat must contain a sufficient prey density, along with vegetative structure conducive to the foraging behavior for a particular species of bat (Lacki et al., chap. 4 in this volume). Habitats chosen for foraging may be distinct from roosting habitat for some species of bats (Broders and Forbes 2004). Indeed, several species of North American bats are known to travel long distances from their roosts to preferred foraging areas (Everette et al. 2001; Pierson 1998). However, the linear distances between roosts and foraging sites may be less than the actual distances traveled (Miller and Russell 2004), because bats may rely on corridors and habitat edges to avoid directly traversing certain landscape elements. The willingness of bats to cross large areas of unsuitable habitat is species dependent (Verboom and Huitema 1997). During foraging and commuting from roosts to foraging areas, many species of bats follow particular structural elements such as woodland edges, waterways, and cliffs (Adam et al. 1994; Brigham 1991; Pavey 1998; Verboom and Huitema 1997; Walsh and Harris 1996b). Clearly, both foraging and commuting habitats need to be maintained for resident populations of bats.

The quality of roosts and foraging patches defines their role in the ecology of the resident populations of bats (Dunning et al. 1992). Low-quality habitat for bats could be a function of low density of resources, degradation of resources, unsuitable vegetative structure, or unsuitable or inhospitable location in the landscape; any of which could lead to reduced fitness of the resident population (Pulliam 1988). For example, if the majority of suitable snag roosts occur in older forest stands located at high elevations that may not typically be occupied by female bats (Arnett and Hayes 2003; Cryan et al. 2000), maternity roosts could be limiting, influencing reproductive success of a population subjected to these conditions. Measures of variation in demographic parameters are needed to assess the impact of variation in resource suitability and associated landscape quality. Unfortunately, we lack standardized techniques for tracking demographic parameters at roosts for most species of bats (O'Shea et al. 2003). The lack of a clear relationship between resource quality and landscape characteristics currently prevents accurate prediction of impacts of landscape change on survival and fitness of bat populations and, thus, should be a focal area of future research.

Surveys comparing multiple, spatially distinct landscapes have found correlations between the presence of bat species and measured landscape variables in forested ecosystems fragmented by an urban and agricultural matrix (Gehrt and Chelsvig 2004; Jaberg and Guisan 2001; Russ and Montgomery 2002; Walsh and Harris 1996a) and in managed forest settings consisting of patches of forest stands (Arnett and Hayes 2003).

Three of these studies used a grid to divide their study regions so that
each cell designated a separate landscape (Jaberg and Guisan 2001; Russ and Montgomery 2002; Walsh and Harris 1996a). Each study measured landscape composition, but not spatial configuration within sampled landscapes. Russ and Montgomery (2002) and Walsh and Harris (1996a) acoustically sampled bats from randomly selected 1-km landscape cells within their respective national boundaries (Ireland and Britain). Jaberg and Guisan (2001) relied on historical records of a variety of sources within 2.5-km landscape cells that had adequate sampling among 786 km^2 in western Switzerland. Gehrt and Chelsvig (2004) randomly selected sites for acoustic monitoring of bats within 15 study areas that served as patches of natural habitat in a 3,500-km^2 suburban matrix, extending from the Chicago metropolitan area. In addition to measuring landscape composition of each study area and the surrounding 2 km, they measured study area size, the distance to Chicago's urban center, and the distance to the nearest undeveloped area. Landscape elements that repeatedly surfaced as correlates of bat species presence in these studies were quantity of woodlands, tree lines, wetlands, residential areas, and arable land (Gehrt and Chelsvig 2004; Jaberg and Guisan 2001; Russ and Montgomery 2002; Walsh and Harris 1996a). Additionally, Jaberg and Guisan (2001) noted a correlation between bat occurrence and elevation, and Gehrt and Chelsvig (2004) found correlations between distance to urban center and the activity of some species of bats.

In a managed forest system in western Oregon, Arnett and Hayes (2003) mist netted at 36 different ponds and modeled capture rates of bats using several landscape-level variables. The landscape was defined as a circle with a 4.8-km radius surrounding each capture site. The most parsimonious model indicated that snag density within each landscape best explained capture rates of bats when species of bats and sex were pooled. However, sex-specific models revealed that capture rates of male bats for all species were positively related to both elevation and snag density, whereas capture rates of females were negatively related to elevation. Snag density was a poor predictor of female presence in their study area. They attributed this finding to snag-rich, late-seral, and old-growth forest stands being situated most frequently at higher elevations.

Given an appropriate sampling design, landscape-level studies can help identify both critical resource configurations for bat populations within a landscape and the scale at which the distribution of resources influences populations of bats. A landscape-level assessment also can help land managers recognize additional landowners with whom they need to interact and coordinate for effective management of bat populations (Kautz and Cox 2001; Larkin et al. 2003; Wigley et al., chap. 11 in this volume). Relationships between landscape patterns and bat populations can be used to predict the impact of land-use change and aid in both the planning and resource allocation of conservation efforts. These relationships can also be linked to other models that predict the availability of a particular resource. For example, availability of roosting habitat could be linked to forest landscape-change models such as LANDIS

(Gustafson et al. 2000; Larson et al. 2003). The remainder of this chapter focuses on linking the information acquired from landscape-level studies to both management decisions and predictive resource models.

MANAGING HABITAT ACROSS BROAD SPATIAL SCALES

The spatial pattern and suitability of roosts, foraging areas, and water are fundamental to the management of landscapes for bat communities in forested ecosystems (Hayes 2003). Landscapes with an abundance and diversity of suitable roosts, adequate foraging habitat, and sources of open water that are well distributed across the landscape should provide conditions necessary to support healthy populations of bats in most forested ecosystems (Hayes 2003). Managers also need to consider the temporal availability of these resources and that the rates of change will depend on the type of resource and species that use them. Because resource suitability is influenced at multiple temporal and spatial scales, managers need to account for these relationships in areas where bat conservation is a goal in forested landscapes. Until more studies further quantify landscape-scale relationships and response of bats to landscape change, the following discussion and suggested strategies seem prudent in regard to landscape-scale management of habitat for bats in forested landscapes.

We concur with Hayes (2003) that maintaining existing roosts and perpetuating the development of suitable roosts through space and time is the cornerstone of management for bats in forested landscapes. Selection of roost structures has been well studied for many species of management concern. In a review of habitat characteristics selected by tree-roosting bats, Lacki and Baker (2003) found that, when compared with randomly located trees in the same stand, crevice-roosting bats typically chose taller roost trees that were larger in diameter and near the height of the surrounding canopy. These trees also were surrounded by lower amounts of canopy cover and were further from the nearest tree of greater height. Unfortunately, the density, distribution, and optimal location of roosts on the landscape are not well understood, and will vary temporally, topographically, and among tree species and forested ecosystems (Hayes 2003).

Despite these gaps in our understanding of roosting requirements of bats, some generalizations can be drawn from existing literature concerning suitable roost location in a landscape context. Within a stand or patch, roost trees and snags are often located close to the edge (Hayes 2003). When compared with randomly located stands, stands containing roost trees typically have a lower overall density of trees, a higher density of snags, and a higher proportion of larger trees. Providing roosts in relatively close proximity to water also appears to be an important consideration in some forested ecosystems, in particular, in western coniferous forests where water can be limiting (Hayes 2003). Day roosts often are found closer to open water and riparian habitat than random structures (Arnett and Hayes 2003; Carter et al. 2002; Evelyn et al. 2004; Waldien et al. 2000; Weller and Zabel 2001). Several studies report, however, that bats in western forests generally choose day roosts in upslope habitats and rarely use available structures immediately within or adjacent to riparian zones for day roosting (Arnett and Hayes 2003; Campbell et al. 1996; Waldien 1998). In forests of the midwestern United States, bottomland and other wetland forests seem to provide important roost locations (Kurta et al. 2002; Whitaker and Gummer 2001, 2003), but selection of bottomland habitats relative to other available forested habitats in the same area has not been tested. Elevation appears to have a profound influence on spatial distribution of bats and their use of habitats (Arnett and Hayes 2003; Brack et al. 2002; Cryan et al. 2000; Grindal et al. 1999). In some regions female bats do not readily occupy higher elevations (Arnett and Hayes 2003; Brack et al. 2002; Cryan et al. 2000; Grindal et al. 1999), even when suitable roosts are abundant and well distributed (Arnett and Hayes 2003). Because many species of bats appear to segregate by sex during the maternity season, distinguishing landscape needs by this criterion will be important for management of many species of bats.

Green-tree and snag-retention strategies vary by forested ecosystem, and with the goals of regulatory mandates imposed by land management agencies. Implementation of these strategies often lacks clear objectives and empirical data to support optimal locations of these resources to meet specific needs of wildlife, including bats (Wigley et al., chap. 11 in this volume). In the Pacific Northwest, a common practice on privately managed forests is to retain green trees and snags in riparian zones or other inoperable areas when attempting to meet the requirements of state laws governing forest practices. While logistically feasible, the ecological relationships of bats must be considered when managing for future roosting habitat. Further study is warranted to resolve questions on the relative quality of roosts located within riparian habitat versus those located in upland sites. In lieu of this research, it is reasonable to assume that maintaining roost structures and replacement trees for bats only in riparian areas, or only in upland areas, is not likely to meet the needs of a diverse species group (Hayes 2003). Furthermore, if older trees and snags provide important sites for night roosts and hibernacula, the locations in a landscape that are most appropriate to provide these functions remain uncertain (Hayes 2003). Regardless of the region and forest type, and in the absence of adequate information (Hayes 2003; Miller et al. 2003), maintaining and perpetuating a diversity of roost types and characteristics across the landscape in a variety of topographic settings is a logical and conservative approach that should provide the broad spectrum of conditions necessary to meet the varying needs of bats.

Within an area of interest, the proportion of different types of forest stands and other habitat patches (landscape composition) may influence the number and species of bats that an area can support, whereas juxtaposition and size of those habitat patches in relation to one another (landscape configuration) determines the amount and complexity of edge habitat and the level of fragmentation of a site (Harris 1984; Hayes 2003). The quality of a particular patch of forest for roosting or foraging is contingent on the availability of suitable roosts and vegetation density (clutter). High levels of bat activity and use of roosts in mature and old-growth forests is thought to be, in part, a consequence of the availability of large, older trees for roosting combined with open, multilayered canopies within these stands (Christy and West 1993; Crampton and Barclay 1998; Jung et al. 1999; Kunz and Lumsden 2003; Thomas 1988). In western coniferous forests, snag density has been shown to increase substantially with stand age (Arnett and Hayes 2003; Ohmann and Waddell 2002; Spies et al. 1988). Stand age is not necessarily the driving force behind resource selection by bats (Hayes 2003), however, and active management can provide the structural conditions conducive for both foraging and roosting by bats in forest patches of various ages. Thinning can accelerate development of large-diameter trees and snags (Carey and Curtis 1996; Hayes 1997) for roosting by bats and reduces clutter, rendering young stands more suitable for foraging by bats (Guldin et al., chap. 7 in this volume; Humes et al. 1999). Small patch cuts are known to receive high levels of use by bats (Grindal and Brigham 1998; Menzel et al. 2002), suggesting that uneven-age silvicultural prescriptions, such as group selection or individual tree selection, may be beneficial for bats (Hayes 2003), although response to such harvest practices will vary with the size of the cut (Menzel et al. 2002) and the species of bat (Jung et al. 1999; Menzel et al. 2002).

Planning for bats at the landscape scale in forested ecosystems appears to fit the seemingly simplistic, yet profound advice of Bunnell (1997), who noted that managers should not "do the same thing on every acre!" Instead, planning efforts should consider providing a shifting mosaic of forest patches of varying age and structural conditions favored by bats for both roosting and foraging, juxtaposed with riparian habitat, open water, and, if available, rock outcrop and karst systems. Retaining remnant patches of structurally diverse older forest, in particular, in intensively managed landscapes, will likely be an important conservation strategy to maintain some local populations of bats until patches within the matrix of young forest develop structural characteristics conducive to roosting and foraging by bats (Jung et al. 1999; Waldien 1998; Zielinski and Gellman 1999). Active forest management can and should be employed, where appropriate, to create and maintain desired structural conditions conducive to roosting and foraging habitats for bats through space and time (Guldin et al., chap. 7 in this volume).

PREDICTING EFFECTS OF LAND MANAGEMENT ON BATS

Balancing new forest policies to meet goals for sustaining biological diversity with social and economic demands of forests has become a major challenge for managers and policymakers throughout North America (McComb et al. 2002; Wiersum 1995). Land management decisions affect wildlife populations by influencing changes in landscapes that provide habitat. Mathematical models can be used to assess the potential effects of management actions on wildlife, including bats, and are important tools for comparing ecological benefit to economic cost. Such models are needed to connect land management plans to changes in landscapes and to link landscape change to the status of wildlife populations. Linkages between landscapes and populations are often made with maps of habitat suitability (Kliskey et al. 1999; Rempel et al. 1997). Habitat maps can be generated by using a variety of methods, including the application of habitat suitability index (HSI) models (U.S. Fish and Wildlife Service 1981), which are based on expert opinion, or resource selection functions based on empirical data (Manly et al. 2002). Linkages between habitat suitability and population viability of bats can be made by assessing the availability of high-quality habitat (Roloff and Haufler 2002), applying Bayesian belief networks (Oliver and Smith 1990), or simulating population growth with habitat-dependent rates of reproduction and mortality. The computers, software, and information for modeling landscapes (e.g., LANDIS, He et al. 1996, 1999; Mladenoff and He 1999; LMS, McCarter 1997; McCarter et al. 1998), habitats, and populations (e.g., RAMAS GIS, Akçakaya 1998, 2000) are now widely available.

From a wildlife perspective, it is useful to begin by considering habitat features, or resources, that are most likely to affect populations of bats. Models of landscape change can incorporate many landcover types and mechanisms causing change, but often few are necessary to capture the critical components of the landscape that directly affect bats. An example of a mathematical model for predicting the effects of land management on bats is the use of Geographic Information Systems (GIS)-based habitat suitability models applied to output from simulations of forest landscape dynamics (Larson et al. 2003).

PREDICTING THE EFFECTS OF LAND MANAGEMENT ON NORTHERN MYOTIS IN SOUTHERN MISSOURI

Consideration of northern myotis (*Myotis septentrionalis*) in forest management planning required the application of a habitat suitability model to the results of a forest landscape simulation. Shifley et al. (2000) used LANDIS to simulate the effects of alternative tree harvest methods on a 71,000-ha tract of the Mark Twain National Forest in the Ozark region of southern Missouri. One of the alternatives was to harvest trees on a 100-year rotation using an equal mixture of even-aged management (i.e., clearcuts) and uneven-aged management (i.e., group selection cuts). The authors calibrated LANDIS with a resolution of 30- × 30-m grid cells, a time step of 10 years, and four tree species groups. LANDIS simulated the reproduction and growth of trees and disturbances due to wind, fire, and tree harvest. Output from this landscape model provided information on composition and age structure of tree species for each grid cell once each decade.

Larson et al. (2003) developed a GIS-based HSI model for northern myotis to apply to the forested landscapes simulated by Shifley et al. (2000). HSI models relate specific environmental characteristics to habitat suitability on an index scale from 0 (i.e., not habitat) to 1 (i.e., habitat of maximum suitability). The Larson et al. (2003) habitat model was based on an extensive literature review. It included variables for roost sites and foraging areas during summer, but not variables for hibernacula in caves used as winter habitat. Northern myotis roost under exfoliating bark or in crevices and cavities of large trees and snags in summer months. Therefore, the first two variables related the age of the oldest trees on a grid cell to suitability of habitat for providing roosting sites (e.g., 0 for cells with trees <40 years old, 1 for cells with trees >100 years old, and a linear relationship in between). The third variable specified that habitat suitability for roosting declined from 1 to 0 between 1 and 2 km from a permanent source of surface water for drinking. The roosting component of the habitat model was, thus, a product of the first three variables (fig. 9.2). The foraging component of the model assigned high-suitability values to cells containing mature forest, especially those within 30 m of a forest opening. Edges between forest and nonforest or very young forest were used to indicate canopy gaps which northern myotis are thought to use for foraging. The last variable in the habitat model identified cells within 2 km of both roosting and foraging habitat, as this was the maximum distance observed between roost sites used by the same bat of this species (Foster and Kurta 1999). The HSI value for a cell was the maximum of the roosting and foraging components, but was zero if either component was not within 2 km (figs. 9.2 and 9.3).

Based on the methods described above, a map of HSI values can be calculated for each time step of a landscape simulation scenario. These HSI maps can be summarized over an entire landscape to facilitate comparisons between different points in time or between management alternatives (fig. 9.4). Viewing the cumulative distribution of HSI values is more informative than basing inference on mean or median HSI values because the distribution may not be symmetrical or continuous (fig. 9.5). In some circumstances it may be useful to analyze characteristics of habitat patches (e.g., distribution of sizes, spatial configuration), but often these characteristics can be incorporated directly into spatially explicit habitat models and, therefore, be reflected in the frequency distribution of resulting habitat suitability values. In other situations it may be useful to go a step further and use results of habitat modeling as inputs for a habitat-based population viability model (Akçakaya et al. 2004; Larson et al. 2004). Population viability, not habitat quality, is often the ultimate concern and the only way to fully quantify uncertainty in how landscape changes affect bats.

Figure 9.2.
Relationships between life requisites and components of an HSI model for northern myotis in southern Missouri.
Adapted from Larson et al. 2003

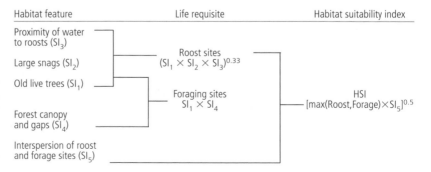

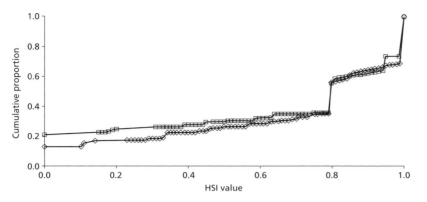

Figure 9.5.
Cumulative frequency distribution of habitat suitability index values for northern myotis in a forested landscape (3,260 ha) in southern Missouri. The landscape experienced simulated tree growth, disturbance, and regeneration. The HSI values declined slightly between years 30 (diamonds) and 100 (squares) of the simulation, mainly because more of the landscape was considered nonhabitat (i.e., 13% vs. 21% of the landscape with HSI = 0).

LIMITATIONS AND BENEFITS OF MODELING

Predictions about the effects of land management on bat populations have limitations that should be considered. Models, by definition, are simplifications of the systems they represent. Therefore, models and their results are limited by the parameters, or factors, that are included and are unlikely to incorporate all the factors that actually influence the desired predictions. Models also are limited by how each factor is incorporated. Uncertainty in the numerical coefficients and the functional forms describing relationships between factors in a model contribute to uncertainty in the resulting predictions. A critical interpretation of modeling results requires knowledge of these limitations. More detailed discussions about interpreting results of habitat models are provided by Starfield (1997), Roloff and Kernohan (1999), and Burgman et al. (2001).

Models, however, offer several benefits that are not diminished by these limitations. Models provide the only rapid and effective means for evaluating wildlife habitat suitability across large landscapes and serve as a convenient framework for comparing land management alternatives. They also summarize the current understanding of a system. The models described in the example above formalized purported wildlife-habitat relationships into testable hypotheses. For example, do bats select roosts in areas identified by the model as high quality roosting habitat, and do measures of bat foraging activity and success in different areas correlate with the predicted suitability of foraging habitat? Conceptual and mathematical models may still be useful even if information about bat populations and how they interact with landscapes is lacking. Models can help document sources of uncertainty in making management decisions and help identify and prioritize research needs.

SUMMARY

The reliance of bats on multiple resources over a large area requires incorporating a landscape perspective in both research and management of bat populations. Unfortunately, landscape-level requirements are poorly understood for most species of bats. In this chapter, we have attempted to facilitate further consideration of landscape-level attributes in both research and management plans by reviewing basic concepts of landscape ecology and summarizing current studies on bats that incorporate multiple landscapes. Although, relationships have been found between populations of bats and overall forest cover and elevation, the more detailed information required by land managers is still lacking.

In the absence of such information, we suggest that managers attempt to incorporate a mosaic of landscape types that encompasses a range of foraging and roosting opportunities for bats. Within that mosaic, older trees and snags at low elevations will likely be a critical roosting resource for many species. We believe that planning for such mosaics should involve assessing different management options by using a habitat suitability index when possible. Habitat models (e.g., habitat suitability index, resource selection functions) summarize the current knowledge of habitat requirements of a species, and can be used to quantitatively assess the quality of different predicted landscapes under varying management scenarios. As new research is compiled these models can be updated to improve future planning efforts for bats in forests.

LITERATURE CITED

Adam, M.D., M.J. Lacki, and T.G. Barnes. 1994. Foraging areas and habitat use of the Virginia big-eared bat in Kentucky. Journal of Wildlife Management 58:462–469.

Addicott, J.F., J.M. Aho, M.F. Antolin, D.K. Padilla, J.S. Richardson, and D.A. Soluk. 1987. Ecological neighborhoods: scaling environmental patterns. Oikos 49:340–346.

Akçakaya, H.R. 1998. RAMAS GIS: linking landscape data with population viability analysis (version 3.0). Applied Biomathematics, Setauket, NY.

———. 2000. Viability analyses with habitat-based metapopulation models. Population Ecology 42:45–53.

Akçakaya, H.R., V.C. Radeloff, D.J. Mladenoff, and H.S. He. 2004. Integrating landscape and metapopulation modeling approaches: viability of the sharp-tailed grouse in a dynamic landscape. Conservation Biology 18:526–537.

Arnett, E.B., and J.P. Hayes. 2003. Presence, relative abundance, and resource selection of bats at multiple spatial scales in western Oregon. Final report prepared for USGS-BRD. Oregon State University, Corvallis, OR.

Augustine, N.H., M.A. Mugglestone, and S.T. Buckland. 1996. An autologistic model for the spatial distribution of wildlife. Journal of Applied Ecology 33:339–347.

Barclay, R.M.R., and R.M. Brigham. 2001. Year-to-year reuse of tree-roosts by California bats (*Myotis californicus*) in southern British Columbia. American Midland Naturalist 146:80–85.

Barclay, R.M.R., and L.D. Harder. 2003. Life histories of bats: life in the slow lane, pp. 209–253, *in* Bat ecology (T.H. Kunz and M.B. Fenton, eds.). University of Chicago Press, Chicago, IL.

Beier, P., and R.F. Noss. 1998. Do habitat corridors provide connectivity? Conservation Biology 12:1241–1252.

Bissonette, J.A. 2003. Linking landscape pattern to biological reality, pp. 15–34, *in* Landscape ecology and resource management: linking theory with practice (J.A. Bissonette and I. Storch, eds.). Island Press, Washington, DC.

Boonman, M. 2000. Roost selection by noctules (*Nyctalus noctula*) and Daubenton's bats (*Myotis daubentonii*). Journal of Zoology (London) 251:385–389.

Brack, V.J., C.W. Stihler, R.J. Reynolds, C.M. Butchkoski, and C.S. Hobson. 2002.

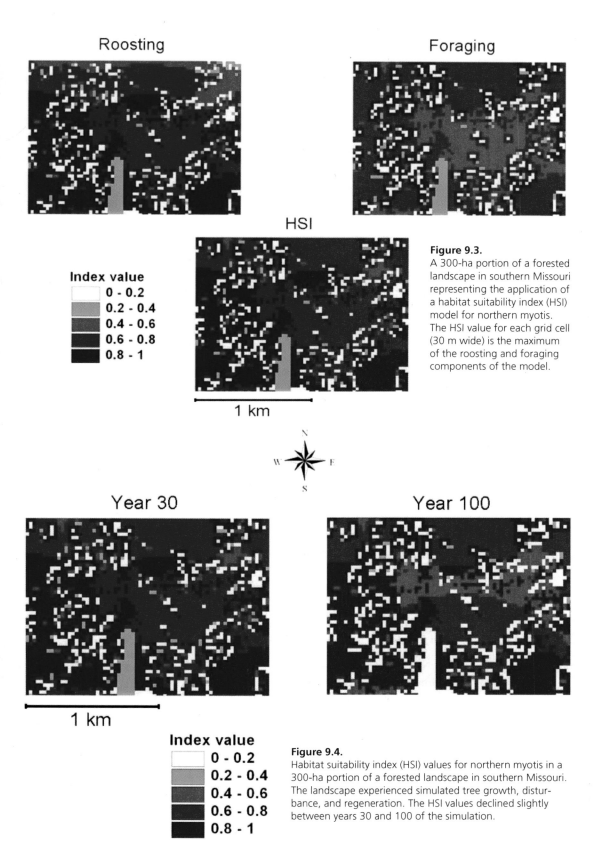

Roosting

Foraging

HSI

Index value
- 0 - 0.2
- 0.2 - 0.4
- 0.4 - 0.6
- 0.6 - 0.8
- 0.8 - 1

1 km

Figure 9.3.
A 300-ha portion of a forested landscape in southern Missouri representing the application of a habitat suitability index (HSI) model for northern myotis. The HSI value for each grid cell (30 m wide) is the maximum of the roosting and foraging components of the model.

Year 30

Year 100

1 km

Index value
- 0 - 0.2
- 0.2 - 0.4
- 0.4 - 0.6
- 0.6 - 0.8
- 0.8 - 1

Figure 9.4.
Habitat suitability index (HSI) values for northern myotis in a 300-ha portion of a forested landscape in southern Missouri. The landscape experienced simulated tree growth, disturbance, and regeneration. The HSI values declined slightly between years 30 and 100 of the simulation.

Duff, A.A. 2004. Predicting bat occurrences in Northern California using land-scape-scale variables. Ball State University, Muncie, IN.

Dunning, J.B., B.J. Danielson, and H.R. Pulliam. 1992. Ecological process that affect populations in complex landscapes. Oikos 65:169–175.

Edwards, T.C., Jr., G.G. Moisen, T.S. Frescino, and J.J. Lawler. 2003. Modeling multiple ecological scales to link landscape theory to wildlife conservation, in Landscape ecology and resource management: linking theory with practice (J.A. Bissonette and I. Storch, eds.). Island Press, Washington, DC.

Ekman, M., and J. de Jong. 1996. Local patterns of distribution and resource utilization of four bat species (*Myotis brandti, Eptesicus nilssoni, Plecotus auritus* and *Pipistrellus pipistrellus*) in patchy and continuous environments. Journal of Zoology (London) 238:571–580.

Elmore, L.W., D.A. Miller, and F.J. Vilella. 2005. Foraging area size and habitat use by red bats (*Lasiurus borealis*) in an intensively managed pine landscape in Mississippi. American Midland Naturalist 153:405–417.

Erickson, J.L., and S.D. West. 1996. Managed forests in the western Cascades: effects of seral stage on bat habitat use patterns, pp. 215–227, *in* Bats and forests symposium (R.M.R. Barclay and R.M. Brigham, eds.). Research Branch, British Columbia Ministry of Forests, Victoria, BC.

Estrada, A., and R. Coates-Estrada. 2001. Bat species richness in live fences and in corridors of residual rain forest vegetation at Los Tuxtas, Mexico. Ecography 24:94–102.

———. 2002. Bats in continuous forest, forest fragments and in an agricultural mosaic habitat-island at Los Tuxtlas, Mexico. Biological Conservation 103:237–245.

Evelyn, M.J., D.A. Stiles, and R.A. Young. 2004. Conservation of bats in suburban landscapes: roost selection by *Myotis yumanensis* in a residential area in California. Biological Conservation 115:463–473.

Everette, A.L., T.J. O'Shea, L.E. Ellison, L.A. Stone, and J.L. McCance. 2001. Bat use of a high-plains urban wildlife refuge. Wildlife Society Bulletin 29:967–973.

Fleming, T.H., and P. Eby. 2003. Ecology of bat migration, pp. 156–208, *in* Bat ecology (T.H. Kunz and M.B. Fenton, eds.). University of Chicago Press, Chicago, IL.

Forman, R.T.T. 1995. Land mosaics: the ecology of landscapes and regions. Cambridge University Press, Cambridge, U.K.

Foster, R.W., and A. Kurta. 1999. Roosting ecology of the northern bat (*Myotis septentrionalis*) and comparisons with the endangered Indiana bat (*Myotis sodalis*). Journal of Mammalogy 80:659–672.

Furlonger, C.L., H. J. Dewar, and M.B. Fenton. 1987. Habitat use by foraging insectivorous bats. Canadian Journal of Zoology 65:284–288.

Gehrt, S.D., and J.F. Chelsvig. 2004. Species-specific patterns of bat activity in an urban landscape. Ecological Applications 14:625–635.

Gorresen, P.M., and M.R. Willig. 2004. Landscape response of bats to habitat fragmentation in Atlantic forest of Paraguay. Journal of Mammalogy 85:688–697.

Grindal, S.D., and R.M. Brigham. 1998. Short-term effects of small-scale habitat disturbance on activity by insectivorous bats. Journal of Wildlife Management 62:996–1003.

Grindal, S.D., J.L. Morissette, and R.M. Brigham. 1999. Concentration of bat activity in riparian habitats over an elevational gradient. Canadian Journal of Zoology 77:972–977.

Gustafson, E.J. 1998. Quantifying landscape spatial pattern: what is the state of the art? Ecosystems 1:143–156.

Gustafson, E.J., S.R. Shifley, D.J. Mladenoff, H.S. He, and K.K. Nimerfro. 2000. Spatial simulation of forest succession and timber harvesting using LANDIS. Canadian Journal of Forest Research 30:32–43.

Haila, Y. 2002. A conceptual genealogy of the fragmentation research: from island biogeography to landscape ecology. Ecological Applications 12:321–334.

Hargrove, W.W., and J. Pickering. 1992. Pseudoreplication: a *sine qua non* for regional ecology. Landscape Ecology 6:251–258.

Harris, L.D. 1984. The fragmented forest: island biogeography theory and the preservation of biotic diversity. University of Chicago Press, Chicago, IL.

Hayes, J.P. 1997. Temporal variation in activity of bats and the design of echolocation-monitoring studies. Journal of Mammalogy 78:514–524.

———. 2003. Habitat ecology and conservation of bats in western coniferous forests, pp. 81–119, *in* Mammal community dynamics in coniferous forests of western North America: management and conservation (C.J. Zabel and R.G. Anthony, eds.). Cambridge University Press, Cambridge, MA.

He, H.S., D.J. Mladenoff, and J. Boeder. 1996. LANDIS: a spatially explicit model of forest landscape change. LANDIS 2.0 User's Guide. Department of Forest Ecology and Management, University of Wisconsin, Madison, WI.

He, H.S., D.J. Mladenoff, and T.R. Crow. 1999. Object-oriented design of LANDIS, a spatially explicit and stochastic landscape model. Ecological Modelling 119:1–19.

Hess, G.R., and R.A. Fischer. 2001. Communicating clearly about conservation corridors. Landscape and Urban Planning 55:195–208.

Holland, J.D., D.G. Bert, and L. Fahrig. 2004. Determining the spatial scale of species' response to habitat. BioScience 54:227–233.

Humes, M.L., J.P. Hayes, and M.W. Collopy. 1999. Bat activity in thinned, unthinned, and old-growth forests in western Oregon. Journal of Wildlife Management 63:553–561.

Humphrey, S.R. 1975. Nursery roosts and community diversity of Nearctic bats. Journal of Mammalogy 56:321–346.

Hurlbert, S.H. 1984. Pseudoreplication and the design of ecological field experiments. Ecological Monographs 54:187–211.

Jaberg, C., and A. Guisan. 2001. Modelling the distribution of bats in relation to landscape structure in a temperate mountain environment. Journal of Applied Ecology 38:1169–1181.

Jung, T.S., I.D. Thompson, R.D. Titman, and A.P. Applejohn. 1999. Habitat selection by forest bats in relation to mixed-wood stand types and structure in central Ontario. Journal of Wildlife Management 63:1306–1319.

Kautz, R.S., and J.A. Cox. 2001. Strategic habitats for biodiversity conservation in Florida. Conservation Biology 15:55–77.

Kliskey, A.D., E.C. Lofroth, W.A. Thompson, S. Brown, and H. Schreier. 1999. Simulating and evaluating alternative resource-use strategies using GIS-based habitat suitability indices. Landscape and Urban Planning 45:163–175.

Knight, T.W., and D.W. Morris. 1996. How many habitats do landscapes contain? Ecology 77:1756–1764.

Kunz, T.H., and L.F. Lumsden. 2003. Ecology of cavity and foliage roosting bats, pp. 3–83, *in* Bat ecology (T. H. Kunz and M. B. Fenton, eds.). University of Chicago Press, Chicago, IL.

Kurta, A., D. King, J.A. Teramino, J.M. Stribley, and K.J. Williams. 1993. Summer roosts of the endangered Indiana bat (*Myotis sodalis*) on the northern edge of its range. American Midland Naturalist 129:132–138.

Kurta, A., S.W. Murray, and D.H. Miller. 2002. Roost selection and movements across the summer landscape, pp. 118–129, *in* The Indiana bat: biology and management of an endangered species (A. Kurta and J. Kennedy, eds.). Bat Conservation International, Austin, TX.

Lacki, M.J., and M.D. Baker. 2003. A prospective power analysis and review of habitat characteristics used in studies of tree-roosting bats. Acta Chiropterologica 5:199–208.

Larkin, J.L., D.S. Maehr, T. Hector, M.A. Orlando, and K. Whitney. 2003. Landscape linkages and conservation planning for the black bear in southwest Florida. Animal Conservation 7:23–24.

Larson, M.A., W.D. Dijak, F.R. Thompson III, and J.J. Millspaugh. 2003. Landscape-level habitat suitability models for twelve wildlife species in southern Missouri. USDA Forest Service, North Central Research Station, General Technical Report NC-233, St. Paul, MN.

Larson, M.A., F.R. Thompson III, J.J. Millspaugh, W.D. Dijak, and S.R. Shifley. 2004. Linking population viability, habitat suitability, and landscape simulation models for conservation planning. Ecological Modelling 180:103–118.

Lewis, S.E. 1995. Roost fidelity of bats: a review. Journal of Mammalogy 76:481–496.

Lewis, T. 1965. The effects of an artificial windbreak on the aerial distribution of flying insects. Annals of Applied Biology 55:502–512.

———. 1970. Patterns of distribution of insects near a windbreak of tall trees. Annals of Applied Biology 65:213–220.

Lichstein, J.W., T.R. Simons, S.A. Shriner, and K.E. Franzreb. 2002. Spatial auto-correlation and autoregressive models in ecology. Ecological Monographs 72:445–463.

Lima, S.L., and P.A. Zollner. 1996. Towards a behavioral ecology of ecological landscapes. Trends in Ecology and Evolution 11:131–135.

Manly, B.F.J., L.L. McDonald, D.L. Thomas, T.L. McDonald, and W.P. Erickson. 2002. Resource selection by animals, 2nd ed. Kluwer Academic Publishers, Boston, MA.

Mayer, A.L., and G.N. Cameron. 2003. Consideration of grain and extent in landscape studies of terrestrial vertebrate ecology. Landscape and Urban Planning 65:201–207.

McCarter, J.B. 1997. Integrating forest inventory, growth and yield, and computer visualization into a landscape management system. General Technical Report, USDA Forest Service, Intermountain Research Station, Ogden, UT.

McCarter, J.B., J.S. Wilson, P.J. Baker, J.L. Moffett, and C.D. Oliver. 1998. Landscape management through integration of existing tools and emerging technologies. Journal of Forestry 96:17–23.

McComb, W.C., M.T. McGrath, T.A. Spies, and D. Vesely. 2002. Models for mapping potential habitat at landscape scales: an example using northern spotted owls. Forest Science 48:203–216.

McGarigal, K., and S.A. Cushman. 2002. Comparative evaluation of experimental approaches to the study of habitat fragmentation effects. Ecological Applications 12:335–345.

McGarigal, K., and B.J. Marks. 1995. Fragstats: spatial pattern analysis program for quantifying landscape structure. General Technical Report PNW-GTR-351. USDA Forest Service, Pacific Northwest Research Station, Portland, OR.

Menzel, M.A., T.C. Carter, J.M. Menzel, W.M. Ford, and B.R. Chapman. 2002. Effects of group selection silviculture in bottomland hardwoods on the spatial activity patterns of bats. Forest Ecology and Management 162:209–218.

Miller, D.A., E.B. Arnett, and M.J. Lacki. 2003. Habitat management for forest-roosting bats of North America: a critical review of habitat studies. Wildlife Society Bulletin 31:30–44.

Miller, M., and R.E. Russell. 2004. Species-specific response to landscape heterogeneity: improving estimates of connectivity, *in* Conserving biodiversity in agricultural landscapes: model-based planning tools (R.K. Swihart and J.E. Moore, eds.). Purdue University Press, West Lafayette, IN.

Mitchell, M.S., and R.A. Powell. 2003. Linking fitness of landscapes with the behavior and distribution of animals, pp. 93–124, *in* Landscape ecology and resource management: linking theory with practice (J.A. Bissonette and I. Storch, eds.). Island Press, Washington, DC.

Mladenoff, D.J., and H.S. He. 1999. Design and behavior of LANDIS, an object-oriented model of forest landscape disturbance and succession, *in* Advances in spatial modelling of forest landscape change: approaches and applications (D.J. Mladenoff and W.L. Baker, eds.). Cambridge University Press, Cambridge, U.K.

Morand, S. 2000. Geographic distance and the role of island area and habitat diversity in the species-area relationships of four Lesser Antillean faunal groups: a complementary note to Ricklefs and Lovette. Journal of Animal Ecology 69:1117–1119.

Morrison, M.L. 2001. A proposed research emphasis to overcome the limits of wildlife-habitat relationship studies. Journal of Wildlife Management 65:613–623.

Murcia, C. 1995. Edge effect in fragmented forests: implications for conservation. TREE 10:58–62.

Ohmann, J.L., and K.L. Waddell. 2002. Regional patterns of dead wood in forested habitats of Oregon and Washington. General Technical Report PSW-GTR-181. USDA Forest Service, Portland, OR.

Oliver, R.M., and J.Q. Smith, eds. 1990. Influence diagrams, belief nets and decision analysis. Wiley, Chichester, U.K.

O'Neil, R.V., C.T. Hunsaker, S.P. Timmins, B.L. Jackson, K.B. Jones, K.H. Riitters, and J.D. Wickham. 1996. Scale problems in reporting landscape pattern at the regional scale. Landscape Ecology 11:169–180.

O'Neil, R.V., J.R. Krummel, R.H. Gardner, G. Suigihara, B. Jackson, D.L. De Angelis, B.T. Milne, M.G. Turner, B. Zygmunt, S. Christensen, V.H. Dale, and R.L. Graham. 1988. Indices of landscape pattern. Landscape Ecology 1:153–162.

Ormsbee, P.A., and W.C. McComb. 1998. Selection of day roosts by female long-legged myotis in the central Oregon Cascade Range. Journal of Wildlife Management 62:596–603.

O'Shea, T.J., M.A. Bogan, and L.E. Ellison. 2003. Monitoring trends in bat populations of the United States and territories: status of the science and recommendations for the future. Wildlife Society Bulletin 31:16–29.

Pavey, C.R. 1998. Habitat use by the eastern horseshoe bat, *Rhinolophus megaphyllus,* in a fragmented woodland mosaic. Wildlife Research 25:489–498.

Pierson, E.D. 1998. Tall trees, deep holes, and scarred landscapes: conservation biology of bats in North America, pp. 309–324, *in* Bat biology and conservation (T.H. Kunz and P. A. Racey, eds.). Smithsonian Institution Press, Washington, DC.

Porter, W.F., and K.E. Church. 1987. Effects of environmental pattern on habitat preference analysis. Journal of Wildlife Management 51:681–685.

Pulliam, H.R. 1988. Sources, sinks, and population regulation. American Naturalist 132:652–661.

Raudenbush, S.W., and A.S. Bryk. 2002. Hierarchical linear models: applications and data analysis methods. Sage Publications, Thousand Oaks, CA.

Rempel, R.S., P.C. Elkie, A.R. Rodgers, and M.J. Gluck. 1997. Timber-management and natural-disturbance effects on moose habitat: landscape evaluation. Journal of Wildlife Management 61:517–524.

Ricketts, T.H. 2001. The matrix matters: effective isolation in fragmented landscapes. The American Naturalist 158:87–99.

Riitters, K.H., R.V. O'Neil, C.T. Hunsaker, J.D. Wickham, D.H. Yankee, S.P. Timmons, K.B. Jones, and B.L. Jackson. 1995. A factor analysis of landscape pattern and structure metrics. Landscape Ecology 10:23–40.

Roloff, G.J., and J.B. Haufler. 2002. Modeling habitat-based viability from organism to population, *in* Predicting species occurrence: issues of accuracy and scale (J.M. Scott, P.J. Heglund, M.L. Morrison, J.B. Haufler, M.G. Raphael, W.A. Wall, and F.B. Samson, eds.). Island Press, Washington, DC.

Roloff, G.J., and B.J. Kernohan. 1999. Evaluating reliability of habitat suitability index models. Wildlife Society Bulletin 27:973–985.

Russ, J.M., and W.I. Montgomery. 2002. Habitat associations of bats in Northern Ireland: implications for conservation. Biological Conservation 108:49–58.

Shifley, S.R., F.R. Thompson III, D.R. Larsen, and W.D. Dijak. 2000. Modeling forest landscape change in the Missouri Ozarks under alternative management practices. Computers and Electronics in Agriculture 27:7–24.

Spies, T.A., J.F. Franklin, and T.B. Thomas. 1988. Coarse woody debris in Douglas-fir forests of western Oregon and Washington. Ecology 69:1689–1702.

Starfield, A.M. 1997. A pragmatic approach to modeling for wildlife management. Journal of Wildlife Management 61:261–270.

Swihart, R.K., T.M. Gehring, M.B. Kolozsvary, and T.E. Nupp. 2003. Responses of "resistant" vertebrates to habitat loss and fragmentation: the importance of niche breadth and range boundaries. Diversity and Distributions 9:1–18.

Thomas, D.W. 1988. The distribution of bats in different ages of Douglas-fir forests. Journal of Wildlife Management 52:619–626.

Turner, M.G., R.H. Gardner, and R.V. O'Neil. 2001. Landscape ecology in theory and practice. Springer-Verlag, New York.

U.S. Fish and Wildlife Service. 1981. Standards for the development of habitat suitability models for use in the habitat evaluation procedure, *in* Ecological service manual 103. Division of Ecological Services, Washington, DC.

Veilleux, J.P., J.O. Whitaker Jr., and S.L. Veilleux. 2003. Tree-roosting ecology of reproductive female eastern pipistrelles, *Pipistrellus subflavus,* in Indiana. Journal of Mammalogy 84:1068–1075.

Verboom, B., and H. Huitema. 1997. The importance of linear landscape elements for the pipistrelle *Pipistrellus pipistrellus* and the serotinus bat *Eptesicus serotinus.* Landscape Ecology 12:117–125.

Verboom, B., and K. Spoelstra. 1999. Effects of food abundance and wind on the use of tree lines by an insectivorous bat, *Pipistrellus pipistrellus.* Canadian Journal of Zoology 77:1393–1401.

Vonhof, M.J., and R.M.R. Barclay. 1996. Roost-site selection and roosting ecology of forest-dwelling bats in southern British Columbia. Canadian Journal of Zoology 74:1797–1805.

———. 1997. Use of tree stumps as roosts by the western long-eared bat. Journal of Wildlife Management 61:674–684.

Waldien, D.L. 1998. Characteristics and spatial relationships of day-roosts and activity areas of female long-eared myotis (*Myotis evotis*) in western Oregon. Ph.D. dissertation. Oregon State University, Corvallis, OR.

Waldien, D.L., J.P. Hayes, and E.B. Arnett. 2000. Day-roosts of female long-eared myotis in western Oregon. Journal of Wildlife Management 64:785–796.

Walsh, A.L., and S. Harris. 1996a. Factors determining the abundance of vespertilionid bats in Britain: geographical, land class and local habitat relationships. Journal of Applied Ecology 33:519–529.

———. 1996b. Foraging habitat preferences of vespertilionid bats in Britain. Journal of Applied Ecology 33:508–519.

Walsh, A.L., and B.A. Mayle. 1991. Bat activity in different habitats in a mixed lowland woodland. Myotis 29:97–104.

Weller, T.J., and C.J. Zabel. 2001. Characteristics of fringed myotis day roosts in northern California. Journal of Wildlife Management 65:489–497.

Whitaker, J.O., Jr., and S.L. Gummer. 2001. Bats of the Wabash and Ohio river basins of southwestern Indiana. Proceedings of the Indiana Academy of Science 110:126–140.

———. 2003. Current status of the evening bat, *Nycticeius humeralis,* in Indiana. Proceedings of the Indiana Academy of Science 112:55–60.

Wiens, J.A. 1989. Spatial scaling in ecology. Functional Ecology 3:385–397.

Wiersum, K.F. 1995. 200 years of sustainable forestry: lessons from history. Environmental Management 19:321–329.

Zielinski, W.J., and S.T. Gellman. 1999. Bat use of remnant old-growth redwood stands. Conservation Biology 13:160–167.

Zimmerman, G.E., and W.E. Glanz. 2000. Habitat use by bats in eastern Maine. Journal of Wildlife Management 64:1032–1040.

ASSESSING POPULATION STATUS OF BATS IN FORESTS: CHALLENGES AND OPPORTUNITIES

10

Theodore J. Weller

Interest in bats has increased during the past two decades in scientific communities, land management agencies, and the general public (Fenton 1997). Growing knowledge of the interdependence between bats and forests in concert with concern over human-induced changes to forested ecosystems has spurred the need for a greater understanding of the ecology of forest-dwelling bats (Barclay and Brigham 1996). Increased understanding has been facilitated by technological advancements (e.g., miniature radiotransmitters and affordable bat detectors) that have allowed exciting insights into many formerly unknown aspects of bat ecology. Prior to these advances, scientific and public interest in bats was focused on roosts where bats aggregated in large numbers and could be easily viewed, such as caves and buildings (Brigham, chap. 1 in this volume). The advent of affordable bat detectors meant an increasing number of people could eavesdrop on the formerly silent world of bats for a variety of purposes ranging from nature walks to in-depth studies of echolocation. Similarly, radiotransmitters small enough to be carried by bats have vastly increased our knowledge of bats that use less conspicuous roosts. Nowhere are these gains more evident than for forest-dwelling bats: bat detectors have been used extensively to compare habitat use among forest types and ages, and the use of radiotelemetry has determined that many bats in forests tend to roost in small groups and switch between a large number of roosts (Barclay and Kurta, chap. 2 in this volume).

With increased interest and recognition come questions regarding the vitality of bat populations. Population declines have been measured for cave-dwelling species in the eastern United States (Tuttle 1979; U.S. Fish and Wildlife Service 1999), but other bat populations may also be declining (Pierson 1998). Concerns about population declines have led to the listing of several species under the U.S. Endangered Species Act, and incorporation on provincial red lists in Canada and other regional lists of sensitive species. As a result, a large number of agency personnel are responsible for managing bats and their habitats (Weller, in prep.).

The challenges of monitoring bat populations are myriad (O'Shea and Bogan 2003). Most of these challenges result because bats are small, nocturnal flying animals that roost secretively and produce calls largely outside the hearing range of humans; thus, bats are difficult to detect without the use of specialized tools such as bat detectors. Our ability to assess the population status of bats varies according to the ecology of the species and their susceptibility to existing inventory methods. Species that roost in large numbers in relatively accessible locations afford the best opportunities to achieve rigorous estimates of conventional population parameters such as abundance and survival (O'Shea and Bogan 2003; Sendor and Simon 2003; Warren and Witter 2002). However, most species of North American bats exhibit several ecological attributes that render their monitoring difficult, including low detectability, aggregation in low densities, unpredictable movement patterns, and poorly known biology (Clarke et al. 2003; O'Shea and Bogan 2003).

Forest-dwelling bats are among the most difficult group to monitor. During the warm season most forest bats roost in cracks, crevices, and cavities, or under the bark of trees (Barclay and Kurta, chap. 2 in this volume; Kunz and Lumsden 2003). These roosts are difficult to observe because, in general, they are inconspicuous and used by relatively few individuals. Fidelity to these roosts is also much less than for bats that roost in more permanent structures (Lewis 1995). As a result, monitoring individual roost sites to assess population status of forest-dwelling bats presents even greater challenges than for bats that roost more conspicuously. Away from roosts, bats are difficult to capture and activity patterns vary greatly in both space and time (Lacki et al., chap. 4 in this volume). Often just determining whether a given species occurs in a particular forested area can require significant effort (Weller, in prep.), and evaluating more conventional population parameters using existing methods, such as abundance levels, is nearly impossible (O'Shea and Bogan 2003).

Despite these obstacles, biologists are asked to assess the status and trend of bat populations, and often for several species simultaneously. Although little specific guidance exists as to how to do this for forest-dwelling bats (Resources Inventory Committee 1998; Vonhof 2000), there are several resources that identify important considerations for conducting inventory and monitoring of wildlife (Goldsmith 1991; Morrison et al. 2001; Thompson et al. 1998; Yoccoz et al. 2001). These sources provide valuable guidance on the fundamentals of this type of work that should be consulted before initiating a program for bats.

The challenges of applying conventional approaches, used to monitor other wildlife populations, to populations of bats were the subject of a 1999 workshop (O'Shea and Bogan 2003). The workshop concluded that forest-dwelling bats were most difficult to study and prospects for monitoring populations of these bats using conventional measures were limited. Since then, analytical techniques have been developed that may improve our ability to monitor populations of bats (MacKenzie et al. 2002; Manley et al. 2004; Tyre et al. 2003). The dearth of information on population status of many bat species, against the backdrop of widespread disturbance to forested ecosystems, requires creative solutions to assess the status of populations of bats. This chapter serves not only to review the challenges facing biologists who conduct inventories for forest-dwelling bats, but also to call attention to new approaches for assessing the population status of bats and highlight the need for additional innovative approaches for monitoring this important group of animals.

INVENTORY AND MONITORING BASICS
DEFINING TERMS

Terms such as inventory, monitoring, and population have been applied in a wide variety of situations and contexts and have different meanings to different individuals. Successful communication requires standard terminology and definitions for concepts (Morrison and Hall 2002). I offer the following definitions for important terms that are used throughout the chapter.

Survey

A survey is a set of qualitative or quantitative observations conducted during a specified time. Ideally, surveys follow a rigorous set of instructions or "survey protocol." Surveys may count the number of bats leaving a roost on a given night or create a list of the species of bats captured in mist nets at a given pond on a given night. In contrast, a census has been defined as a complete count of all individuals and can be distinguished from a survey, because a survey typically results in only a partial count (Thompson et al. 1998). Owing to the cryptic nature of forest-dwelling bats and the imperfect tools available to assess their presence and abundance, censuses are generally not possible.

Inventory

An inventory describes the state of a system at a given point in time. The intensity of effort will influence the accuracy and precision of an inventory. Because of the spatiotemporal variability with which bats use their habitats, an accurate inventory may require a series of surveys across space and time. For instance, creating an accurate list of species that occur in a hypothetical recreation area may require x surveys at each of y sites. In turn, a precise estimate of the number of bats using a roost may require several counts.

Monitoring

Monitoring assesses the change, or trend, in abundance, composition, or distribution of a resource over time. For instance, an investigator may be interested in assessing a change in the number of bats using a roost (Chung-MacCoubrey 2003) or the number of sample units occupied by a species across its range (Manley et al. 2004). Monitoring is distinguished from an inventory by including a temporal component with the collection of data and by the purpose of its implementation (Hellawell 1991). Where an inventory seeks to describe what is present at a given point in time, monitoring seeks to determine the extent and direction of change from some established norm over time. In an inventory, surveys are conducted without any preconception of what will be found, but monitoring presupposes that the investigator has an idea, however vague, of what will be found (Hellawell 1991). Indeed, frequently a collection of surveys will be used to understand the norm and variation expected (i.e., baseline inventory or monitoring), such that targets can be set and a rigorous monitoring program designed (Noon 2003). Although monitoring programs can be developed without baseline information, understanding inherent levels of variation allows for creation of a more efficient and relevant monitoring scheme (Hellawell 1991; Pollock et al. 2002).

Population

A population is an aggregation of individuals of the same species in a particular place at a particular time (Morrison and Hall 2002; O'Shea and Bogan 2003). It is up to the investigator to define a population, both in terms of taxonomy and size of study area (Morrison and Hall 2002). For instance, population status can be assessed across the range of the species or within a single watershed. Selecting the appropriate spatial scale at which to assess bat populations is difficult because neither their daily nor seasonal movements are limited to an easily defined parcel of land. Despite these challenges, investigators should define the target population, or the group of animals about which inference is desired, as explicitly as possible in terms of the area of interest and time period to which the inference applies (Morrison and Hall 2002). It is also important to articulate the assumptions used in defining the population.

Colony

Colony has conventionally been the term used to describe a group of bats. Implicit in this description is the expectation that bats that roost together are part of a cohesive social group. Even for relatively permanent roost sites, though, the collection of individuals using a roost may change according to time of day (day roost versus night roost), day (roost switching), season, and among seasons (summer versus winter use) (Barclay and Kurta, chap. 2 in this volume; Ormsbee et al., chap. 5 in this volume). Forest-dwelling bats often do not move between roosts as a cohesive group (Brigham et al. 1997; Kerth and Reckardt 2003; Weller and Zabel 2001), but a group of bats may use a series of tree roosts over the course of a season, coming in and out of contact with individuals within the group (Kerth and Reckardt 2003). Like the term population, a colony should be defined by the investigator with the assumptions clearly stated. Because both colony and population refer to a user-defined group of bats, they may often be synonymous.

ANSWERING IMPORTANT QUESTIONS

Successful inventory and monitoring programs address basic questions about the program prior to implementation (Hellawell 1991; Morrison et al. 2001; Noon 2003). Fundamental to these programs is a clear understanding of the reason for undertaking an inventory or monitoring program, and the goal one hopes to achieve.

Why?

The decision to conduct an inventory may be motivated by, among other things, the need to meet legal requirements, determine the effects of a stressor on the population of bats, or satisfy some biological curiosity. Whatever the motivation, it is important to define as clearly as possible the reason for undertaking an inventory. A clear definition will guide answers to other critical questions such as whether the inventory targets one or multiple species and whether data on absolute abundance, relative abundance, or species occurrence is required. It will also indicate whether a "snapshot" of the state of the system is desired or whether the inventory will be repeated in the future. Clearly defining and remaining mindful of the motivation for the inventory will help to provide accurate and meaningful answers to other important questions.

What?

The question of what to inventory has two components: the species of interest and the metrics used to describe the species of interest. The first component is usually to address one of two objectives: listing the species that occur in an area or assessing the status of a single species at a particular location (Weller, in prep.). Once the target species has been selected, the investigator must determine which metrics can reliably be used to assess the status of the species. Investigators should choose metrics that both satisfy the goals of the project and can be reliably measured with available inventory methods. Defining target species and metrics will often lead directly to answers to other important questions. For example, well-defined objectives such as estimating the number of bats that roost in a particular tree or listing the species that occur at a particular pond between June and September, provide specifics as to where and when the inventory will be conducted.

Where?

The study area also should be defined as explicitly as possible. Often this is determined by administrative boundaries, but in other cases the investigator defines the limits of the area in question. To improve the accuracy of the inventory, investigators must decide whether it is necessary to expand the study area beyond the specific area of interest. For instance, forest-dwelling bats frequently roost in upland habitats, but may be difficult to detect there. In this case, even though the area of interest is upland habitats, it may also be beneficial to survey adjacent water sources to more accurately document the species of bats using the upland habitats nearby.

When?

Decisions as to when surveys should take place depend on the objectives of the study. For instance, if the objective includes documenting the presence of juveniles, then surveys need to be conducted between the time of parturition and the end of the summer season. To date, most inventories of bat communities have been conducted during summer, so this is where I will focus my attention. Note, however, that with increasing interest and number of sightings of bats outside the summer months (Cryan and Veilleux, chap. 6 in this volume), there may be additional need to specifically define the season of interest and design inventories accordingly.

Often the more difficult question with regard to timing is how often to survey? Given the spatiotemporal variability with which bats use their habitat, additional surveys will almost always increase precision and reliability of an inventory. Unfortunately, resources are usually limited, so the goal is for inventories to be as efficient as possible. The appropriate number and allocation of surveys depend on the goals and scale of the project and will be discussed in greater detail later in the chapter.

How?

The final consideration is to determine the inventory methods that will be used to evaluate the metric of interest (e.g., abundance) for the target species at the places and times specified. For instance, acoustic surveys should only be used if the echolocation calls of the target species can be detected and accurately identified to species using bat detectors. When existing inventory methods do not provide a reliable means of evaluating the desired measure of the population, it may be necessary to revisit project goals or the choice of population metrics used.

ADDRESSING SPATIOTEMPORAL VARIATION

Spatial variation and detectability, which incorporates temporal variability, are the two largest concerns that need to be addressed in designing a monitoring program for any animal species (Pollock et al. 2002; Yoccoz et al. 2001). Spatial variation occurs when animals use multiple sites, and where observers can only monitor a subset of the population during the survey. For instance, bats often use multiple roosts, but rarely is it possible to survey all roosts simultaneously. The issue of detectability refers to the inability to detect all individuals or species of interest during a given survey. This may be due to limitations of inventory methods or variability in the number of individuals of a species at a particular location among surveys. Regardless of the methods used, it will be necessary to account for both temporal and spatial variation in habitat use of bats.

Temporal variation in bat activity occurs at multiple scales and has received greater attention than spatial variation (Hayes and Loeb, chap. 8 in this volume). Presence of a particular species and number of individuals at a site can vary seasonally (Kuenzi and Morrison 2003; O'Donnell 2000), nightly (Hayes 1997), or by time of night (Hayes 2000; Kuenzi and Morrison 2003; Miller 2003). Night-to-night variation is often the largest concern for inventory design and is illustrated by a hypothetical example (fig. 10.1). In the example, species A was only available for detection at the site during surveys 1 and 2, and failure to detect species A on surveys 3 through 6 was attributed to temporal variation in the use of the site. Clearly, sampling only between the dates of surveys 3 through 6 would yield a different outcome than completing all six surveys.

Evaluation of spatial variation in habitat use by bats has received far less attention than temporal variation (Duchamp et al., chap. 9 in this volume; Hayes 2000), but may be of equal or greater importance when

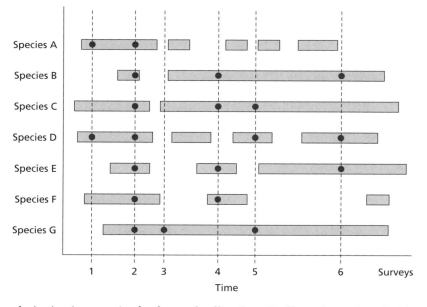

Species A

Species B

Species C

Species D

Species E

Species F

Species G

1 2 3 4 5 6 Surveys

Time

Figure 10.1.
Schematic illustrating the challenges in detecting bats in forests. Horizontal bars indicate periods when a species was present at the particular habitat feature surveyed, dashed vertical lines indicate points in time when surveys were conducted, and filled circles indicate when a species was detected by the inventory methods employed.

designing inventories for forest-dwelling bats (Mills et al. 1996). Individuals of the same species often use multiple roosts and foraging areas within an area (Barclay and Kurta, chap. 2 in this volume; Lacki et al., chap. 4 in this volume). As a result, it is difficult to select survey locations with any certainty of where a particular species or number of individuals will be present on a given night.

Whether such variation is categorized as spatial or temporal variation depends on the scale of interest and is a matter of semantics. For instance, if inference about the site depicted in figure 10.1 is desired, the absence of species A during surveys 3 through 6 is reasonably interpreted as temporal variation, but absence of species A on those nights also means that the species was somewhere else within a larger area. Therefore, if the scale of inference actually covers a much larger area than that contained by the site illustrated in figure 10.1, the movement of species A will be interpreted as spatial variation. For this reason, variation in habitat use is best referred to as spatiotemporal variation.

Spatiotemporal variation is unpredictable and best addressed by replicating surveys at the appropriate scale to address the objectives. For instance, multiple surveys of a site will improve the accuracy of the species list for that site (fig. 10.1). However, if the site is one of many within a larger area of interest, an identical number of surveys may be more effective if distributed among additional sites (Field et al. 2005; Link et al. 1994; Thompson et al. 2002). Although inventory accuracy will always improve with increased replication, resources for such work are often limited and means of improving inventory efficiency are needed.

INVENTORY TECHNIQUES

Inventories for bats are primarily conducted using three methods: capture surveys, acoustic surveys, and roost surveys. Each inventory method

Table 10.1. Measures of population status derived from common survey methods for bats

Type of survey	Measures of population status		
	Species occurrence	Abundance	Demographic parameters (e.g., survival)
Capture	*Yes*	*No* Probability of capturing individual animals has not been estimated.	*Limited* Index to basic measures (e.g., sex, juvenile ratios) Recapture rates too low in most situations
Acoustic	*Limited* All species not equally detectable or identifiable	*No* Individuals cannot be identified.	*No*
Tree roost	*No* Most species cannot be seen in roosts or visually identified in flight upon exit	*Limited* Generally from exit counts. Inference is to individual roosts unless most alternate roosts known	*No* Fidelity to individual roosts and recapture rates too low in most cases
Structure roost (e.g., mine or bridge)	*Yes* Species identification visual or by capture at roost	*Yes* Internal or exit counts Strongest inference where individuals can be marked/recaptured Inference is to individual roosts unless most alternate roosts known	*Yes* Where individuals can be marked/recaptured

yields data that can be used to evaluate one or more population metrics (table 10.1, fig. 10.2), and each of these metrics can be used to describe population status at a particular level of resolution. For instance, abundance is a more direct measure of population status than is spatial distribution, yet both are valid population metrics that depend on the type of data available to evaluate the particular metric (Thompson et al. 1998). The urgent need for reliable assessments of the status of populations of forest-dwelling bats requires that we understand both the potential and limitations of current inventory methods. Below I review the contributions that existing inventory methods make in evaluating population metrics of bats.

CAPTURE SURVEYS

Capture surveys are an essential technique for determining the species that occur in an area and assessing presence of bats in particular sex, age, or reproductive classes (table 10.1). Although bats are often captured at their roosts or hibernacula to meet a variety of objectives (Kunz et al. 1996a; Sedgeley and O'Donnell 1999; Sendor and Simon 2003), most capture surveys of bats are directed at free-flying individuals away from their roosts, so I will focus on this application.

Reliable estimates of abundance or demographic parameters are not currently possible using capture surveys away from roosts because recapture rates are low and the probability of recapturing an individual bat

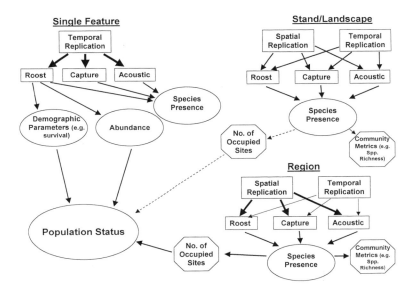

Figure 10.2.
Relationships between inventory methods, spatial and temporal replication, and metrics used to describe the status of bat populations at three spatial scales. Strength of relationship is indicated by thickness of arrow. Single feature refers to a discrete location where surveys are conducted (e.g., pond). Landscapes contain multiple features and regions contain multiple landscapes.

is unknown (O'Shea and Bogan 2003). For abundance to be estimated from capture surveys, the probability of detecting an individual bat must be determined. The best estimates would come from mark-recapture studies, but this is not possible in most situations because of poor rates of recapture, especially at locations away from roosts. Even when large numbers of bats have been marked at hibernacula, recapture rates away from the roost are low (O'Shea et al. 2004). Capture surveys of forest bats typically occur away from roosts, count an unknown proportion of the individuals present, and do not specify the area that was counted (Sauer 2003). As an alternative, a strong index to abundance could be achieved if it was possible to estimate the proportion of individuals in an area that were captured. The prospect of developing a correction factor by combining capture methods with other observational techniques, such as infrared video (Lang et al. 2004) or thermal imaging (Frank et al. 2003; Sabol and Hudson 1995), deserves future consideration.

Bats are also captured to obtain demographic information such as sex ratios (Miller 2003; Schulz 1999), juvenile ratios (Miller 2003), proportion of reproductively active individuals (Barclay et al. 2004), and timing of reproduction (Grindal et al. 1992; Lewis 1993; Miller 2003). Such measures represent an index to the true measure of the population and contain the implicit assumption that capture probabilities are equal among the classes compared (e.g., probability of capturing a juvenile is the same as an adult). Variation in capture probability according to sex, age, or reproductive class is poorly understood but may be biased against certain classes of individuals. Annual comparisons of such ratios are somewhat less susceptible to these biases, in particular, for large sample sizes and for comparisons within the same study area by the same investigator. For example, the ability to capture juveniles of a particular species probably does not vary greatly on an annual basis. Nevertheless, such comparisons should clearly state assumptions of the index. Where it is important to simply document areas where a particular sex is present, or where reproduction is occurring, capture surveys are the only method of obtaining such information.

Probably the most important use of capture surveys is to document species presence. Capture surveys often detect species not recorded with acoustic methods (Duffy et al. 2000; Kuenzi and Morrison 1998; O'Farrell and Gannon 1999), and captured individuals can often be reliably identified using external morphologic characters. In contrast, certain pairs of species can be difficult to distinguish from each other using morphologic features, but new molecular techniques are likely to change how we view and study closely related species of bats (McCracken et al. 2000; Rodriguez and Ammerman 2004; Zinck et al. 2004). Whenever possible, voucher or genetic samples should be taken from captured individuals so that species identity can be accurately confirmed.

The two most prevalent techniques for capturing bats away from their roosts are mist nets and harp traps (Kunz et al. 1996a). Comparisons in Australia and Malaysia have indicated that harp traps were more effective than mist nets for capturing small species and resulted in a larger number of individuals and species captured (Francis 1989; Kingston et al. 2003; Tidemann and Woodside 1978). Use of harp traps, as opposed to mist nets, has also been credited with rediscovery of the golden tipped bat (*Kerivoula papuensis*) in Australia after it was thought to be extinct (Schulz 1999). In North America, the nearly universal use of mist nets (Weller, in prep.) probably reflects their perceived effectiveness. Comparisons of the efficacy of harp traps and mist nets for capturing North American species in different situations may indicate species-specific biases of each device which could then be used to improve capture rates. Similarly, studies that evaluate the number, size, and configuration of mist nets to maximize inventory effectiveness would be helpful.

Regardless of the device employed, the success of capture surveys is limited by the ability of an investigator to intercept a flying bat with a device that represents only a tiny fraction of the three-dimensional space within the home range of a bat (O'Farrell and Gannon 1999). Therefore, capture of bats is most effective in situations where bats tend to aggregate and are accessible, such as water sources or travel corridors along roads and trails (Kunz et al. 1996a). Bats have also been captured along the edge of clearcuts (Brigham et al. 1997) and in forest gaps (Crampton and Barclay 1998, Tibbels and Kurta 2003); however, comparisons of capture rates at such features relative to more conventional sites, such as trails or water sources, have not been made. Placing nets in the canopy has been effective for capturing Indiana myotis (*Myotis sodalis;* Gardner et al. 1989) and the New Zealand long-tailed bat (*Chalinolobus tuberculatus;* Sedgeley and O'Donnell 1999). Mist nets within the forest interior were more effective at capturing northern myotis (*Myotis septentrionalis*) than those set at more conventional sites, such as streams and roads (Carroll et al. 2002). Other species-specific recommendations have not been made for North American bats, but capture success may be improved through an understanding of the interaction between physical characteristics of a site and the ecomorphology of the species to be detected (Aldridge and Rautenbach 1987; Crome and Richards 1988). For example, capture surveys along trails or small streams are more likely to capture species that tend to fly relatively low and slow in more cluttered situations, than species that tend to fly high and fast across open habitats. The habitat feature selected for a survey will influence the number and species of bats captured, and the most complete list of bat species will likely result from sampling across different types of habitat features (e.g., canopy, pond, forest interior) where capture devices are deployed.

ACOUSTIC SURVEYS

Acoustic surveys, which use bat detectors to monitor echolocation calls of bats, have been used extensively in recent years to meet a variety of objectives. They offer a number of advantages for use in field studies because they can be used to survey bats in places where capture methods are ineffective and can be configured to collect data automatically without the necessity of on-site field personnel (Corben and Fellers 2001; Hayes and Hounihan 1994). However, the age, sex, or reproductive condition of free-flying bats cannot be determined using existing acoustic devices. In laboratory situations, echolocation call characteristics have been demonstrated to vary according to sex (Jones et al. 1992), age (Kazial et al. 2001; Masters et al 1995), and reproductive status (Grilliot et al. 2004) of the individual bat producing the call. Similar studies have not been conducted in forested habitats, and it is unlikely in the near future that sex or age differences will be able to be distinguished from other sources of intraspecific or interspecific variation in echolocation call structure of free-flying bats.

Abundance cannot be determined from echolocation data because detection of multiple bats, each recorded once, cannot be distinguished from a single bat recorded multiple times (table 10.1). Hence acoustic data may be useful to describe use of an area by a particular species, but not the number of bats using that area (Hayes 2000; Miller et al. 2003). Future developments that allow calibration of the number echolocation calls against a more reliable measure of abundance would improve the prospect for use of bat detectors to assess populations, in particular, at the local scale.

This leaves documenting species occurrence as the only useful metric that can be derived from acoustic surveys. In this respect, acoustic monitoring offers a number of potential advantages over other inventory methods. A much wider variety of habitats and habitat features can be surveyed with bat detectors than capture devices. For instance, open meadows and lakes are ideally suited to acoustic inventories because they often have high levels of bat activity, but capturing bats in such locations is difficult. The ability to conduct acoustic surveys in a wider variety of situations, and thus survey additional habitat niches, may partly explain why acoustic surveys detect species that are often not captured (O'Farrell and Gannon 1999). Additionally, acoustic inventories can be conducted with little disruption to the normal behavior of bats. Analogous to capture surveys, acoustic surveys are limited by the need to "intercept" the call of a bat in the course of flight. Bats must pass relatively near to the detector microphone to be recorded. The distance and angle between the detector microphone and the bat affects call quality and, consequently, the ability of an investigator to identify the species of bat that produced the call (Corben and Fellers 2001; O'Farrell et al. 1999a; Waters and Gannon 2004).

When acoustic surveys are used to document species occurrence it is important to remember that echolocation calls of different species are not equally detectable (Fenton 2003; O'Farrell and Gannon 1999). Different species echolocate at different intensities and those with greater intensities will be recorded more easily and over greater distances. In general, larger, fast-flying species (e.g., hoary bats, *Lasiurus cinereus*) echolocate at higher intensity, whereas smaller, slow-flying species (e.g., northern myotis) echolocate at lower intensity (Aldridge and Rautenbach 1987). Moreover, species of bats may vary the intensity of their echolocation calls to accomplish different tasks (Fenton 2003). Therefore, acoustic surveys will not necessarily result in detection of all of the species present at the survey location, rendering interspecific comparisons of activity problematic (Hayes 2000).

Perhaps the greatest limitation with the use of acoustic surveys to document species occurrence is the uncertainty surrounding species identification based on echolocation call characteristics (Barclay 1999; Hayes 2000). Many North American species have been putatively identified by their echolocation calls (Fenton et al. 1983; O'Farrell et al. 1999b), but concern over reliability of acoustic identification means that recorded calls are often not accepted as proof of species presence. For instance, in the Indiana myotis draft recovery plan, echolocation calls are not acceptable for establishing the presence of Indiana myotis (U.S. Fish and Wildlife Service 1999). Gannon et al. (2003, 57) identified species of free-flying bats from their echolocation calls to assess habitat use, but provided the caveat that "any study that produces an extension of the range of a species, suggests management needs, or even documents new aspects of natural history should not be based on acoustic data alone without some form of verification or elaboration." This sentiment portends a lack of confidence in species identification with use of echolocation data.

Several factors influence the ability to identify bats from their echolocation calls, including choice of detection system, intraspecies variability in call design, and difficulty in obtaining representative calls under controlled situations (Fenton 2003; Parsons and Jones 2000). Contributions to intraspecific variation include the age and sex of bats recorded (Jones et al. 1992; Kazial et al. 2001; Masters et al 1995) and the presence of conspecifics (Obrist 1995). Geographic variation in call structure across the range of a species may also contribute to intraspecific variation (Barclay 1999; Thomas et al. 1987), and its importance to the study of bats is the subject of debate (Murray et al. 2001; O'Farrell et al. 2000). The sur-

roundings in which an individual bat is flying may be the largest con-
tributor to intraspecific variation and the greatest impediment to identification of bats using echolocation. Bats alter their calls as they move
among areas with varying degrees of habitat clutter (Broders et al. 2004;
Obrist 1995), and the magnitude of this variation may be sufficient to
mask differences that would allow discrimination among species (Tibbels 1999). Therefore, species identification from echolocation calls requires not only an understanding of intraspecific variation of the species
of interest, but also that of sympatric species with which it may be con-
fused (Broders et al. 2004). Investigators must realize the limitations to
such work and be willing to forgo identification to the species level when
necessary (Barclay 1999; Hayes 2000).

Creating call libraries

Understanding the variation in the call repertoire of a given species requires a large sample of calls from individuals that have been identified
by other means (e.g., capture, observation). A collection of such calls for
the species in a particular area is commonly referred to as a reference call
library. Constructing such a call library has a number of considerations
(Waters and Gannon 2004). Limitations have been noted for most of the
common methods used to obtain reference calls including recordings of
bats flying in rooms or other enclosed spaces (Barclay 1999; Mukhida et
al. 2004), hand-released bats (Barclay 1999; O'Farrell et al.1999a), and
bats leaving roosts (Jones et al. 2000; Murray et al. 1999). Attaching
chemiluminescent tags to bats and recording their calls as long as possible after release, within habitats where the species is likely to be encountered, is currently the best means of recording "natural" calls (Broders et
al 2004; Murray et al. 1999). Nevertheless, the low proportion of individuals from which echolocation calls are recorded using this method often prompts reevaluation of the costs, in terms of money, personnel, and
stress to the bat, relative to the benefits gained in obtaining representative calls with its application. Innovative approaches for reliably collecting representative reference calls are needed to advance progress in identification of species using echolocation calls.

In general, it is recommended that reference calls be recorded from
bats in the study area because of the potential for geographic variation
(Barclay 1999). Confidence in the description of intraspecific variation
will increase with sample size for each species. Duffy et al. (2000) quantitatively established that 38 reference calls were necessary to capture
intraspecific variation in characteristic frequency and duration of *Chalinolobus morio* at their study site in Australia. Murray et al. (1999) recommended a sample of 30–100 individuals from a small geographic area,
and Parsons and Jones (2000) used 12–96 samples for each of 15 species
in Britain.

The effort necessary to build a call library for each project, with robust sample sizes for each species across a variety of habitat and clutter
conditions, is clearly an obstacle to documenting species occurrence using acoustics. The necessity of local call libraries is founded on the as-

sumption that geographic variation is a significant component of intraspecific variation. If geographic variation is demonstrated to be a minor component of intraspecific variation (Murray et al. 1999; O'Farrell et al. 2000), then calls from a wider geographic area could be pooled to create more comprehensive reference libraries (Duffy et al. 2000). The need for a repository of echolocation calls recorded from individuals that have been identified by other methods (e.g., capture) has been recognized for quite some time and there have been several attempts to establish such a repository for North American species (Waters and Gannon 2004). Relatively few bat biologists have deposited "known calls" in such repositories, however, and maintenance and quality control of these repositories requires a commitment of time and funding that has so far not been sustained. Nevertheless, the benefits of a rigorously screened, consistently updated, regional repository should not be dismissed; in particular, when compared with the level of effort necessary for individual investigators to create reliable libraries for specific study areas.

Identifying species

Creating a representative library of calls is only an interim step toward the ultimate goal of identification of free-flying bats via a comparison with that library. A number of methods have been proposed to conduct such comparisons: qualitative identifications that include visual cues observed in the field (Ahlén and Baagøe 1999; O'Farrell et al. 1999b), use of dichotomous keys (Beck and Peterson 2003; Fenton et al. 1983), and quantitative or statistical methods (Murray et al. 1999; Parsons 2001; Parsons and Jones 2000; Vaughan et al. 1997). Reliable information can only be obtained when the process used to identify the call is repeatable (Barclay 1999).

Dichotomous keys use quantitative information measured from the calls (e.g., minimum and maximum frequency, duration) to define differences among species in a repeatable manner (Beck and Peterson 2003; Fenton et al. 1983). Use of such keys also allows for reevaluation of identifications as more information about echolocation in a given species becomes available. Depending on the species composition of an area, some species of bats will be easily identified, whereas call parameters for other species may overlap widely making it difficult to assign calls to a species. Automated filters that assign calls to species in a repeatable manner have been discussed (Corben and Fellers 2001) but are not in widespread use.

Methods such as discriminant function analysis (Gehrt and Chelsvig 2004; Krusic and Neefus 1996; Tibbels and Kurta 2003; Vaughan et al. 1997) and artificial neural networks (Broders et al. 2004; Parsons 2001; Parsons and Jones 2000) have been tested and allow assignment of echolocation calls to species for individuals that have been identified by other methods (e.g., capture). Both techniques assign a probability that a given echolocation call was produced by a certain species and have been used to identify free-flying bats (Gehrt and Chelsvig 2004; Tibbels and Kurta 2003; Vaughan et al. 1997; Wickramsinghe et al. 2003). An artificial neural network has improved performance over discriminant function analysis in the ability to discriminate among species of British bats (Parsons and Jones 2000). Such quantitative approaches are rigorous and repeatable methods for identifying species and should be encouraged.

Some species will be easier to discriminate from others and have correspondingly higher classification rates (Murray et al. 1999; Parsons and Jones 2000). Wickramasinghe et al. (2003) assigned calls to individual species when confidence levels were at least 85%, but individual investigators will need to establish the level of confidence with which they are comfortable. The selected standard should depend on the cost of misidentifying the species (Barclay 1999). Note that quantitative tools were developed by individual researchers working in specific areas and are applicable only to those species and areas. Development of additional models based on large samples of known calls collected at regional scales would greatly increase the applicability of these techniques.

ROOST SURVEYS

Several important inventory and monitoring objectives for bats can be achieved using surveys at roosts, including estimates of demographic parameters, abundance, and documenting species occurrence (O'Shea and Bogan 2003; Sendor and Simon 2003; Warren and Witter 2002). Such techniques are most effective at roosts where bats are easy to both capture and recapture, the number of individual bats is low to moderate, and the number of alternate roosts used by the colony are few and well known (Kunz 2003; Warren and Witter 2002). Most forest-dwelling bats roost in relatively inaccessible places, such as cavities, cracks, or beneath the bark of tall trees and snags, and they move frequently among alternate roosts (Barclay and Kurta, chap. 2 in this volume; Kunz and Lumsden 2003). Therefore, mark-recapture techniques necessary to estimate demographic parameters or achieve precise abundance estimates are currently not feasible for forest-dwelling bats in most situations (Kerth and Reckardt 2003).

Tree roosts

Because bats generally cannot be directly observed in tree roosts, abundance estimates are instead based on counts of individuals exiting the roost (Barclay and Brigham 2001; Chung-MacCoubrey 2003). The number of bats exiting an individual tree roost can vary widely because of low fidelity to both individual trees and conspecific roost mates (Brigham et al. 1997; Chung-MacCoubrey 2003; Weller and Zabel 2001). Although bats may return to the same roost area annually (Kurta and Murray 2002), and in some cases reuse individual trees (Chung-MacCoubrey 2003; Willis et al. 2003), the chance of seeing bats emerge from a given tree roost on a given night is low in many areas, even when the tree has been previously confirmed as a roost (Barclay and Brigham 2001). Furthermore, observation of a single roost only provides information on use of that roost. Estimation of abundance requires knowledge of the location of the majority of the roosts, such that all roosts could be chosen for survey with a specified probability of selection (O'Shea et al. 2004). The number of roosts used by an individual bat, even over the course of a single summer, is unknown for most species (Kerth and Reckardt 2003; O'Donnell and Sedgeley 1999), making simultaneous roost counts untenable in most situations.

Surveys at tree roosts are also not an effective means of documenting species presence in an area (table 10.1). Most tree roosts are located by radiotracking a previously captured and identified bat to its roost. If a roost is located without radiotelemetry, identifying the species exiting the roost is difficult in most situations (Kunz 2003). The notable exception is where bats roost in the basal hollows of large trees (e.g., bottomland hardwoods, redwoods), and where species and number of individuals can often be determined by direct observation or the presence of guano (Gellman and Zielinski 1996; Lance et al. 2001; Mazurek 2004). Because of their relative accessibility within the roost, basal hollows may also represent one of the few opportunities for application of mark-recapture techniques in tree-roosting species.

Other roosts

Although they are not unique to forests, structures such as bridges, buildings, caves, and mines occur within forests and are used as both day and night roosts by some species of bats (Ormsbee et al., chap. 5 in this volume). Structures offer opportunities to estimate abundance and sometimes the demographic parameters of bat populations, because bats that use them as roosts are generally easier to observe (Kunz 2003) and exhibit greater fidelity to them than to roosts in trees (Lewis 1995). These structures can also be a valuable resource for establishing the presence of species that would require much greater effort using capture or acoustic surveys (e.g., Rafinesque's big-eared bat, *Corynorhinus rafinesquii;* Lance et al. 2001), especially when species can be identified from guano (Zinck et al. 2004). Therefore, when roosts in these structures occur within forested landscapes they should be used to enhance the capability for assessing bat populations.

Demographic parameters (e.g., survival, longevity) have been estimated at roosts where animals can be marked and recaptured. For example, survival has been estimated for bats hibernating in buildings, caves, and mine tunnels (Hitchcock et al. 1984; Keen and Hitchcock 1980; Sendor and Simon 2003), and for bats roosting in buildings (Neubaum et al. 2005; O'Shea et al. 2004). A recent review evaluated the field methods and analytical approaches used to evaluate survival of bats (O'Shea et al. 2004). Historically, bats have been individually marked using forearm bands. However, bands can have deleterious effects on bats and should be used with extreme caution (Baker et al. 2001; O'Shea et al. 2004). Freeze branding may represent an improved alternative to forearm bands (Sherwin et al. 2002), but this technique has not been widely tested. Passive integrated transponder (PIT) tags were successfully used in several recent studies (Kerth and Reckardt 2003; O'Shea et al. 2004) and show great promise for marking bats at roosts. PIT tags involve subdermal injection of small transponders that are activated only when the tag passes a reader at very close range. Reports have indicated that PIT tags did not have deleterious effects on bats and tag malfunction rates were low (O'Shea et al. 2004). Perhaps the greatest benefit to the use of PIT tags is that bats can be "resighted" without having to capture and handle the bat. One limitation to the use of PIT tags is that a bat must pass in close proximity to the reader. Readers are most effective when roost entrances are small, because bats are forced to pass near to the reader. Reliable estimates of abundance or survival require relatively high resighting rates, so the number of roosts and roost entrances must be relatively few for this technique to be cost-effective.

Abundance estimates can also be achieved where bats can be reliably counted. Counts can occur either within the roost or as bats emerge from the roost. Emergence counts at 79 maternity roosts in buildings were used to estimate the population of lesser horseshoe bats (*Rhinolophus hipposideros*) in Wales, U.K. (Warren and Witter 2002), and both internal and external counts at 184 colonies in buildings were used to estimate population sizes for three species of bats in Scotland (Speakman et al. 1991). Use of roost counts to estimate abundance requires that bats show high fidelity to individual roosts and the majority of roosts for the population are known (Warren and Witter 2002). In general, these assumptions are more applicable to the large roosts in structures (e.g., buildings, caves) than for the smaller, more ephemeral roosts often used by forest-dwelling bats.

When there are a known number of occupied roost structures and an unknown number of undiscovered roost structures, the dual-frame method represents a means of estimating abundance across the area of interest (Haines and Pollock 1998). The dual-frame method uses two sampling frames: the list frame, which in the case of bats would contain all previously known roosts, and the area frame, which is used to describe the geographic boundaries of the area of interest. Random sampling is conducted in both sampling frames. In the list frame previously known roosts are surveyed to estimate the number of individuals, whereas the area frame would be sampled to locate and survey additional roosts, not included in the list frame. Separate population estimates are generated for each sampling frame and summed to estimate the total population. In subsequent years, roosts located in the area frame are included in the new list frame and the area frame is again used to locate new roosts. Although the dual-frame method was demonstrated to be a robust and efficient option for estimating the population size of bald eagles (*Haliaeetus leucocephalus;* Haines and Pollock 1998), there are several caveats to using it with forest-dwelling bats. First, the approach is designed for use on highly visible aggregations of animals with high year-to-year fidelity to roosts (Haines and Pollock 1998). Hence, the dual-frame method is likely to be most applicable to structure roosts rather than to tree roosts. However, it may be worth exploring the use of the method where tree roosts are limited and year-to-year reuse of individual trees has been noted. Second, status of bald eagle nests can be established reliably with relatively little effort in comparison with the effort necessary to achieve a reliable count of bats using a roost, especially given the temporal variability with which bats use even relatively permanent roost structures (Sherwin et al. 2003).

COMBINING TECHNIQUES

Species occurrence can be documented using most of the common inventory methods for bats (table 10.1, fig. 10.1). Although certain inventory methods may be highly effective and reliable for detecting particular species, the use of a single technique will usually not be effective for detecting all the species in an area. Several studies have compared the utility of acoustic and capture surveys for detecting species presence in an area and, without exception, have concluded that a combination of methods is the best way to assess species richness (Duffy et al. 2000; Kuenzi and Morrison 1998; Murray et al. 1999; O'Farrell and Gannon 1999). Inclusion of roost searches when conspicuous roosts are available should increase the efficiency of the inventory (Kunz et al. 1996b).

TECHNOLOGICAL ADVANCEMENTS

One of the common themes of the 1999 workshop on monitoring bat populations (O'Shea and Bogan 2003) was the hope that future technological advancements would increase monitoring capabilities in the way that bat detectors and miniature radiotransmitters have helped expand the knowledge of the ecology of bats. Chief among the needs were improvements in the ability to estimate abundance. Roosts are the easiest place to improve abundance estimates, as the time and place of exit can be reliably predicted and observed. PIT tags appear to be a safer alternative to banding bats (O'Shea et al. 2004) and their expanded use will likely increase the number of studies that employ modern mark-recapture approaches to obtain reliable estimates of capture probability, abundance, and survival. Reductions in cost of PIT tag readers would expand the ability to monitor bats that use multiple tree roosts. Improvements in PIT technology that allow tags to be read at greater distances, such that free-flying bats might be "resighted" without having to be captured, would improve prospects for estimating abundance in specific areas.

Improvement and cost reduction of infrared thermal-imaging technology (Frank et al. 2003; Kunz 2003; Sabol and Hudson 1995) will likely result in expanded use of this technique in the study of bats, in particular, at roosts. Similarly, radar has been used to track the movements of individual birds and bats (Reynolds et al. 1997). Both of these techniques are most effective where animals are depicted as high-contrast objects against a homogenous background (e.g., the sky), a condition that can be difficult to achieve in forested habitats. Expanded use of these technologies may improve the ability to estimate number of individuals in an area or serve as correction factors that calibrate the proportion of individuals captured or detected acoustically. However, like other technological advancements, they may not be the "silver bullets" they were originally perceived to be (Fenton 2003). Technological advancements are often eye catching and costly but may not address the root problem (Sauer 2003). A clear assessment of the benefits and limitations of such tools relative to inventory goals should be made before their investment or implementation.

SPECIES OCCURRENCE AS A POPULATION PARAMETER

This review of currently available inventory methods supports the conclusion from the 1999 workshop on monitoring bat populations (O'Shea and Bogan 2003) that, in most circumstances, species occurrence is the most reliable measure of population status that can be achieved for forest-dwelling bats. Presence data can be used to meet a variety of objectives: assessment of community metrics such as species richness and diversity (Moreno and Halffter 2000), modeling habitat associations (Jaberg and Guisan 2001), and assessment of population status, in particular, at large scales (Manley et al. 2004). Although species occurrence is not a direct estimate of population size it can be used to assess changes in distribution, which is a valid measure of population status (Telfer et al. 2002, Thompson et al. 1998).

Importantly for forest-dwelling bats, the information required (species occurrence) to assess population status with these approaches can be collected using currently available inventory methods. Given the limitations of available methods to provide more intensive measures of population status at the local level, such an approach has been recommended as an "early warning system" to monitor bat populations at large spatial scales (Ahlén and Baagøe 1999; O'Shea and Bogan 2003; Weller et al. 2002). Some of the challenges associated with monitoring large areas over extended periods, including conducting surveys in a wide variety of habitat types and use of a large number of personnel with varying skill levels, are ameliorated by the requirement for relatively simple information such as species presence (Usher 1991). Some researchers are concerned, however, that reliance on presence data would indicate potential rather than actual population trends (O'Shea and Bogan 2003).

Since the 1999 workshop on monitoring bat populations (O'Shea and Bogan 2003), modern analytical techniques have been applied to other taxa that use presence/not-detected data to estimate site occupancy across large areas (Gu and Swihart 2004; MacKenzie et al. 2002; Tyre et al. 2003). These analytical approaches provide sound inference because they come with strong theoretical underpinnings and allow assessment of precision for estimates. Hence, these approaches provide an alternative means of rigorously assessing population status that avoids the cost and difficulties associated with estimating abundance, in particular, at large scales (MacKenzie et al. 2003; Manley et al. 2004; Royle and Nichols 2003).

The parameter of interest in these approaches is the proportion of sites occupied by an individual species (Gu and Swihart 2004; MacKenzie et al. 2002; Tyre et al. 2003). Because of the spatiotemporal variability with which bats use their habitat and the limitations of current inventory methods, species present during a survey often go undetected (fig. 10.1). Negative surveys are therefore more appropriately termed "not-detected" rather than "absent." The issue of nondetection during a single survey is addressed by conducting multiple surveys at individual sites to create a capture history for each species at each site. Maximum likelihood estimation is then used to simultaneously estimate the probability that a site was occupied and the probability that the species was detected given that it was occupied (MacKenzie et al. 2002; Tyre et al. 2003). In other words, these analytical approaches extend our ability to estimate the number of sites occupied beyond those where a species was observed in the field (MacKenzie et al. 2002; Tyre et al. 2003).

Application of these methods requires at least two repeat surveys of individual sites, but increasing the number of surveys improves estimates of detection probabilities (Field et al. 2005). Previous work focused largely on developing the analytical methods and application was demonstrated on a limited number of species whose biology was well understood and for which well-developed survey protocols existed (MacKenzie et al. 2002, 2003). Where such an approach was applied to multiple species, meaningful estimates were achieved for only the 19 most frequently detected woodland birds and 2 of 14 species of forest frogs, because of the low number of sites with detections (Tyre et al. 2003).

The feasibility of this approach for monitoring a change in the number of sites occupied by bats was evaluated in central Sierra Nevada (Weller et al. 2002). Ninety-six sites were surveyed two to eight times each during 2001 and 2002 by using mist nets to detect the presence of multiple bat species. Eight of eleven species were captured frequently enough to estimate probabilities of site occupancy and detection. Estimates for proportion of occupied sites were generally two or more times higher than the proportion of sites where a species was captured in the field. Probability of detection for the eight species with a single visit to a single site ranged from 0.07 to 0.544 and increased as $1 - (1 - p)^n$, where p was the conditional probability of detection given that it occurred at a site and n was the number of surveys. For example, with three surveys the probability of detecting the most frequently detected species improved to 0.91. Although this particular effort was limited to mist net results, it is equally applicable when species are detected by acoustic surveys. If capture and acoustic methods were combined, detection probabilities would undoubtedly increase for species that can be identified acoustically.

Results from the Sierra Nevada study also illustrate one of the limitations inherent in these approaches: estimators perform poorly for species with low probabilities of detection (Tyre et al. 2003). Although this poses problems for monitoring rare species that are often the subject of conservation and monitoring efforts (Manley et al. 2004; Pierson and Rainey 1998; Schulz 1999), there is a growing interest in monitoring common species (Agosta 2002; Manley et al. 2004), which would benefit greatly from the application of these methods. Where rare or difficult to detect species are of interest, conducting additional surveys at habitat elements where the species is most likely to be found will improve estimates of detection probability (Manley et al. 2004).

Another disadvantage of these approaches is that their application is limited to the large-scale monitoring contexts for which they were designed (MacKenzie et al. 2002). Even at large scales, multiple surveys of individual sites and of a large number of sites will be necessary to achieve reliable estimates, in particular, for species with low-detection probabilities (MacKenzie et al. 2002; Tyre et al. 2003). However, the difficulties of assessing population status at smaller scales, the potential that changes in local populations could be offset by counter trends in nearby areas, and the urgent need for some measure of the population status of forest-dwelling bats (O'Shea and Bogan 2003), suggest these approaches deserve further attention.

ALLOCATING EFFORT AT MULTIPLE SPATIAL SCALES

Decisions regarding allocation of effort between spatial and temporal replicates are dependent on the scale of interest. For instance, when a single habitat feature (e.g., roost, pond) is under consideration, spatial replication is not applicable and precision of the estimate improves solely as a function of temporal replication (fig. 10.3a). However, when the scale of interest is an area that contains multiple habitat features, surveys should also be conducted at multiple habitat features (i.e., spatial replication). Several studies of other taxa have concluded that increasing the number of sites surveyed is preferable to increasing the number of surveys at each site (Colwell and Coddington 1994; Field et al. 2005; Link et al. 1994; Thompson et al. 2002), and this has held true for bats (Mills et al. 1996; Weller et al. 2002). Thus, for very large areas, spatial replication should be emphasized at the expense of temporal replication. Because such sampling schemes will inevitably be carried out over multiple nights they will indirectly incorporate temporal replication as well. Allocation of spatial and temporal replicates at the extremes of spatial scale is relatively straightforward: single features can only accommodate temporal replicates and large regional areas should emphasize spatial replicates. However, the scale of most projects (e.g., landscape) will fall somewhere between these extremes and the appropriate balance between spatial and temporal replication is often unclear. The appropriate balance in such cases will likely be influenced by more practical considerations such as budget, availability of equipment and personnel, and the number of habitat features in the area that are amenable to bat surveys (Field et al. 2005).

Decisions regarding the allocation of survey effort are also strongly influenced by the goals and scale of the inventory. Conducting one to a few visits at individual habitat features may be the most appropriate sampling scheme to characterize a large area. However, such surveys are unlikely to allow reliable inference about the individual habitat feature surveyed, due to inadequate temporal replication at the feature (fig. 10.3b). Likewise an inventory of the species present at a particular habitat feature, regardless of temporal replication, will provide limited inference about the surrounding landscape or region. In addition to strength of inference, cost efficiency should be weighed when allocating survey effort among spatial and temporal replicates. The cost of surveying new sites is generally higher than additional surveys of existing sites (Field et al. 2005; Link et al. 1994).

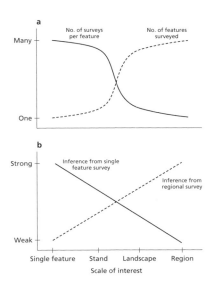

Figure 10.3.
Relationships among spatial replication, temporal replication, and inference at multiple spatial scales: (a) number of surveys at a single habitat feature (temporal replication) and number of habitat features surveyed (spatial replication) to maximize accuracy and efficiency of an inventory at a given spatial scale; (b) strength of inference from an inventory at one spatial scale that can be drawn to other spatial scales, assuming that an appropriate level of spatio-temporal replication has been applied during the original inventory.

CONCLUSIONS

Technological advancements such as miniature radiotransmitters and portable, inexpensive bat detectors have resulted in a rapid expansion of investigations into the ecology and habitat requirements of bats. Similar gains have not been achieved in understanding the status of their populations. One of the reasons for this is that metrics used to monitor populations of other taxa (e.g., abundance) cannot currently be estimated for most bats (O'Shea and Bogan 2003). For instance, except in rare situations that allow bats to be marked and regularly recaptured or resighted (Kerth and Reckardt 2003; O'Donnell and Sedgeley 1999; U.S. Fish and Wildlife Service 1999), assessment of abundance or vital rates for most populations of bats is generally not feasible. Nevertheless, many investigators have applied the available inventory methods with the vague hope that information obtained will yield a better understanding of local population status. Although such practices have recently been discouraged (Fenton 2003; Hayes 2000; Miller et al. 2003; O'Shea and Bogan 2003), a viable alternative has not been offered to the well-intentioned biologist. It seems that many bat biologists are resigned to wait for new or improved technologies that will allow the monitoring of bat populations to begin in earnest. In the meantime, decisions that affect both bats and their habitats are made in the absence of even the most basic information on the status of bat populations at any scale.

The major impediment to applying existing inventory methods to estimate abundance of bats is that currently no means of estimating detection probabilities for individual animals are available. Recent innovations that use presence data to assess population status provide an alternative that may allow progress in our ability to monitor the status of bat populations. Such techniques use multiple surveys to estimate detection probabilities for individual species. This approach is a trade-off that favors increased statistical rigor using a lower-resolution metric (species occurrence) over more conventional metrics (e.g., abundance) that are difficult to assess for forest-dwelling bats. The result is a rigorous estimate of the number of occupied sites and the ability to monitor changes in that proportion over time across large areas. Although it is unlikely that a single individual or institution would have the resources to apply such an approach at the largest scales, numerous inventories in forested habitats using similar inventory tools are conducted each year to address local needs. If the protocols for these inventories could be standardized to an extent that would allow results to be pooled at larger spatial scales, they could be used effectively to monitor population status of bats. The need for standardization of methods and centralized databases has been recognized (O'Shea and Bogan 2003), but there has been little progress toward achieving these ends. Standardization does not necessarily mean rigid guidelines that do not allow individual investigators the flexibility to adapt to local conditions or achieve individual objectives. For instance, the Indiana myotis mist-netting guidelines are only two pages long (U.S. Fish and Wildlife Service 1999), but simply prescribing the number of nets per kilometer of stream and the minimum level of effort to expend is enough to allow pooling of data among sites and studies. Although the standards prescribed in the Indiana myotis survey protocol may not have been based on empirical evidence or data analysis, simply having a mutually agreed upon, or legally mandated, standard has resulted in data that can be pooled for analysis.

Clearly, administrative, institutional, and even interpersonal hurdles will be associated with the development of standardized methods and centralized databases. Nevertheless, surmounting these relatively well-defined obstacles to obtain the information necessary to assess status of bat populations may be imminently more achievable than waiting for technological advancements that allow more proximate population metrics to be estimated. Pooling inventory data at large scales would likely identify only large magnitudes of population change or suggest areas where more intensive effort should be applied (O'Shea and Bogan 2003). Metrics that provide more proximate measures of the status of local populations of bats are clearly desirable and additional energy should be directed toward such innovation; however, development and implementation of new technologies or innovative approaches will occur at unpredictable intervals and will likely require several years to fully implement. Use of presence data at large spatial scales represents an interim solution that can be applied until methods are developed that provide more proximate measures of the status of populations of forest-dwelling bats. The urgent need for information on the population status of bats demands that the current "wait and see" approach be supplemented by more active measures that make full use of the information available from existing inventory methods.

LITERATURE CITED

Agosta, S.J. 2002. Habitat use, diet and roost selection by the big brown bat (*Eptesicus fuscus*) in North America: a case for conserving an abundant species. Mammal Review 32:179–198.

Ahlén, I., and H.J. Baagøe. 1999. Use of ultrasound detectors for bat studies in Europe: experiences from field identification, surveys, and monitoring. Acta Chiropterologica 1:137–150.

Aldridge, H.D.J.N., and I.L. Rautenbach 1987. Morphology, echolocation, and resource partitioning in insectivorous bats. Journal of Animal Ecology 56:763–778.

Baker, G.B., L.F. Lumsden, E.B. Dettman, N.K. Schedvin, M. Schulz, D. Watkins, and L. Jansen. 2001. The effect of forearm bands on insectivorous bats (Microchiroptera) in Australia. Wildlife Research 28:229–237.

Barclay, R.M.R. 1999. Bats are not birds—a cautionary note on using echolocation calls to identify bats: a comment. Journal of Mammalogy 80:290–296.

Barclay, R.M.R., and R.M. Brigham, eds. 1996. Bats and forests symposium. Research Branch, British Columbia Ministry of Forests, Victoria, BC.

———. 2001. Year-to-year reuse of tree roosts by California bats (*Myotis californicus*) in southern British Columbia. American Midland Naturalist 146:80–85.

Barclay R.M.R., J. Ulmer, C.J.A. MacKenzie, M.S. Thompson, L. Olson, J. McCool, E. Cropley, and G. Poll. 2004. Variation in reproductive rate of bats. Canadian Journal of Zoology 82:688–693.

Beck, J., and C. Peterson. 2003. Determining bat species composition using call analysis. Bat Research News 44:126.

Brigham, R.M., M.J. Vonhof, R.M.R. Barclay, and J.C. Gwilliam. 1997. Roosting behavior and roost-site preferences of forest-dwelling California bats (*Myotis californicus*). Journal of Mammalogy 78:1231–1239.

Broders, H.G., C.S. Findlay, and L. Zheng. 2004. Effects of clutter on echolocation call structure of *Myotis septentrionalis* and *M. lucifugus*. Journal of Mammalogy 85:273–281.

Carroll, S.K., T.C. Carter, and G.A. Feldhamer. 2002. Placement of nets for bats: effects on perceived fauna. Southeastern Naturalist 1:193–198.

Chung-MacCoubrey, A.L. 2003. Monitoring long-term reuse of trees by bats in pinyon-juniper woodlands of New Mexico. Wildlife Society Bulletin 31:73–79.

Clarke, R.H., D.L. Oliver, R.L. Boulton, R.P. Cassey, and M.F. Clarke. 2003. Assessing programs for monitoring threatened species-a tale of three honeyeaters (Meliphagidae). Wildlife Research 30:427–435.

Colwell, R.K., and J.A. Coddington. 1994. Estimating terrestrial biodiversity through extrapolation. Philosophical Transactions of the Royal Society of London B 345:101–118.

Corben, C., and G.M. Fellers. 2001. Choosing the 'correct' bat detector-a reply. Acta Chiropterologica 2:253–256.

Crampton, L.H., and R.M.R. Barclay. 1998. Selection of roosting and foraging habitat by bats in different-aged aspen mixedwood stands. Conservation Biology 12:1347–1358.

Crome, F.H.J., and G.C. Richards. 1988. Bats and gaps: microchiropteran community structure in a Queensland rain forest. Ecology 69:1960–1969.

Duffy, A.M., L.F. Lumsden, C.R. Caddle, R.R. Chick, and G.R. Newell. 2000. The efficacy of Anabat ultrasonic detectors and harp traps for surveying microchiropterans in south-east Australia. Acta Chiropterologica 2:127–144.

Fenton, M.B. 1997. Science and the conservation of bats. Journal of Mammalogy 78:1–14.

———. 2003. Science and the conservation of bats: where to next? Wildlife Society Bulletin 31:6–15.

Fenton, M.B., H.G. Merriam, and G.L. Holroyd. 1983. Bats of Kootenay, Glacier, and Mount Revelstoke national parks in Canada: identification by echolocation calls, distribution and biology. Canadian Journal of Zoology 61:2503–2508.

Field, S.A., A.J. Tyre, and H.P. Possingham. 2005. Optimizing landscape-scale monitoring under economic and observational constraints. Journal of Wildlife Management 69:473–482.

Francis, C.M. 1989. A comparison of mist nets and two designs of harp traps for capturing bats. Journal of Mammalogy 70:865–870.

Frank, J.D., T.H. Kunz, J. Horn, C. Cleveland, and S. Petronio. 2003. Advanced infrared detection and image processing for automated bat censusing, pp. 261–271, *in* Infrared technology and applications XXIX (B.G. Andersen and G.F. Fulop, eds.), vol. 5074. Proceedings of SPIE, International Society for Optical Engineering, Bellingham, WA.

Gannon, W.L., R.E. Sherwin, and S. Haymond. 2003. On the importance of articulating assumptions when conducting acoustic studies of habitat use by bats. Wildlife Society Bulletin 31:45–61.

Gardner, J.E., J.D. Garner, and J.E. Hofmann. 1989. A portable mist netting system for capturing bats with emphasis on *Myotis sodalis* (Indiana bat). Bat Research News 30:1–8.

Gehrt, S.D., and J.E. Chelsvig. 2004. Species-specific patterns of bat activity in an urban landscape. Ecological Applications 14:625–635.

Gellman, S.T., and W.J. Zielinski. 1996. Use by bats of old-growth redwood hollows on the north coast of California. Journal of Mammalogy 77:255–265.

Goldsmith, F.B., ed. 1991. Monitoring for conservation and ecology. Chapman and Hall, London, U.K.

Grilliot, M.E., S.C. Burnett, and M.T. Mendonça. 2004. Sex and seasonal differences in echolocation signals of *Eptesicus fuscus*. Bat Research News 45:223.

Grindal, S.D., T.S. Collard, R.M. Brigham, and R.M.R. Barclay. 1992. The influence of precipitation on reproduction by Myotis bats in British Columbia. American Midland Naturalist 128:339–344.

Gu, W., and R.K. Swihart. 2004. Absent or undetected? Effects of non-detection on species occurrence on wildlife-habitat models. Biological Conservation 116:195–203.

Haines, D.E., and K.H. Pollock. 1998. Estimating the number of active and successful bald eagle nests: an application of the dual frame method. Environmental and Ecological Statistics 5:245–256.

Hayes, J.P. 1997. Temporal variation in activity of bats and the design of echolocation-monitoring studies. Journal of Mammalogy 78:514–524.

———. 2000. Assumptions and practical considerations in the design and interpretation of echolocation-monitoring studies. Acta Chiropterologica 2:225–236.

Hayes, J.P., and P. Hounihan. 1994. Field use of the Anabat II bat-detector system to monitor bat activity. Bat Research News 35:1–3.

Hellawell, J.M. 1991. Development of a rationale for monitoring, pp. 1–14, *in* Monitoring for conservation and ecology (F.B. Goldsmith, ed.). Chapman and Hall, London, U.K.

Hitchcock, H.B., R. Keen, and A. Kurta. 1984. Survival rates of *Myotis leibii* and *Eptesicus fuscus* in southeastern Ontario. Journal of Mammalogy 65:126–130.

Jaberg, C., and A. Guisan. 2001. Modelling the distribution of bats in relation to landscape in a temperate mountain environment. Journal of Applied Ecology 38:1169–1181.

Jones, G., T. Gordon, and J. Nightingale. 1992. Sex and age differences in the echolocation calls of the lesser horseshoe bat, *Rhinolophus hipposideros*. Mammalia 56:189–193.

Jones, G., N. Vaughan, and S. Parsons. 2000. Acoustic identification of bats from directly sampled and time expanded recordings of vocalizations. Acta Chiropterologica 2:155–170.

Kazial, K.A., S.C. Burnett, and W.M. Masters. 2001. Individual and group variation in echolocation calls of big brown bats, *Eptesicus fuscus* (Chiroptera: Vespertilionidae). Journal of Mammalogy 82:339–351.

Keen, R., and H.B. Hitchcock. 1980. Survival and longevity of the little brown bat (*Myotis lucifugus*) in southeastern Ontario. Journal of Mammalogy 61:1–7.

Kerth, G., and K. Reckardt. 2003. Information transfer about roosts in female Bechstein's bats: an experimental field study. Proceedings of the Royal Society of London B 270:511–515.

Kingston, T., C.F. Francis, Z. Akbar, and T.H. Kunz. 2003. Species richness in an insectivorous bat assemblage from Malaysia. Journal of Tropical Ecology 19:67–79.

Krusic, R.A., and C.D. Neefus. 1996. Habitat associations of bat species in the White Mountain National Forest, pp. 185–198, *in* Bats and forests symposium (R.M.R. Barclay and R.M. Brigham, eds.). Research Branch, British Columbia Ministry of Forests, Victoria, BC.

Kuenzi, A.J., and M.L. Morrison. 1998. Detection of bats by mist-nets and ultrasonic sensors. Wildlife Society Bulletin 26:307–311.

———. 2003. Temporal patterns of bat activity in southern Arizona. Journal of Wildlife Management 67:52–64.

Kunz, T.H. 2003. Censusing bats: challenges, solutions, and sampling biases, pp. 9–19, *in* Monitoring bat populations of the U. S. and territories: problems and prospects (T.J. O'Shea, and M.A. Bogan, eds.). U.S. Geological Survey Information and Technology Report ITR-2003-0003.

Kunz, T.H., and L.F. Lumsden. 2003. Ecology of cavity and foliage roosting bats, pp. 3–89, *in* Bat ecology (T.H. Kunz and M.B. Fenton, eds.). University of Chicago Press, Chicago, IL.

Kunz, T.H., G.R. Richards, and C.R. Tidemann. 1996a. Capturing small volant mammals, pp.157–164, *in* Measuring and monitoring biological diversity (D.E. Wilson, J. Nichols, R. Rudrin, R. Cole, and M. Foster, eds.). Smithsonian Institution Press, Washington, DC.

Kunz, T.H., D.W. Thomas, G.R. Richards, C.D. Tidemann, E.D. Pierson, and P.A. Racey. 1996b. Observational techniques for bats, pp. 105–114, *in* Measuring and monitoring biological diversity (D.E. Wilson, J. Nichols, R. Rudran, R. Cole, and M. Foster, eds.). Smithsonian Institution Press, Washington, DC.

Kurta, A., and S.W. Murray. 2002. Philopatry and migration of banded Indiana bats (*Myotis sodalis*) and effects of radio transmitters. Journal of Mammalogy 83:585–589.

Lance, R.F., B.T. Hardcastle, A. Talley, and P.L. Leberg. 2001. Day-roost selection by Rafinesque's big-eared bats (*Corynorhinus rafinesquii*) in Louisiana forests. Journal of Mammalogy 82:166–172.

Lang, A.B., C.D. Weise, E.K.V. Kalko, and H. Roemer. 2004. The bias of bat netting. Bat Research News 45:235–236.

Lewis, S.E. 1993. Effect of climatic variation on reproduction by pallid bats (*Antrozous pallidus*). Canadian Journal of Zoology 71:1429–1433.

———. 1995. Roost fidelity of bats: a review. Journal of Mammalogy 76:481–496.

Link, W.A., R.J. Barker, J.R. Sauer, and S. Droege. 1994. Within-site variability in surveys of wildlife populations. Ecology 75:1097–1108.

MacKenzie, D.I., J.D. Nichols, J.E. Hines, M.G. Knutson, and A.B. Franklin. 2003. Estimating site occupancy, colonization, and local extinction when a species is detected imperfectly. Ecology 84:2200–2207.

MacKenzie, D.I., J.D. Nichols, G.B. Lachman, S. Droege, J.A. Royle, and C.A. Langtimm. 2002. Estimating site occupancy rates when detection probabilities are less than one. Ecology 83:2248–2255.

Manley, P.N., W.J. Zielinski, M.D. Schlesinger, and S.R. Mori. 2004. Evaluation of a multiple-species approach to monitoring species at the ecoregional scale. Ecological Applications 14:296–310.

Masters, W.M., K.A.S. Raver, and K.A. Kazial. 1995. Sonar signals of big brown bats, *Eptesicus fuscus,* contain information about individual identity, age, and family affiliation. Animal Behavior 50:1243–1260.

Mazurek, M.J. 2004. A maternity roost of Townsend's big-eared bats (*Corynorhinus townsendii*) in coast redwood basal hollows in northwestern California. Northwest Naturalist 85:60–62.

McCracken, G.F., L.R. Saidak, and R.C. Currie. 2000. Genetic analysis of the species status of the Indiana bat, *Myotis sodalis*. Bat Research News 41:129.

Miller, D.A. 2003. Species diversity, reproduction, and sex ratios of bats in managed pine forest landscapes of Mississippi. Southeastern Naturalist 2:59–72.

Miller, D.A., E.B. Arnett, and M.J. Lacki. 2003. Habitat management for forest-roosting bats of North America: a critical review of habitat studies. Wildlife Society Bulletin 31:30–44.

Mills, D.J., T.W. Norton, H.E. Parnaby, R.B. Cunningham, and H.A. Nix. 1996. Designing surveys for microchiropteran bats in complex forest landscapes-a pilot study from south-east Australia. Forest Ecology and Management 85:149–161.

Moreno, C.E., and G. Halffter. 2000. Assessing completeness of bat biodiversity inventories using species accumulation curves. Journal of Applied Ecology 37:149–158.

Morrison, M.L., W.M. Block, M.D. Strickland, and W.L. Kendall. 2001. Wildlife study design. Springer-Verlag, New York.

Morrison, M.L., and L.S. Hall. 2002. Standard terminology: toward a common language to advance ecological understanding and application, pp.43–52, *in* Predicting species occurrences: issues of accuracy and scale (J.M. Scott, P.J. Heglund, and M.L. Morrison, eds.). Island Press, Washington, DC.

Mukhida, M., J. Orprecio, and M.B. Fenton. 2004. Echolocation calls of *Myotis lucifugus* and *M. leibii* (Vespertilionidae) flying inside a room and outside. Acta Chiropterologica 6:91–97.

Murray, K.L., E.R. Britzke, B.M. Hadley, and L.W. Robbins. 1999. Surveying bat communities: a comparison between mist nets and the Anabat II bat detector system. Acta Chiropterologica 1:105–112.

Murray, K.L., E.R. Britzke, and L.W. Robbins. 2001. Variation in search-phase calls of bats. Journal of Mammalogy 82:728–737.

Neubaum, D.J., M.A. Neubaum, L.E. Ellison, and T.J. O'Shea. 2005. Survival and condition of big brown bats (*Eptesicus fuscus*) after radiotagging. Journal of Mammalogy 36:95–98.

Noon, B.R. 2003. Conceptual issues in monitoring ecological resources, pp. 27–71, *in* Monitoring ecosystems: interdisciplinary approaches for evaluating ecoregional initiatives (D. E. Busch and J. C. Trexler, eds.). Island Press, Washington DC.

Obrist, M.K. 1995. Flexible bat echolocation: the influence of individual, habitat, and conspecifics on sonar signal design. Behavioral Ecology and Sociobiology 36:207–219.

O'Donnell, C.F.J. 2000. Influence of season, habitat, temperature, and invertebrate availability on nocturnal activity of the New Zealand long-tailed bat (*Chalinolobus tuberculatus*). New Zealand Journal of Zoology 27:207–221.

O'Donnell, C.F.J., and J.A. Sedgeley 1999. Use of roosts by the long-tailed bat, *Chalinolobus tuberculatus*, in temperate rainforest in New Zealand. Journal of Mammalogy 80:913–923.

O'Farrell, M.J., C. Corben, and W.L. Gannon. 2000. Geographic variation in the echolocation calls of the hoary bat (*Lasiurus cinereus*). Acta Chiropterologica 2:185–196.

O'Farrell, M. J., C. Corben, W. L. Gannon, and B. W. Miller. 1999a. Confronting the dogma: a reply. Journal of Mammalogy 80:297–302.

O'Farrell, M.J., and W.L. Gannon. 1999. A comparison of acoustic versus capture techniques for the inventory of bats. Journal of Mammalogy 80:24–30.

O'Farrell, M.J., B.W. Miller, and W.L. Gannon. 1999b. Qualitative identification of free-flying bats using the Anabat detector. Journal of Mammalogy 80:11–23.

O'Shea, T.J., and M.A. Bogan, eds. 2003. Monitoring bat populations of the U.S. and territories: problems and prospects. U.S. Geological Survey Information and Technology Report ITR-2003-0003.

O'Shea, T.J., L.E. Ellison, and T.R. Stanley. 2004. Survival estimation in bats: historical review, critical appraisal, and suggestions for new approaches, pp. 297–336, *in* Sampling rare or elusive species: concepts, designs, and techniques for estimating population parameters (W. L. Thompson, ed.). Island Press, Washington, DC.

Parsons, S. 2001. Identification of New Zealand bats (*Chalinolobus tuberculatus* and *Mystacina tuberculata*) in flight from analysis of echolocation calls by artificial neural networks. Journal of Zoology (London) 253:447–456.

Parsons, S., and G. Jones. 2000. Acoustic identification of twelve species of echolocating bat by discriminant function analysis and artificial neural networks. Journal of Experimental Biology 203:2641–2656.

Pierson, E.D. 1998. Tall trees, deep holes, and scarred landscapes, conservation of biology of North American bats, pp. 309–325, *in* Bat conservation and ecology (T.H. Kunz and P.A. Racey, eds.). Smithsonian Institution Press, Washington, DC.

Pierson, E.D., and W.E. Rainey. 1998. Distribution of the spotted bat, *Euderma maculatum,* in California. Journal of Mammalogy 79:1296–1305.

Pollock, K.H., J.D. Nichols, T.R. Simons, G.L. Farnsworth, L.L. Bailey, and J.R. Sauer. 2002. Large scale wildlife monitoring studies: statistical methods for design and analysis. Environmetrics 13:105–119.

Resources Inventory Committee. 1998. Inventory and monitoring methods for bats. Standards and components of British Columbia's Biodiversity No. 20. Ministry of Land Environment and Parks, Victoria, BC.

Reynolds, M.H., B.Z. Cooper, and R.H. Day. 1997. Radar study of seabirds and bats on windward Hawai'i. Pacific Science 51:97–106.

Rodriguez, R.M., and L.K. Ammerman. 2004. Mitochondrial DNA divergence does not reflect morphological differences between *Myotis californicus* and *Myotis ciliolabrum.* Journal of Mammalogy 85:842–851.

Royle, J.A., and J.D. Nichols. 2003. Estimating abundance from repeated presence-absence data or point counts. Ecology 84:777–790.

Sabol, B.M., and K. Hudson. 1995. Technique using thermal infrared-imaging for estimating populations of gray bats. Journal of Mammalogy 76:1242–1248.

Sauer, J.R. 2003. A critical look at national monitoring programs for birds and other wildlife species, pp. 119–126, *in* Monitoring bat populations of the U.S. and territories: problems and prospects (T.J. O'Shea, and M.A. Bogan, eds.). U.S. Geological Survey Information and Technology Report ITR-2003-0003.

Schulz, M. 1999. Relative abundance and other aspects of the natural history of the rare golden-tipped bat, *Kerivoula papuensis* (Chiroptera: Vespertilionidae). Acta Chiropterologica 1:165–178.

Sedgeley, J.A., and C.F.J. O'Donnell. 1999. Roost selection by the long-tailed bat, *Chalinolobus tuberculatus,* in temperate New Zealand rainforest and its implications for the conservation of bats in managed forests. Biological Conservation 88:261–276.

Sendor, T., and M. Simon. 2003. Population dynamics of the pipistrelle bat: effects of sex, age, and winter weather on seasonal survival. Journal of Animal Ecology 72:308–320.

Sherwin, R.E., W.L. Gannon, and J.S. Altenbach. 2003. Managing complex systems simply: understanding inherent variation in the use of roosts by Townsend's big-eared bat. Wildlife Society Bulletin 31:62–72.

Sherwin, R.E., S. Haymond, D. Stricklan, and R. Olsen. 2002. Freeze-branding to permanently mark bats. Wildlife Society Bulletin 30:97–100.

Speakman, J.R., P.A. Racey, C.M. Catto, P.I. Webb, S.M. Swift, and A.M. Burnett. 1991. Minimum summer populations and densities of bats in N.E. Scotland, near the northern borders of their distributions. Journal of Zoology (London) 225:327–345.

Telfer, M.G., C.D. Preston, and P. Rothery. 2002. A general method for measuring relative change in range size from biological atlas data. Biological Conservation 107:99–109.

Thomas, D.W., G.P. Bell, and M.B. Fenton. 1987. Variation in echolocation call frequencies recorded from North American vespertilionid bats: a cautionary note. Journal of Mammalogy 68:842–847.

Thompson, F.R., III, D.E. Burnhans, and B. Root. 2002. Effects of point count protocol on bird abundance and variability estimates and power to detect population trends. Journal of Field Ornithology 73:141–150.

Thompson, W.L., G.C. White, and C. Gowan. 1998. Monitoring vertebrate populations. Academic Press, San Diego, CA.

Tibbels, A. 1999. Do call libraries reflect reality? Bat Research News 40:153–155.

Tibbels, A.E., and A. Kurta. 2003. Bat activity is low in thinned and unthinned stands of red pine. Canadian Journal of Forest Research 33:2436–2442.

Tidemann, C.R., and D.P. Woodside. 1978. A collapsible bat-trap and a comparison of results obtained with the trap and with mist-nets. Australian Wildlife Research 5:355–362.

Tuttle, M.D. 1979. Status, causes of decline, and management of endangered gray bats. Journal of Wildlife Management 43:1–17.

Tyre, A.J., B. Renhumberg, S.A. Field, D. Niejalke, K. Parris, and H.P. Possingham. 2003. Improving precision and reducing bias in biological surveys: estimating false-negative error rates. Ecological Applications 13:1790–1801.

U.S. Fish and Wildlife Service. 1999. Agency draft Indiana bat (*Myotis sodalis*) revised recovery plan. Fort Snelling, MN.

Usher, M.B. 1991. Scientific requirements of a monitoring programme, pp. 15–32, *in* Monitoring for conservation and ecology (F.B. Goldsmith, ed.). Chapman and Hall, London, U.K.

Vaughan, N., G. Jones, and S. Harris. 1997. Habitat use by bats (Chiroptera) assessed by means of a broad-band acoustic method. Journal of Applied Ecology 34:716–730.

Vonhof, M.J. 2000. Handbook of inventory methods and standard protocols for surveying bats in Alberta. Report prepared for Alberta Environment, Edmonton, Alberta.

Warren, R.D., and M.S. Witter. 2002. Monitoring trends in bat populations through roost surveys: methods and data from *Rhinolophus hipposideros*. Biological Conservation 105:255–261.

Waters, D.A., and W.L. Gannon. 2004. Bat call libraries: management and potential use, pp. 150–157, *in* Bat echolocation research: tools, techniques, and analysis (R.M. Brigham, E.K.V. Kalko, G. Jones, S. Parsons, and H.J.G.A. Limpens, eds.). Bat Conservation International, Austin, TX.

Weller, T.J., P.N. Manley, J.A. Baldwin, and M.M. McKenzie. 2002. Designing regional-scale monitoring for free-flying bats: incorporation of detectability estimates. Bat Research News 43:191.

Weller, T.J., and C.J. Zabel. 2001. Characteristics of fringed myotis day roosts in northern California. Journal of Wildlife Management 65:489–497.

Wickramasinghe, L.P., S. Harris, G. Jones, and N. Vaughan. 2003. Bat activity and species richness on organic and conventional farms: impact of agricultural intensification. Journal of Applied Ecology 40:984–993.

Willis, C.K.R., K.A. Kolar, A.L. Karst, M.C. Kalcounis-Rüppell, and R.M. Brigham. 2003. Medium-and long-term reuse of trembling aspen cavities as roost by big brown bats (*Eptesicus fuscus*). Acta Chiropterologica 5:85–90.

Yoccoz, N.G., J.D. Nichols, and T. Boulinier. 2001. Monitoring of biological diversity in space and time. Trends in Ecology and Evolution 16:446–453.

Zinck, J.M., D.A. Duffield, and P.C. Ormsbee. 2004. Primers for identification and polymorphism assessment of Vespertilionid bats in the Pacific Northwest. Molecular Ecology Notes 4:239–242.

PLANNING FOR BATS ON FOREST INDUSTRY LANDS IN NORTH AMERICA

11

T. Bently Wigley, Darren A. Miller, and Greg K. Yarrow

Natural disturbances (e.g., fire, wind throw, and ice storms) and associated successional changes and conversion of forests to alternative land uses (e.g., roads, agriculture, and urban uses) have obvious influences on habitat for bats in North America. Additionally, management of North American forests to provide wood products has a widespread and often pervasive influence on ecosystem function. Practices such as site preparation and stand establishment, thinning, timber harvest, and vegetation control clearly affect current and future forest structure and composition (Guldin et al., chap. 7 in this volume). Implications of these forestry practices for biological diversity, including bats, are complicated and depend on the spatial and temporal scales being considered (Miller et al. 2003; Sallabanks et al. 2001).

Silvicultural practices on forest industry lands are used to increase forest growth and yield and produce a sustainable flow of wood products, but these forests also have been shown to contribute to conservation of biological diversity (Wigley et al. 2000). Recent findings have indicated that the full range of species of bats can likely be maintained on intensively managed forest landscapes (Campbell et al. 1996; Duchamp et al., chap. 9 in this volume; Hayes 2003; Miller 2003), in particular, if considerations for bats are made during forest planning (e.g., provision of legacy trees and snags in the Pacific Northwest). Maintaining a balance between sustainable economic value and sustainable ecological objectives can be realized best when industry managers carefully plan for both.

In this chapter, we describe our perspectives on the realities of providing habitat for forest-dwelling bats within forests whose primary function is providing wood products. We focus on coniferous forests of Canada, the Pacific Northwest, and the southeastern United States because of their prevalence in wood production. Our objective is to describe: (1) current characteristics of the industry and its forests; (2) the paradigm of sustainable forestry and aspects of sustainable forestry certification programs, including provisions related to biological diversity; (3) a sample process that captures how many companies plan for biological diversity; (4) examples of biodiversity planning by individual companies, with a special focus on bats; and (5) considerations for including bats within management plans on industrial ownership.

CHARACTERISTICS OF THE FOREST PRODUCTS INDUSTRY AND ITS LANDS

The forest products industry consists of companies and individuals that operate primary wood-using facilities and manage forests they own or control primarily for wood products (Helms 1998). In the United States, the forest products industry directly employs about 1.7 million people in wood and paper production, or about 1.1% of the U.S. workforce (American Forest and Paper Association and Clemson University 2001). For every job that is directly forestry related, another two jobs are related indirectly (e.g., transportation, distribution, and sales), meaning that about 5.7 million jobs in the United States are linked to the forest products industry. This industry can be vital to rural economies, such as in Mississippi where 10% of all jobs in the state are forestry related and, during 2002, forestry contributed $11.4 billion dollars to the Mississippi economy (http://www.msforestry.net/pdf/Forest_Facts.pdf). The forest products industry also is one of Canada's largest employers with approximately 376,300 individuals employed in wood product and paper manufacturing, logging, and forestry services in 2003 (www.nrcan-rncan.gc.ca/cfs-scf/national/what-quoi/sof/latest_e.html). Also during 2003, the forest products industry contributed $33.7 billion to Canada's gross domestic product and $29.7 billion to Canada's balance of trade, and it was responsible for $3.3 billion in new capital investments. The economic importance of wood products cannot be overlooked when addressing ecological issues related to forest management.

The considerable complexity of the industry is due to the variable size, character, and objectives of its constituent companies. This includes ownership (individuals/families versus stockholders) and source of wood supply (extent to which timber is purchased from public or nonindustrial private forests). Over the past several decades, ownership of industry land has become even more complex as many forest products companies have sold lands to organizations that manage timberlands on behalf of institutional (e.g., pension funds, foundations, and endowments) and other types of investors. Known as timber investment management organizations (TIMOs) or real estate investment trusts (REITs) (Ravenel et al. 2002; Stanturf et al. 2003), some of these ventures seek to optimize economic return within a much shorter time frame (10–15 years) than forest products companies and may include nontimber objectives, such as real estate sales, as a primary motivation.

Differences among companies, intermingling of company lands with nonindustrial private and public forests, and age and forest type diversity within industrial ownerships usually result in diverse habitat conditions across landscapes that vary in their ability to provide roosting and foraging habitat for bats. Companies that require wood for solid products often manage with uneven-aged systems or even-aged systems, with rotations of 30 years or more in the southern United States to more than 60 years in Canada and the Pacific Northwest, using thinnings to remove wood for paper products as part of the silvicultural system to achieve larger, higher-quality trees (Guldin et al., chap. 7 in this volume). In general, companies that primarily require wood for paper products favor

even-aged systems with short rotations (<15 years in the southern United States). Variability in management approaches among companies and resulting forest structure make it difficult to apply broad guidelines for bat conservation across all industry ownerships because site-specific conditions and habitat opportunities differ.

The character of industrial forests differs between the United States and Canada. In Canada, 93% of the 401.9 million ha of forest and wooded land is under provincial, territorial, or federal jurisdictions (table 11.1). Only 7% is privately owned, primarily (80%) occurring in New Brunswick, Nova Scotia, and Prince Edwards Island. Although only 7% of Canada's forests are privately owned, those lands alone would rank Canada among the top 10 countries in the world in terms of forest area and production of roundwood (Abusow 2004).

Because of this ownership pattern, in general, forest products companies in Canada manage Crown lands under license or tenure from the provincial governments through a collaborative process involving government, the public, nongovernmental organizations, and companies. However, only about 144.6 million ha (36%) of Canada's forests are considered accessible and likely to be managed, and <1 million ha (0.3% of all Canada's forests) are actually harvested annually (www.nrcan.gc.ca/cfs-scf/national/what-quoi/sof/latest_e.html). In 2002, 96% of industrial roundwood and 88% of harvested land was in British Columbia, Quebec, Ontario, Alberta, and New Brunswick (table 11.1).

In the United States, over 57% of forests are privately owned with

Table 11.1. Area of forest and other woodlands (million ha), industrial harvest of roundwood (m³/ha), area of harvested land (million ha), and forest ownership (% forest area by owner class) in 2002 by Canadian province/territory

Province/territory	Forest area*	Harvest		Forest ownership[†]		
		Volume	Area	Provincial	Federal	Private
Quebec	84.6	39.6	309,195	89	0	11
Ontario	68.3	26.3	184,322	91	1	8
British Columbia	64.1	81.5	189,277	96	1	3
Alberta	36.4	24.6	68,430	89	8	3
Manitoba	36.3	2	15,042	95	2	3
Northwest Territories	33.3	0.003	50	0	100	0
Saskatchewan	24.3	4.3	23,222	90	4	6
Yukon	22.8	0.007	42	0	100	0
Newfoundland/Labrador	20.1	2.1	22,027	99	0	1
New Brunswick	6.2	10.1	105,834	48	2	50
Nova Scotia	4.3	6	49,959	29	3	68
Nunavut	0.9	0	0	0	100	0
Prince Edward Island	0.3	0.4	4,903	8	1	91
Total	401.9	189.2	972,303	77	16	7

Source: www.nrcan.gc.ca/cfs-scf/national/what-quoi/sof/latest_e.html
*Million ha of forest and other woodland.
[†]Percent of 401.9 million ha of forest and wooded land in Canada.

Table 11.2. Percent area of forestland by owner class and Renewable Resource Planning Act region, 2002

Owner class	USA	North	South	Rocky Mountain	Pacific Coast
Total forest land (million ha)	303.09	68.67	86.85	58.42	89.15
Public land					
National forest	19.82	6.72	5.76	50.93	23.23
Other federal	13.08	1.41	3.36	18.42	28.05
State	8.43	11.94	2.36	4.04	14.52
County/municipality	1.28	4.31	0.53	0.13	0.43
Total public	42.62	24.38	12.00	73.53	66.23
Private land					
Forest industry	8.86	8.74	16.74	2.03	5.77
Nonindustrial private	48.52	66.88	71.26	24.44	28.00
Total private	57.38	75.62	88.00	26.47	33.77

Source: From Smith et al. 2004.

about 26.9 million ha of land owned by the forest industry, comprising about 9% of total forest ownership (Smith et al. 2004) (table 11.2). Most industry ownership (14.5 million ha) is in the USDA Forest Service's Southern region where industry owns about three times the area of national forests and almost 1.5 times that in all public ownerships (Smith et al. 2004). In the Rocky Mountain and Pacific Coast regions combined, industry owns about 4% of all forestland with public entities owning about 15 times more.

Contributions of different ownerships and regions in the United States to overall wood supply also differ (American Forest and Paper Association and Clemson University 2001). In 2001, the South supplied 58% of America's total roundwood production, the North about 21.3%, the Pacific Coast about 16.6%, and the Rocky Mountains only 4% (Smith et al. 2004). In 2001, industrial forests provided 29% of the U.S. timber harvest and private forests in total provided 92% (Smith et al. 2004). Therefore, the current trend is for increasing wood production from private lands in the southern United States, demonstrated by an anticipated increase in intensively managed pine (*Pinus* spp.) forests in the South from 12.9 million ha in 1999 to 21.8 million ha by 2040 (Wear and Greis 2002).

A growing human population and associated urban development is significantly influencing North American forests, in particular, in the eastern United States (Egan and Luloff 2000; Wear and Greis 2002). For example, about 30% of U.S. forests (34% in the East and 25.2% in the West) are located in counties with urban centers of >20,000 persons (Smith et al. 2004). Development emanating from these expanding urban centers often results in parcelization of forest ownerships (Mehmood and Zhang 2001) complicating landscape-level planning and leading to increased fragmentation.

Increased production of wood fiber on forest industry lands could ease pressure on nonintensively managed forests and likely will be a primary component of strategies for meeting needs of future global wood supply on a globally decreasing forest base (Wagner et al. 2004). And, because industrial forests have conservation value over alternative land uses (e.g., agriculture, urban), intensive management of some forests is likely to be a key element in future strategies to conserve biological diversity (Wagner et al. 2004). This provides a tremendous challenge, and a worthwhile opportunity, to integrate conservation of bats within provisions for forest products.

SUSTAINABLE FORESTRY PROGRAMS

Forest products companies rarely plan exclusively for individual species or taxonomic groups (e.g., forest-dwelling bats) except those formally identified as being of high conservation concern (table 11.3). Rather, forest products companies generally plan for overall biological diversity under the auspices of sustainable forestry programs. The principal certification programs in North America that contain provisions related to biological diversity and are applicable to forest products companies are the Sustainable Forestry Initiative® (SFI), Forest Stewardship Council (FSC), and the Canadian Standards Association for Sustainable Forest Management (CAN/CSA) (Guynn et al. 2004a). These certification programs, in general, seek to balance the economic, ecological, and social benefits of forestland managed to yield a sustainable supply of wood products. These three programs have approximately 61.5 (SFI), 32.9 (CSA), and 10.5 million ha (FSC) enrolled in North America, meaning a significant portion of industrial forestlands in North America are oper-

Table 11.3. Bat species in North America designated as critically imperiled or imperiled by NatureServe, threatened or endangered under the U.S. Endangered Species Act, or threatened under Canada's Species at Risk Act, January 2005

Scientific name	Common name	NatureServe Global*	United States ESA[†]	Canada COSEWIC[‡]	Distribution
Leptonycteris curasoae yerbabuenae	southern long-nosed bat	G4T3T4	E		USA: AZ, NM
Leptonycteris nivalis	Mexican long-nosed bat	G3	E		USA: NM, TX
Antrozous pallidus	pallid bat	G5		T	CAN: BC USA: AZ, CA, CO, ID, KS, MT, NM, NV, OK, OR, TX, UT, WA, WY
Corynorhinus townsendii ingens	Ozark big-eared bat	G4T1	E		USA: AR, MO , OK
Corynorhinus townsendii virginianus	Virginia big-eared bat	G4T2	E		USA: KY, NC, VA, WV
Myotis grisescens	gray myotis	G3	E		USA: AL, AR, FL, GA, IL, IN, KS, KY, MO, OK, SC, TN, VA
Myotis keenii	Keen's myotis	G2G3		DD	CAN: BC USA: AK, WA
Myotis sodalis	Indiana myotis	G2	E		USA: AL, AR, CT, IA, IL, IN, KY, MA, MD, MI, MO, NC, NJ, NY, OH, OK, PA, SC, TN, VA, VT, WV

Source: www.natureserve.org/explorer/
DD, data deficient.

*Status ranks reported for NatureServe are critically imperiled (G1 or T1), imperiled (G2 or T2), G3 (vulnerable), G4 (apparently secure), and G5 (secure). G ranks are for full species, rangewide and T ranks are for subspecies, varieties, and populations rangewide. Two G or T ranks for the same element indicate a range of potential scores.
[†]E, listed as endangered under the U.S. Endangered Species Act (ESA).
[‡]T, listed as threatened under Canada's Species at Risk Act by the Committee on the Status of Endangered Wildlife in Canada (COSEWIC).

ating within a sustainability paradigm. Sustainable forestry certification programs contain many provisions that serve to enhance biological diversity, several of which are particularly relevant to bats. These considerations include: (1) minimal and prudent use of forest chemicals, (2) protection of water quality, (3) protection of imperiled species and communities, (4) provisions for habitat diversity, and (5) and support of research.

MINIMAL AND PRUDENT USE OF FOREST CHEMICALS

Certification programs generally seek to ensure that silvicultural chemicals are used minimally and prudently. Because forest products companies rarely use insecticides as part of silvicultural operations, the potential for widespread impact on bats from use of insecticides in forestry is not a concern. The most common chemicals used by the forest products industry are herbicides, which are used to control competing vegetation during site preparation, immediately after stand establishment, and sometimes later in the rotation (Shepard et al. 2004; Wagner et al. 2004). Therefore, herbicides usually are applied only a few times over the typical rotation of an even-aged forest (~25–50 years depending on ecosystem and forest type).

Direct toxicity from herbicides to wildlife, including bats and their insect prey, is not a concern as forest herbicides do not persist in the environment, are nontoxic, do not bioaccumulate, and only affect biochemical processes unique to plants (Tatum 2004). Recent technological advances (sophisticated spray-drift models) also have helped managers avoid drift and unintended ecological impact from chemical applications (Teske and Ice 2002). The primary effect of herbicides on wildlife communities is through alterations to forest structure and plant species composition (Guynn et al. 2004b). These effects in North American forests are generally short term ($<$5 years) and response by wildlife is very species specific (Guynn et al. 2004b; Lautenschlager and Sullivan 2004; Miller and Wigley 2004). Herbicides may increase habitat quality for some bat species by decreasing structural complexity (i.e., clutter) within mature forest stands and helping to establish a diverse herbaceous plant community (Burger 2003; Guynn et al. 2004b), which may result in a more diverse insect community (Haddad et al. 2001; Siemann et al. 1998).

PROTECTION OF WATER QUALITY

Another common expectation of certification programs is that participants protect water quality. To accomplish this, most companies follow state or provincial regulations or Best Management Practices to retain areas of undisturbed or lightly harvested forest vegetation in the form of streamside (SMZs) or riparian management zones (RMZs) (Blinn and Kilgore 2001). These riparian forests often extend into headwater areas and can provide a mature forest component on managed forest landscapes, ensuring a long-term supply of snags and larger trees. However, the value of these riparian forests for bats is not entirely clear (Hayes and Loeb, chap. 8 in this volume).

PROTECTION OF IMPERILED SPECIES AND COMMUNITIES

Forest plans of companies participating in certification programs typically provide for protection of imperiled and critically imperiled species and communities (Brown et al. 2004). Companies usually acquire data sets from NatureServe or its member programs (state heritage programs or conservation data centers) that describe government-listed species or imperiled species or communities associated with company lands and their known locations (element occurrences). Companies then compare known locations of these species or communities with operations information (e.g., harvest unit scheduling) to identify priorities for conservation and protection. Company personnel may then work in partnership with scientists within NatureServe member programs to verify status and viability of element occurrences, address any apparent discrepancies between ground surveys and the acquired dataset, and develop strategies to conserve viable occurrences of imperiled species or communities.

PROVISIONS FOR HABITAT DIVERSITY

Under sustainable forestry programs, companies generally are expected to provide stand- and landscape-level habitat diversity. At the stand level, the forest plans of many companies contain criteria for within-stand wildlife habitat elements to be retained (e.g., snags, down woody debris, den trees). At the landscape level, companies may retain patches of legacy forest or ensure that a diversity of stand ages and, where possible, forest types occurs. Many companies have adopted limits on the size of clearcut harvests and "green-up" or "adjacency" policies that require stands to be a certain height or age before adjacent stands are harvested. These constraints directly affect landscape-level diversity because, as these policies are implemented, landowners often disperse relatively small harvest units across their ownership.

SUPPORT OF RESEARCH

Certification programs expect participating companies to implement scientifically based management practices. Unfortunately, the availability of reliable data upon which forest managers can make decisions regarding some aspects of biological diversity (including bats) is limited, in particular, with regard to silvicultural prescriptions and landscape-habitat relationships (Miller et al. 2003). Thus, companies participating in some certification programs also are expected to support research. Companies may meet this expectation through activities of their own staff, through financial contributions to research institutions or universities, or by participating in cooperative research through organizations such as the National Council for Air and Stream Improvement, Inc. (NCASI; www.ncasi.org).

For example, NCASI and its member forest products companies are involved in the Northwest Bat Cooperative (NBC; Arnett and Haufler 2003), which may be a model for other regions in the United States. The NBC has a wide array of partners (e.g., industry, state agencies, federal agencies, and nongovernment organizations) that have collaboratively prioritized research topics. By pooling their fiscal resources, the partners have been able to gather enough data to provide useful results and to maintain studies for relatively long periods. Participating forest products companies have helped secure study sites, implemented treatments, and contributed to other aspects of planned research.

BIODIVERSITY PLANNING ON INDUSTRY LANDS

To address conservation of biological diversity, in compliance with environmental policies or programs, managers often use a formal planning process. Yarrow (2005) have proposed a sample planning process that uses basic principles of adaptive management (Walters and Holling 1990) and provides a general outline of how companies may address biological diversity on their lands. The key steps in the process include: (1) conducting an assessment, (2) establishing broad resource goals and objectives, (3) developing and implementing a plan, and (4) monitoring and adjusting the plan. This adaptive management approach should result in opportunities to continually improve management plans. Details about each of the four steps are provided below.

CONDUCTING AN ASSESSMENT

Companies often establish baseline data that describe the biological and physical features on their land, including an assessment of available habitats such as intensively managed stands, natural stands, unique sites, and inoperable sites. Companies may also consult outside experts to document known or predicted occurrences of sensitive species or communities. Landowners may already have much of the inventory or map-based information needed for an assessment (e.g., ecological classification, cover or habitat type, age and composition of forest stands, stand history and condition, and silvicultural practices). However, some landowners may want to conduct a field assessment to confirm the presence of some species or communities, ground truth remotely sensed data, or document other unique ecological characteristics on their land. For bats, this may include availability and juxtaposition of foraging habitat, roost sites (both appropriately aged stands and specific roost structures such as large snags), bodies of water, and other features. These baseline data are useful for identifying opportunities to conserve biological diversity and serve as a starting point for documenting progress toward selected habitat and biodiversity goals.

ESTABLISHING BROAD RESOURCE GOALS AND OBJECTIVES

Under this approach, landowners should consider establishing broad resource goals and objectives, which may include provision of habitat components specifically for bats. Although Yarrow (2005) recommend this as a second step some landowners will already have these resource goals in place as part of corporate environmental policies or personal objectives. Landowners often express their commitment to conservation of biological diversity in writing to help initially guide what baseline information needs to be collected.

Once broad goals have been established and captured in writing, landowners should consider prioritizing considerations for biodiversity and identifying explicit short- and long-term resource goals. Landowners interested specifically in bats could consult the Strategic Plan of the North American Bat Conservation Partnership (www.batcon.org/nabcp/newsite/index.html), information available through regional bat working groups (e.g., Southeast Bat Diversity Network), biologists with Bat Conservation International (www.batcon.org), or other biologists knowledgeable about habitat management for forest-dwelling bats. Landowners should develop objectives (specific actions aimed at accomplishing goals) that are measurable and can be used as a foundation for selecting appropriate metrics that can later be used to evaluate progress.

DEVELOPING AND IMPLEMENTING A PLAN

The third step is development and implementation of a detailed plan outlining specific actions and management prescriptions to meet biodiversity goals and objectives. Although management prescriptions for bats may be linked to land classification units and intended for application across the landscape, they likely will need to be modified to reflect site-specific considerations (e.g., vegetation, soil type and capabilities, unique areas, and other features). The management plan should contain a time-table and schedule for implementation of management recommendations to help landowners gauge progress. The plan should also consider economic and operational constraints, and overall biodiversity objectives, as habitat management options for bats need to align as closely as possible with other land use objectives.

As part of the plan, landowners should identify metrics for measuring and monitoring progress and for identifying opportunities for improvement. Thus, the selected metrics should be cost-effective, related to the landowner objectives, and appropriately measure forest conditions relative to biodiversity objectives (Guynn et al. 2004a). Process-oriented metrics should focus on key management issues, while outcome-oriented metrics should focus on a limited number of important and measurable features of the forest (Guynn et al. 2004a).

MONITORING AND ADJUSTING THE PLAN

It usually is appropriate for landowners to monitor progress and make adjustments to their management plan, the critical component of an adaptive management process (Halbert 1993). Monitoring, of course, requires gathering data related to metrics identified in the plan and contrasting those data with metric targets and expectations for success. Landowners may want to establish readily achievable ranges or thresholds for metrics early in the biodiversity planning process then narrow these ranges over time. For example, a landowner may initially propose

pine trees are considered for retention within TMZs and corridors because they have been used as roost sites by evening bats (Hein et al. 2003; D.A. Miller, unpubl. data; Temple et al. 2004).

MeadWestvaco cooperates on research projects investigating forestry operations and implementation of their EBF approach. Recent and ongoing studies have documented species composition of bat assemblages in different habitats, roosting and foraging habitat use of selected species of bats, and use of forested corridors (Hein et al. 2003, 2004, 2005; Menzel 2003). Investigators have documented 11 of the expected 12 species of bats in the lower Coastal Plain on MeadWestvaco's managed forests and several species of concern including southeastern myotis (*Myotis austroriparius*), Rafinesque's big-eared bat, northern yellow bat (*Lasiurus intermedius*), and hoary bat (*L. cinereus*) (Hein et al. 2003, 2004).

PORT BLAKELY TREE FARMS (WASHINGTON)

A significant portion of lands owned by Port Blakely Tree Farms, L.P. in the State of Washington consists of relatively small tracts surrounded by other land uses. Thus, Port Blakely Tree Farms plans for biological diversity using a formal "harvest unit review process." Prior to harvest operations, Port Blakely screens the unit to determine whether there is any known federally endangered or threatened species within 1.6 km of the property. They also query Washington Department of Fisheries and Wildlife's Priority Habitats or Species database, the Natural Heritage database, and other databases such as the national wetlands inventory for state-listed species, and species and habitats that are considered at risk or high priority. Next they query an internal database to identify streams and other aquatic habitats occurring on the unit. If the unit has not been surveyed previously it will be scheduled for surveys. If they have not already done so, Port Blakely staff will map fish distribution and identify any other wildlife habitat issues or concerns on the property. Once these assessments are complete, Port Blakely biologists provide recommendations to the forester who completes the unit layout and applies correct buffers along streams and other sensitive sites as required by state law. Within this process, any conservation issues pertaining specifically to bats would be addressed through the state priority habitats or species database. If species of concern were known to occur in the vicinity of the harvest unit, Port Blakely biologists would conduct field assessments and determine how to accommodate any special habitat structures such as snags and wildlife trees.

PLUM CREEK TIMBER COMPANY
(MONTANA AND WASHINGTON)

Management provisions used by Plum Creek Timber Company in Montana and Washington to enhance biological diversity, including bats, differ somewhat among states. In Montana, Plum Creek Timber Company has instituted structure retention guidelines for down wood, snags, and green reserve trees. Management for snags and down wood is primarily driven by stand-level guidance on minimum density per stand (with preference given to large snags) and the need to distribute these features across topography. Management for green reserve trees is based on guidelines for stand-level density that are implemented within the context of landscape-scale considerations, such as past and future management, habitat conditions and management direction on adjacent ownerships, and characteristics of natural features. When reserve trees are selected, priority is given to large hollow trees.

In Washington, Plum Creek has a habitat conservation plan that addresses the needs of bats. The Plan requires retention of an average of 7.4 snags per ha harvested in even-aged units and retention of additional green recruitment trees where snags are lacking. Where snags are not limited, Plum Creek seeks to retain 7.4 green recruitment trees per ha that are either dominants or codominants. Large and hollow snags are given priority when selecting snags for retention. When new caves are discovered, the company notifies the U.S. Fish and Wildlife Service, coordinates with the Service and the state wildlife agency on strategies to maintain the integrity of the cave, and designates a forested buffer around each cave. Plum Creek also uses a guild approach based on the breeding and feeding habitat preferences of vertebrate species, including bats, to model and evaluate suitable habitat conditions throughout a 50-year period based on forest-growth and harvesting assumptions.

Additionally, Plum Creek's habitat conservation plan specifies retention of habitat reserves to provide older mature forest conditions with increased snag and legacy tree abundance for northern spotted owls (*Strix occidentalis caurina*); these reserves should provide foraging and roosting resources for bats. The plan also establishes riparian reserves that help maintain water quality, insect-breeding areas, and older mature trees that may serve as roosting and foraging habitat for bats, and contains conservation measures for wetlands, seeps, springs, and talus slopes that increase the probability that larger trees and snags will be retained. Finally, active forest management within the plan area creates early-successional habitats and forest edges providing foraging areas for many species of bats.

CANADIAN FOREST PRODUCTS, LTD., COASTAL OPERATIONS (BRITISH COLUMBIA)

In British Columbia, Canadian Forest Products, Ltd., Coastal Operations (Canfor) plans for biological diversity under the auspices of their sustainable forest management plan, which was developed to obtain certification under the CAN/CSA Z809-6 Standard. The plan identifies 53 objectives and associated, quantifiable indicators related to values and goals of sustainable forest management. For example, seral stage representation is an objective because it relates to landscape and species diversity, health, resiliency, and productivity of forest ecosystems, ecological cycles, and multiple forest benefits. The objective for this indicator is to achieve the desired seral stage representation ($\pm 10\%$) within three rotations with an initial focus on old seral stages.

Many of the 53 objectives and associated indicators such as old growth, wildlife tree and forest patch retention, seral stage distribution, forest interior, and patch size representation relate to forest stand and landscape structure. Another objective is to establish management areas for and mitigate potential impacts on caves and karst features (e.g., by relocating roads and cutblock boundaries, establishing reserves, and employing alternative harvest systems). These provisions will generally enhance stand and landscape diversity and protect important hibernacula. Canfor's sustainable forest management plan also has one objective specific to Keen's myotis (*Myotis keenii*): to establish management areas around hibernacula and maternity sites within one month of discovery. The company also inventories caves systems that have the potential to support Keen's myotis, and is featuring management for this species in certain areas that were identified by using a predictive habitat relationship model.

Canfor uses compliance and effectiveness monitoring to gauge progress toward the objectives within the SFM Plan. For example, the SFM Plan seeks to achieve a diverse landscape by maintaining a variety of communities and ecosystems. Thus, Canfor monitors the number of trees per ha, number and area of forest patches retained, and other variables as part of compliance monitoring; it monitors the response of key indicator species over time and space as part of effectiveness monitoring. As part of effectiveness monitoring, Canfor documents use by Keen's myotis of established wildlife habitat areas every five years and continues to study habitat relationships for the species.

BAT CONSERVATION ON INDUSTRY LANDS

Although fiber production and bat conservation are not mutually exclusive (Hayes 2003), there are significant challenges for integration of these two goals within forest plans. The first and foremost challenge is to understand the effects of forest management at multiple spatial and temporal scales on bats (Hayes 2003; Miller et al. 2003). In any discussion about conservation of bats, the lack of empirical data linking viability of bat communities to specific forest features must be kept in mind. Also, the literature provides conflicting views on compatibility of intensive forest management with bat conservation (Miller et al. 2003). This is because individual bat species respond differently to forest management (Patriquin and Barclay 2003), in particular, generalists (Elmore et al. 2004) versus habitat or roost-specific species (Ormsbee and McComb 1998), and because unreliable generalizations of research results are made across different forest landscapes (Miller et al. 2003).

In this section we discuss issues related to bat conservation that should be considered when planning for bats on industry lands. Because forest certification systems generally address habitat conservation at stand and landscape scales (Sustainable Forestry Board 2005), we focus our discussion on those scales. Likewise, we limit our discussion to even-aged management systems because most forest products companies use them to cost-effectively produce wood.

Forest products companies generally operate on their lands at the stand level. That is, a block of land will be treated as a management unit throughout the rotation. Although management unit sizes may vary, companies usually limit clearcut size based on company policy, certification standards, or government regulations, which in turn generally drive stand size. For example, SFI participants have a maximum average clearcut size of 296 ha except in response to forest health emergencies (Sustainable Forestry Board 2005). In most cases, habitat structure available to bats (e.g., characteristics and availability of roost trees and structural diversity) will be dictated by silvicultural treatments at the stand level.

At the stand level, even-aged management creates challenges to providing within-stand structural diversity likely needed by bats (Humes et al. 1999; Jung et al. 1999; Kunz 1982; Kunz and Lumsden 2003; Thomas 1988). Some species may need a diversity of tree species within a stand (Brigham et al. 1997b; Campbell et al. 1996; Kalcounis et al. 1999); this is generally lacking in intensively managed forest stands. Of greatest concern, however, is provision of roost trees, in particular, larger snags and trees. It is well documented that many bat species require or prefer large snags for roosting (Barclay and Kurta, chap. 2 in this volume; Hayes 2003; Menzel et al. 2000; Miller et al. 2003), especially in western and northern North America. Additionally, bats have been found to select roost trees (snags or live trees) that are taller, have a larger diameter, are in particular decay classes, and have more open space around them than other trees in the same stand (Brigham et al. 1997b; Campbell et al. 1996; Menzel et al. 2000; Ormsbee and McComb 1998; Waldien et al. 2000; Weller and Zabel 2001). Because even-aged stands in general contain equally spaced trees that are of the same size and because most industrial timberlands have short rotation lengths, the opportunity to provide trees and snags of different sizes within individual stands using standard silvicultural regimes is limited. Forest plans should include provisions to conserve roosting structures, in particular, large snags and leave trees.

Another consideration for roosts at the stand level is that most species of bats frequently switch roosts during relatively short periods, sometimes every few days, with alternate roosts often located near one another (Elmore et al. 2004; Grindal 1998; Menzel et al. 2000; Vonhof and Barclay 1996). Therefore, forest patches that are conserved specifically for roosting structure likely need to contain several available roosts nearby. This also indicates that stand- or landscape-level cues may dictate roost site selection (Elmore et al. 2004). Until we understand these cues, the most prudent approach is to provide roosts in a variety of stands contained within different landscape contexts (Duchamp et al., chap. 9 in this volume).

Despite potential limitations for bat habitat on industrial ownerships, there are opportunities to improve habitat by providing missing habitat components through timber harvesting, vegetation management, and timber stand improvement (Hayes 2003; Hayes and Loeb, chap. 8 in this volume; Ormsbee and McComb 1998). This may even include creation of snags using herbicides (Conner et al. 1981; Dickson et al. 1995) or by "topping" live trees (Bull et al. 1981; Lewis 1998) in stands or on landscapes lacking this important habitat feature. Additionally, structural composition of forest stands may be more important than age with respect to foraging and roosting requirements of some species of bats (Elmore et al. 2004; Humes et al. 1999; Menzel et al. 2002b), in particular, if roosts are provided or retained (Betts 1998; Humes et al. 1999; Ormsbee and McComb 1998; Waldien et al. 2000). This provides opportunities to integrate stand-level habitat needs for bats within short-rotation silvicultural systems.

Intensively managed stands undergo changes during the rotation, progressing through three primary developmental stages: (1) young, open-canopy stands; (2) closed canopy stands; and (3) older, open-canopy stands (Miller et al. 2004). Young, open-canopy stands, such as regenerating clearcuts and forest gaps created through timber harvest, have been shown to provide foraging opportunities for several bat species (Brigham et al. 1997b; Grindal and Brigham 1998; Krusic et al. 1996; Menzel et al. 2001b, 2002a). However, bats have limited roosting opportunities within this seral stage (Crampton and Barclay 1998; Jung et al. 1999), although some species of bats roost in residual snags and in stumps (Vonhof and Barclay 1997; Waldien et al. 2000, 2003). Additionally, the high diversity of plant species within young, open-canopy stands likely improves insect diversity and abundance (Haddad et al. 2001; Humphrey et al. 1999; Siemann et al. 1998). In turn, diversity of plant species depends on stand establishment techniques with more intensive techniques and higher planting densities resulting in lower plant diversity (Edwards 2004). Lower tree-planting densities (e.g., <1,200 trees per ha in the southeastern United States) and less intensive stand establishment treatments would likely benefit bats through a more diverse invertebrate community and by extending the window of time a stand spends in an open-canopy condition.

Closed-canopy forest, characterized by high-stem density and little to no understory development, is the least desirable forest stage for bats (Humes et al. 1999; Thomas 1988). Trees in these stands typically are not large enough for roosts and the high degree of clutter makes these stands undesirable as foraging habitat for bats (Brigham et al. 1997a, b; Patriquin and Barclay 2003). The lack of a well-established understory also likely limits invertebrate production (Humphrey et al. 1999; Siemann et al. 1998). As with most other wildlife species, the best management strategy is to minimize the time that stands spend in this stage. This can be accomplished by lower planting densities (requires longer for the stand to close canopy) and by thinning stands as early as possible.

Older open-canopy stands, which primarily occur under sawtimber rotations with commercial thinnings to reduce tree density, likely provide the greatest opportunity to provide habitat structure for bats at the stand level. Thinning densely stocked forest increases tree growth and produces a forest structure conducive to foraging, commuting, and roosting by bats (Elmore et al. 2004; Hayes 2003; Humes et al. 1999; Miles et al. 2004).

In general, thinning creates tree mortality (Paxton et al. 2004), thus, potentially providing roosting snags for some species (Humes et al. 1999). In some cases, thinned stands can be managed to emulate some of the desirable forest structure characteristics of older stands (Barbour et al. 1997; Carey and Curtis 1996; Guldin et al., chap. 7 in this volume; Hayes et al. 1997), providing opportunities to improve habitat conditions within operational and economic constraints. As these stands are scheduled for harvest, opportunities exist to mark green trees and snags to begin the process of providing larger snags, and thus roosting structure, for bats into the next rotation (Betts 1998; Guldin et al., chap. 7 in this volume; Humes et al. 1999; Ormsbee and McComb 1998; Waldien et al. 2000).

Intensively managed landscapes also contain varying amounts of riparian areas or SMZs, inoperable sites, unique ecological communities, or other areas that are not under intensive management. These "natural" stands can greatly contribute to overall habitat diversity and provide habitat structure characteristic of older forests, such as roost trees (Gorresen and Willig 2004; Hogberg et al. 2002; Jung et al. 1999; Kunz 1982; Kunz and Lumsden 2003; Swystun et al. 2001; Waldien et al. 2000; Zielinski and Gellman 1999). However, highly isolated habitat patches may not be available to bats, limiting their usefulness for habitat conservation (Swystun et al. 2001). In the southeastern United States, where thermoregulation of roosts may not be as important as in northern and western North America (Menzel et al. 2002b), provision of roosts within SMZs may be a reasonable and economically viable approach because bats have often been documented roosting in these types of stands (Elmore et al. 2004; Menzel et al. 2001a; Miles et al. 2004; D. A. Miller, unpubl. data; but see Hutchinson and Lacki 2000). These include two bats of conservation concern, the southeastern myotis and Rafinesque's big-eared bat (Hoffman et al. 1999; Lance et al. 2001). In the western and northern parts of North America, however, it is critical to provide roosts in some upland stands (Hayes 2003; Ormsbee and McComb 1998; Perkins and Cross 1988; Waldien et al. 2000; Weller and Zabel 2001). Conservation of forest patches should contain upland and riparian areas (Campbell et al. 1996).

CONSIDERATIONS AT THE LANDSCAPE SCALE

Although management activities on industrial lands occur at the stand scale, industry plans for biological diversity should consider bat habitat relationships at multiple scales, including at the landscape scale (Duchamp et al., chap. 9 in this volume; Jung et al. 1999; Wilhere 2003). Recently, Duchamp et al. (chap. 9 in this volume) suggested that the foundation of habitat conservation for bats is provision of adequate roosting structure with consideration for placement of these structures on the landscape. In the western United States, including the Pacific Northwest, and in western Canada, conservation of adequate roosts does appear fundamental to bat conservation (Campbell et al. 1996; Hayes 2003; Hayes and Loeb, chap. 8 in this volume; Miller et al. 2003; Rabe et al. 1998). This is based on the overriding importance of roosts for bats (Kunz 1982; Kunz and Lumsden 2003) and previous findings that show preference for roost location depend on the species of bat and landscape context, including location of other potentially limiting landscape features such as water resources (Campbell et al. 1996; Waldien et al. 2000). Consideration for provision of foraging habitat and interactions between foraging habitat and roosts are also important (Hutchinson and Lacki 2000; Miller et al. 2003). This need to consider both roosting and foraging habitat has led some to recommend provision of a mosaic of habitat types and forest ages across landscapes to meet the habitat needs of bats on managed landscapes (Jung et al. 1999; Krusic et al. 1996; Patriquin and Barclay 2003).

Provision of such a habitat mosaic via timber harvest provides not only possible benefits for bats, but also possible detriments. Benefits may include provision of foraging habitat (Brigham et al. 1997b; Krusic et al. 1996; Menzel et al. 2002a), easier access to roosts in mature stands (Grindal 1998), the possible benefit of edge habitats to bats (Crampton and Barclay 1998; Menzel et al. 2002a; Swystun et al. 2001), and a continuous supply of different-aged forests through time. However, others have suggested forest fragmentation may negatively affect bats through such factors as reduction in number and distribution of roosts and possibly increased commuting distances (Grindal and Brigham 1998; Hutchinson and Lacki 1999, 2000). Because landscape patterns likely affect bat species differently, and perhaps even within the same species, it will be difficult to determine the "optimum" mixture of disturbance versus nondisturbance for bat communities on forested landscapes. On industrial landscapes, large areas of unharvested or late-rotation forests are likely not an option. Thus, understanding how bat assemblages respond to landscape pattern, and how potential negative effects can be mitigated, may be important for their conservation on many industry landscapes.

The number of roosts required to maintain bat populations is poorly understood (Hayes and Loeb, chap. 8 in this volume). However, if roosts are limiting for some species of bats on industrial forest landscapes (Fenton 1997), in particular, for species requiring large snags, and if habitat selection is strongly influenced by roost location (Kunz 1982), the best solution is likely provision of snags in a variety of landscape contexts (Duchamp et al., chap. 9 in this volume). This presents a significant challenge on intensively managed landscapes. For example, economics dictate that industrial landowners derive as much income as possible from each acre within reasonable constraints. Leaving large, valuable trees to become snags carries real operational costs beyond the value of the individual tree (Guldin et al., chap. 7 in this volume). These costs include working around leave tree areas, reduced productivity from additional encumbrances on the land, and potential safety issues. Additionally, due mostly to past and current forestry practices, size of snags and snag density are often lower on private forestlands than on public land, obviously impacting available roost resources for bats.

When planning for biological diversity, therefore, most forest products companies emphasize the retention of snags, den trees, and large leave trees in inoperable areas or areas that are lightly managed, such as SMZs and legacy forest patches. But placing all or most of these structures in very similar stands and within similar landscape contexts may not be an appropriate conservation strategy for forest-dwelling bats (Duchamp et al., chap. 9 in this volume; Hayes 2003). This means that significant effort may be needed on some lands to improve roosting structures for bats across the entire ownership. Therefore, the challenge becomes understanding how to optimize placement of roosting structures for bats while minimizing the impact on the economic and operational aspects of privately owned forests.

Regional differences in the provision of snags and leave trees across landscapes must also be recognized. In the eastern United States, conservation of large snags across forested landscapes may not be as critical as it is in the western United States, the Pacific Northwest of the United States, or western Canada. This is due to the ability of many eastern bat species to roost in a variety of structures (Barbour and Davis 1969) and in riparian stands that are often retained on the landscape. We are not suggesting that conservation of snags on industrial forests in the eastern United States is not important. Snags are vitally important to many species in addition to bats (Dickson and Wigley 2001; Hunter 1990; USDA Forest Service 1983), and should be part of forest management plans. Additionally, some bat species of conservation concern have specialized roosting requirements (e.g, Indiana myotis, *Myotis sodalis;* Whitaker and Hamilton 1998) that must be considered. However, given the warmer climate, a general lack of high mountains, and the relative abundance of riparian forests in the eastern United States, placement of snags may not be as critical as in other regions. On the other hand, snags and coarse woody debris do not last nearly as long in the eastern United States, particularly in the southeastern United States, as they do in the West (Zimmerman 2004). Therefore, snags are a much more ephemeral roosting resource, adding to the difficulty of providing this resource on managed landscapes across time. For example, a Douglas-fir (*Pseudotsuga menziesii*) snag may last more than 100 years in the Pacific Northwest, but a southern oak (*Quercus* spp.) snag may only last 30 years in the southeastern United States.

CONCLUSIONS

Significant challenges are associated with planning for bats within industrial forest landscapes. The first challenge is integration of management for bats within silvicultural and economic constraints. Opportunities to do this will vary by company, and perhaps even within companies, based on land management objectives. Therefore, management activities proposed by plans will be constrained by landowner objectives, current forest conditions, and economic realities (Yarrow 2005). Second, large-ownership landowners often address bats within a general plan for conservation of biodiversity, which usually directs efforts toward provision of habitat diversity and protection of unique ecological communities. Thus, plans must integrate the needs of bats with the needs of other forest species (Guldin et al., chap. 7 in this volume).

Despite these challenges, there are also opportunities. The greatest opportunity exists simply because the ability to produce income from industrial forests provides a strong incentive to keep forested lands in forest cover, a key conservation benefit of intensively managed forest lands (Hayes et al. 2005). Although some characteristics of these forests may not be ideal from a bat conservation standpoint, they are decidedly more desirable for forest-dwelling bats than alternative land uses, such as urbanization and agriculture. Opportunities also exist to provide unique habitat features and diversity, including older forest patches, across landscapes via compliance to biodiversity measures within forest certification systems (Sustainable Forestry Board 2005). Such measures to retain legacy forest stands on the landscape for overall enhancement of biological diversity should also be beneficial for bats (Duchamp et al., chap. 9 in this volume; Jung et al. 1999; Waldien et al. 2000).

Another opportunity to further the goals of bat conservation on private lands is consideration of bats as a part of broader biodiversity measures. Bats are sometimes considered to be indicators of forest health (Fenton 2003), and reasonable and meaningful biodiversity indicators for sustainable forestry are still under development (Guynn et al. 2004a). Perhaps examination of bat communities on industrial landscapes can serve as one set of indicators. More work is needed, however, to understand what bat species should occur on a given landscape (Humphrey 1975; Miller 2003) and how we can effectively sample these communities (Weller, chap. 10 in this volume)

Forest industry lands, like any other forested landscape, cannot provide everything for all species. Industrial lands will not have a large preponderance of older forests, but perhaps this is not required if the existing forest is managed for the benefit of bats (Hayes 2003; Humes et al. 1999; Menzel et al. 2002b). Creative solutions and realistic expectations will be needed if industry plans are to integrate needs for bat habitat within operational and economic constraints, and within broader efforts to conserve biological diversity.

Finally, although we do not have a thorough understanding of bat-habitat relationships, especially within the available range of forest management scenarios, we contend this should not discourage landowners from including practices in forest plans that currently are thought to be beneficial for bats. We suggest that forest plans follow an adaptive management approach (Halbert 1993; Irwin and Wigley 1993; Lancia et al. 1993; Walters and Holling 1990). Adaptive management provides a framework in which managers make habitat management recommendations and goals based on the current and best available information. Approaches to managing for bats then can be modified based on success of implemented actions and the gathering of new information. This prevents managers from being caught in a "research paralysis" in which we

continue to wait for more information before implementing conservation measures.

ACKNOWLEDGMENTS

We thank NCASI and Weyerhaeuser Company for their support, and
E. B. Arnett, J. P. Hayes, and C. L. Chambers for comments they provided
on an earlier draft of this manuscript. We especially thank J. A. Deal,
G. C. Muckenfuss, K. L. Risenhoover, H. C. Stabins, and D. E. Wood for
descriptions they provided of planning and management processes and
guidelines used by their respective companies. We also gratefully acknowledge the assistance of K. Abusow in identifying appropriate forest
statistics for Canada.

LITERATURE CITED

Abusow, K. 2004. Private woodlot certification: challenges and opportunities. For-
    estry Chronicle 80:658–660.
American Forest and Paper Association and Clemson University. 2001. U.S. forests
    facts & figures, 2001. American Forest and Paper Association, Washington,
    DC.
Arnett, E.B., and J.B. Haufler. 2003. A customer-based framework for funding pri-
    ority research on bats and their habitats. Wildlife Society Bulletin 31:98–
    103.
Barbour, R.J., S. Johnston, J.P. Hayes, and G.F. Tucker. 1997. Simulated stand char-
    acteristics and wood product yields of Douglas-fir forests managed for
    ecosystem objectives. Forest Ecology and Management 91:205–219.
Barbour, R.W., and W.H. Davis. 1969. Bats of America. University Press of Ken-
    tucky, Lexington, KY.
Betts, B.J. 1998. Roosts used by maternity colonies of silver-haired bats in north-
    eastern Oregon. Journal of Mammalogy 79:643–650.
Blinn, C.R., and M.A. Kilgore. 2001. Riparian management practices, a summary
    of state guidelines. Journal of Forestry 99:11–17.
Brigham, R.M., S.D. Grindal, M.C. Firman, and J.L. Morissette. 1997a. The influ-
    ence of structural clutter on activity patterns of insectivorous bats. Cana-
    dian Journal of Zoology 75:131–136.
Brigham, R.M., M.J. Vonhof, R.M.R. Barclay, and J.C. Gwilliam. 1997b. Roosting
    behavior and roost-site preferences of forest-dwelling California bats (*My-
    otis californicus*). Journal of Mammalogy 78:1231–1239.
Brown, N., L. Master, D. Faber-Langendoen, P. Comer, K. Maybury, M. Robles,
    J. Nichols, and T.B. Wigley. 2004. Managing elements of biodiversity in sus-
    tainable forestry programs: Status and utility of NatureServe's information
    resources to forest managers. NCASI Technical Bulletin No. 885. National
    Council for Air and Stream Improvement, Inc., Durham, NC.
Bull, E.L., A.D. Partridge, and W.G. Williams. 1981. Creating snags with explosives.
    USDA Forest Service Research Note PNW-393, Portland, OR.
Burger, L.W., Jr. 2003. Roles and limitations of herbicides in habitat restoration,
    p. 88, *in* The Wildlife Society Tenth Annual Conference. The Wildlife Soci-
    ety, Bethesda, MD.
Campbell, L.A., J.G. Hallett, and M.A. O'Connell. 1996. Conservation of bats in
    managed forests: use of roosts by *Lasionycteris noctivagans*. Journal of Mam-
    malogy 77:976–984.
Carey, A.B., and R.O. Curtis. 1996. Conservation of biodiversity: a useful paradigm
    for ecosystem management. Wildlife Society Bulletin 24:610–620.
Conner, R.N, J.G. Dickson, and B.A. Locke. 1981. Herbicide-killed trees infected by

fungi: potential cavity sites for woodpeckers. Wildlife Society Bulletin 9:308–311.

Crampton, L.H., and R.M.R. Barclay. 1998. Selection of roosting and foraging habitat by bats in different-aged aspen mixedwoods stands. Conservation Biology 12:1347–1358.

Dickson, J.G., and T.B. Wigley. 2001. Managing forests for wildlife, pp. 83–94, *in* Wildlife of southern forests, habitat and management (J.G. Dickson, ed.). Hancock House Publishers, Surrey, BC.

Dickson, J.G., J.H. Williamson, and R.N. Conner. 1995. Longevity and bird use of hardwood snags created by herbicides. Proceedings of the Annual Conference of the Southeastern Association of Fish and Wildlife Agencies 49:332–339.

Edwards, S.L. 2004. Effects of intensive pine plantation management on wildlife habitat quality in southern Mississippi. M.S. thesis, Mississippi State University, Mississippi State, MS.

Egan, A.F., and A.E. Luloff. 2000. The exurbanization of America's forests: research in rural social science. Journal of Forestry 98:26–30.

Elmore, L.W., D.A. Miller, and F.J. Vilella. 2004. Selection of diurnal roost sites by red bats (*Lasiurus borealis*) in an intensively managed pine forest in Mississippi. Forest Ecology and Management 199:11–20.

———. 2005. Foraging area size and habitat use by red bats (Lasiurus borealis) in an intensively managed pine landscape in Mississippi. American Midland Naturalist 157:405–417.

Fenton, M.B. 1997. Science and the conservation of bats. Journal of Mammalogy 78:1–14.

———. 2003. Science and the conservation of bats: where to next? Wildlife Society Bulletin 31:6–15.

Gorresen, P.M., and M.R. Willig. 2004. Landscape responses of bats to habitat fragmentation in Atlantic forests of Paraguay. Journal of Mammalogy 85:688–697.

Grindal, S.D. 1998. Habitat use by bats, *Myotis* spp., in western Newfoundland. Canadian Field-Naturalist 113:258–263.

Grindal, S.D., and R.M. Brigham. 1998. Short-term effects of small-scale habitat disturbance on activity by insectivorous bats. Journal of Wildlife Management 62:996–1003.

Guynn, D.C., S.T. Guynn, P.A. Layton, and T.B. Wigley. 2004a. Biodiversity metrics in sustainable forestry programs. Journal of Forestry 102:46–52.

Guynn, D.C., S.T. Guynn, T.B. Wigley, and D.A. Miller. 2004b. Herbicides and forest biodiversity: what do we know and where do we go from here? Wildlife Society Bulletin 32:1085–1092.

Haddad, N.M., D. Tilman, J. Haarstad, M. Ritchie, and J.M.H. Knops. 2001. Contrasting effects of plant richness and composition on insect communities: a field experiment. American Naturalist 158:17–35.

Halbert, C.L. 1993. How adaptive is adaptive management? Implementing adaptive management in Washington State and British Columbia. Reviews in Fisheries Science 1:261–283.

Hayes, J.P. 2003. Habitat ecology and conservation of bats in western coniferous forests, pp. 81–119, *in* Mammal community dynamics in coniferous forests of western North America: management and conservation (C.J. Zabel and R.G.Anthony, eds.). Cambridge University Press, Cambridge, MA.

Hayes, J.P., S. Chan, W.H. Emmingham, J. Tappeiner, L. Kellogg, and J.D. Bailey. 1997. Wildlife response to thinning young forests in the Pacific Northwest. Journal of Forestry 95:28–33.

Hayes, J.P., S.H Schoenholtz, M.J. Hartley, G. Murphy, R.F. Powers, D. Berg, and S.R. Radosevich. 2005. Environmental consequences of intensively managed forest plantations in the Pacific Northwest. Journal of Forestry 103:83–87.

Hein, C.D., S.B. Castleberry, and K.V. Miller. 2003. Preliminary analysis on the importance of forested corridors on roost-site selection of Seminole (*Lasiurus seminolus*) and evening (*Nycticeius humeralis*) bats in managed forests. Bat Research News 44:141.

———. 2004. Roost-site selection of Rafinesque's big-eared bats and Southeastern myotis on a managed pine forest in the lower Coastal Plain, South Carolina. Bat Research News 45:225.

———. In press. Winter roost-site selection by Seminole bats in the lower Coastal Plain of South Carolina. Southeastern Naturalist.

Helms, J.A., ed. 1998. The dictionary of forestry. Society of American Foresters. Bethesda, MD.

Hoffman, J.E., J.E. Gardner, J.K. Krejca, and J.D. Garner. 1999. Summer records and a maternity roost of the southeastern myotis (*Myotis austroriparius*) in Illinois. Transactions of the Illinois State Academy of Science 92:95–107.

Hogberg, L.K., K.J. Patriquin, and R.M.R. Barclay. 2002. Use by bats of patches of residual trees in logged areas of the boreal forest. American Midland Naturalist 148:282–288.

Humes, M.L., J.P. Hayes, and M.W. Collopy. 1999. Bat activity in thinned, unthinned, and old-growth forests in western Oregon. Journal of Wildlife Management 63:553–561.

Humphrey, J.W., C. Hawes, A.J. Peace, R. Ferris-Kaan, and M.R. Jukes. 1999. Relationships between insect diversity and habitat characteristics in plantation forests. Forest Ecology and Management 113:11–21.

Humphrey, S.R. 1975. Nursery roosts and community diversity of Nearctic bats. Journal of Mammalogy 56:321–346.

Hunter, M.L., Jr. 1990. Wildlife forests, and forestry: principles of managing forests for biological diversity. Prentice-Hall, Inc., Englewood Cliffs, NJ.

Hutchinson, J.T., and M.J. Lacki. 1999. Foraging behavior and habitat use of red bats in mixed mesophytic forests of the Cumberland Plateau, Kentucky, pp.171–177, *in* Proceedings, 12th Central Hardwoods Forest Conference (J.W. Stringer and D.L. Loftis, eds.). USDA Forest Service, General Technical Report SRS-24.

———. 2000. Selection of day roosts by red bats in mixed mesophytic forests. Journal of Wildlife Management 64:87–94.

Irwin, L.L., and T.B. Wigley. 1993. Toward an experimental basis for protecting forest wildlife. Ecological Applications 3:213–217.

Jung, T.S., I.D. Thompson, R.D. Titman, and A.P. Applejohn. 1999. Habitat selection by forest-roosting bats in relation to mixed-wood stand types and structure in central Ontario. Journal of Wildlife Management 63:1306–1319.

Kalcounis, M.C., K.A. Hobson, R.M. Brigham, and K.R. Hecker. 1999. Bat activity in the boreal forest: importance of stand type and vertical strata. Journal of Mammalogy 80:673–682.

Krusic, R.A., M. Yamaski, C.D. Neefus, and P.J. Pekins. 1996. Bat habitat use in White Mountain National Forest. Journal of Wildlife Management 60:625–631.

Kunz, T.H., ed. 1982. Ecology of bats. Plenum Publishing, New York.

Kunz, T.H., and L.F. Lumsden. 2003. Ecology of cavity and foliage roosting bats, pp. 3–89, *in* Bat ecology (T.H. Kunz and M.B. Fenton, eds.). University of Chicago Press, Chicago, IL.

Laerm, J., W.M. Ford, and B.R. Chapman. 2000. Conservation status of terrestrial mammals of the Southeastern United States, pp. 4–16, *in* Fourth colloquium on conservation of mammals in the Southeastern United States (B.R. Chapman, ed.). Occasional Papers of the North Carolina, Museum of Natural Sciences, Raleigh, NC.

Lance, R.F., B.T. Hardcastle, A. Talley, and P.L. Leberg. 2001. Day-roost selection by

Rafinesque's big-eared bats (*Corynorhinus rafinesquii*) in Louisiana forests. Journal of Mammalogy 82:166–172.

Lancia, R.A., T.D. Nudds, and M.L. Morrison. 1993. Opening comments: slaying slippery shibboleths. Transactions of the North American Wildlife and Natural Resources Conference 58:505–508.

Lautenschlager, R.A., and T.P. Sullivan. 2004. Improving research into effects of forest herbicide use on biota in northern ecosystems. Wildlife Society Bulletin 32:1061–1070.

Lewis, J.C. 1998. Creating snags and wildlife trees in commercial forest landscapes. Western Journal of Applied Forestry 13:97–101.

Mehmood, S.R., and D. Zhang. 2001. Forest parcelization in the United States: a study of contributing factors. Journal of Forestry 99:30–34.

Menzel, M.A. 2003. An examination of factors influencing the spatial distribution of foraging bats in pine stands in the southeastern United States. Ph.D. dissertation, West Virginia University, Morgantown, WV.

Menzel, M.A., T.C. Carter, W.M. Ford, and B.R. Chapman. 2001a. Tree-roost characteristics of subadult and female adult evening bats (*Nycticeius humeralis*) in the Upper Coastal Plain of South Carolina. American Midland Naturalist 145:112–119.

Menzel, M.A., T.C. Carter, W.M. Ford, B.R. Chapman, and J. Ozier. 2000. Summer roost tree selection by eastern red, Seminole, and evening bats in the upper coastal plain of South Carolina. Proceedings of the Annual Conference of the Southeastern Association of Fish and Wildlife Agencies 54:304–313.

Menzel, M.A., T.C. Carter, J.M. Menzel, W.M. Ford, and B.R. Chapman. 2002a. Effects of group selection silviculture in bottomland hardwoods on the spatial activity patterns of bats. Forest Ecology and Management 162:209–218.

Menzel, M.A., J.M. Menzel, W.M. Ford, J.W. Edwards, T.C. Carter, J.B. Churchill, and C. Kilgo. 2001b. Home range and habitat use of male Rafinesque's big-eared bats (*Corynorhinus rafinesquii*). American Midland Naturalist 145:402–408.

Menzel, M.A., S.F. Owen, W.M. Ford, J.W. Edwards, P.B. Wood, B.R. Chapman, and K.V. Miller. 2002b. Roost tree selection by northern long-eared bat (*Myotis septentrionalis*) maternity colonies in an industrial forest of the central Appalachian Mountains. Forest Ecology and Management 155:107–114.

Miles, A.C., S.B. Castleberry, D.A. Miller, and L.M. Conner. 2004. Evening bat day roost selection relative to forest management in southwest Georgia. Bat Research News 45:64–65.

Miller, D.A. 2003. Species diversity, reproduction, and sex ratios of bats in managed pine landscapes of Mississippi. Southeastern Naturalist 2:59–72.

Miller, D.A., E.B. Arnett, and M.J. Lacki. 2003. Habitat management for forest-roosting bats in North America: a critical review of habitat studies. Wildlife Society Bulletin 31:30–44.

Miller, D.A., R.E. Thill, M.A. Melchiors, T.B. Wigley, and P.A. Tappe. 2004. Small mammal communities of streamside management zones in intensively managed pine forests of Arkansas. Forest Ecology and Management 203:381–393.

Miller, D.A., and T.B. Wigley. 2004. Introduction: herbicides and forest biodiversity. Wildlife Society Bulletin 32:1016–1019.

Ormsbee, P.C., and W.C. McComb. 1998. Selection of day roosts by female long-legged myotis in the central Oregon Cascade Range. Journal of Wildlife Management 62:596–603.

Patriquin, K.J., and R.M.R. Barclay. 2003. Foraging by bats in cleared, thinned and unharvested boreal forest. Journal of Applied Ecology 40:646–657.

Paxton, B.J., M.D. Wilson, and B.D. Watts. 2004. Relationships between standing dead wood dynamics and bird communities within North Carolina pine plantations. Center for Conservation Biology Technical Report Series, CCBTR-04-08. College of William and Mary, Williamsburg, VA.

Perkins, J.M., and S.P. Cross. 1988. Differential use of some coniferous forest habitats by hoary and silver-haired bats in Oregon. Murrelet 69:21–24.

Rabe, M.J., T.E. Morrell, H. Green, J.C. DeVos Jr., and C.R. Miller. 1998. Characteristics of ponderosa pine snag roosts used by reproductive bats in northern Arizona. Journal of Wildlife Management 62:612–621.

Ravenel, R., M. Tyrrell, and R. Mendelsohn. 2002. Institutional timberland investment: A summary of a forum exploring changing ownership patterns and the implications for conservation of environmental values. Global Institute of Sustainable Forestry, School of Forestry and Environmental Studies, YFF Review 5(2). Yale University, New Haven, CT.

Sallabanks, R., E.B. Arnett, T.B. Wigley, and L. Irwin. 2001. Accommodating birds in managed forests of North America: a review of bird-forestry relationships. NCASI Technical Bulletin No. 822. National Council for Air and Stream Improvement, Inc., Research Triangle Park, NC.

Shepard, J.P., J. Creighton, and H. Duzan. 2004. Forestry herbicides in the United States: an overview. Wildlife Society Bulletin 32:1020–1027.

Siemann, E., D. Tilman, J. Haarstad, and M. Ritchie. 1998. Experimental tests of the dependence of arthropod diversity on plant diversity. American Naturalist 152:738–750.

Smith, W.B., P.D. Miles, J.S. Vissage, and S.A. Pugh. 2004. Forest resources of the United States, 2002. USDA Forest Service General Technical Report NC-241. USDA Forest Service, North Central Research Station, St. Paul, MN.

Stanturf, J.A., R.C. Kellison, F.S. Broerman, and S.B. Jones. 2003. Productivity of southern pine plantations: where are we and how did we get here? Journal of Forestry 101:26–31.

Sustainable Forestry Board. 2005. Sustainable Forestry Initiative® (SFI) Standard 2005–2009 edition. Sustainable Forestry Board, Arlington, VA.

Swystun, M.B., J.M. Psyllarkis, and R.M. Brigham. 2001. The influence of residual tree patch isolation on habitat use by bats in central British Columbia. Acta Chiropterologica 3:197–207.

Tatum, V.L. 2004. Toxicity, transport, and fate of forest herbicides. Wildlife Society Bulletin 32:1042–1048.

Temple, D.L., A.C. Miles, S.B. Castleberry, D.A. Miller, and L.M. Conner. 2004. Evening bat (*Nycticeius humeralis*) use of forked-topped trees: a potential tool for conserving bat roosts in managed pine plantations of the Southeast, *in* Proceedings of the 14th Colloquium on Conservation of Mammals in the Southeastern United States, Helen, GA.

Teske, M.E., and G.G. Ice. 2002. A one-dimensional model for aerial spray assessment in forest streams. Journal of Forestry 100:40–45.

Thomas, D.W. 1988. The distribution of bats in different ages of Douglas-fir forests. Journal of Wildlife Management 52:619–626.

USDA Forest Service. 1983. Snag habitat management: proceedings of the symposium. USDA General Technical Report RM-99, Fort Collins, CO.

Vonhof, M.J., and R.M.R. Barclay. 1996. Roost-site selection and roosting ecology of forest-dwelling bats in southern British Columbia. Canadian Journal of Zoology 74:1797–1805.

———. 1997. Use of tree stumps as roosts by western long-eared bat. Journal of Wildlife Management 61:674–684.

Wagner, R.G., M. Newton, E.C. Cole, J.H. Miller, and B.D. Shiver. 2004. The role of herbicides for enhancing forest productivity and conserving land for biodiversity in North America. Wildlife Society Bulletin 32:1028–1041.

Waldien, D.L., J.P. Hayes, and E.B. Arnett. 2000. Day-roosts of female long-eared myotis in western Oregon. Journal of Wildlife Management 64:785–796.

Waldien, D.L., J.P. Hayes, and B.E. Wright. 2003. Use of conifer stumps in clearcuts by bats and other vertebrates. Northwest Science 77:64–71.

Walters, C.J., and C.S. Holling. 1990. Large-scale management experiments and learning by doing. Ecology 71:2060–2068.

Wear, D.N., and J.G. Greis. 2002. Southern forest resource assessment: summary report. USDA Forest Service General Technical Report SRS-54.

Weller, T.J., and C.J. Zabel. 2001. Characteristics of fringed myotis day roosts in northern California. Journal of Wildlife Management 65:489–497.

Whitaker, J.O., Jr., and W.J. Hamilton Jr. 1998. Mammals of the eastern United States, 3rd ed. Cornell University Press, Ithaca, NY.

Wigley, T.B., W.M. Baughman, M.E. Dorcas, J.A. Gerwin, J.W. Gibbons, D.C. Guynn, Jr., R.A. Lancia, Y.A. Leiden, M.S. Mitchell, and K.R. Russell. 2000. Contributions of intensively managed forests to the sustainability of wildlife communities in the South, *in* Sustaining southern forests: the science of forest assessment. USDA Forest Service, Southern Forest Resource Assessment, Durham, NC.

Wilhere, G.F. 2003. Simulations of snag dynamics in an industrial Douglas-fir forest. Forest Ecology and Management 174:521–539.

Yarrow, G. 2005. Biological diversity and wildlife habitat considerations in managed forests. American Forest and Paper Association, Washington, D.C.

Zielinski, W.J., and S.T. Gellman. 1999. Bat use of remnant old-growth redwood stands. Conservation Biology 13:160–167.

Zimmerman, G. 2004. Dynamics of coarse woody debris in North American forests: a literature review. NCASI Technical Bulletin No. 877. National Council for Air and Stream Improvement, Inc., Research Triangle Park, NC.

AUTHOR INDEX

SPECIES INDEX

SUBJECT INDEX

fire (*continued*)
190–92, 200, 207, 213, 215, 217, 219;
stand replacement, 212–13; suppression
of, 181, 219; timing of burns, 169; top-
killing, 179; wildfire, 189, 192, 196, 213,
219
fission-fusion societies, 44–45
fitness, 8, 83, 211, 245–46; individual, 166;
maximize, 17; physical, 131; due to roost
selection, 3, 33, 145
FOCUS, 244
foliage-roosting bats, 3, 33, 61–77, 83
foraging: activity, 4, 87, 90, 253; areas, 5, 39,
45, 47, 89, 100, 103, 105–6, 220; opportunistically, 5, 90, 160, 164; optimal, 6;
patterns, 5, 97; selectively, 5; specialists,
90; strategy, 84, 87–88
forest gaps, 86, 215, 217, 272, 308
forest management strategies, 177
forest planning, 293
forest products: companies, 294, 297, 311;
industry, 294–96, 298
forest-roosting bats, 8
forestry certification programs, 293, 299;
sustainable, 298, 302, 305
FragStats, 242
freeze branding, 278
frequency-modulated components, 87
functional morphology, 84

geographic distribution, 62, 69–70, 109,
155, 157
geographic information systems, 252
geographic variation, 1, 48, 211, 219, 275–
76
Geometroidea, 88
girdling, 179, 197–98
gleaning, 1, 86, 88, 93
global climatic change, 8
global wood supply, 296
green-tree retention, 145, 213, 218, 249,
304–5, 309
group selection harvest, 109, 183, 187–88,
193, 199, 201

habitat: diversity, 309, 312; fragmentation
of, 48–49, 107, 112, 224, 237, 239–40,
243, 249; patch, 107, 239, 241, 249, 252,
309; riparian, 74, 248, 250; selection, 214
habitat suitability index model, 251–52
hard-bodied insects, 84
hardwood forest, 6, 74, 108, 209, 218, 223
harp traps, 272
Hemiptera, 90–91
herbicides, 179, 190, 192, 198, 200, 220,
298, 308
hibernation, 2, 8, 17, 43, 47, 61, 63–64, 70,
83, 153–54, 156–57, 164, 213
hierarchical linear models, 244
high-frequency, broad-bandwidth calls, 199
high-graded forest, 188
high-resolution aerial photography, 242

home range, 39, 226; adaptive kernel
method, 101–2; bivariate ellipse
method, 100–101; convex polygon
method, 101–2; definition of, 100; estimates, 100; fixed kernel method, 101–2;
Fourier series estimates, 101; harmonic
means method, 101; of a colony, 38, 47,
102; overlap of, 102; size of, 49, 243
Homoptera, 90–91
human-made structures, 1, 8, 136, 140,
167, 209–10
Hymenoptera, 90
hypothesis testing, 2, 31, 33, 44–45, 61, 94

imperiled species, 297–99
improvement cutting, 191–92, 200
industrial forests, 178, 295–97, 310–12
industry lands, 209, 302, 309, 312
information-theoretic approach, 2
insect-breeding areas, 305
insect-eating bats, 6
insecticides, 220, 298
insectivorous bats, 84–85, 90–93, 153
insect pests, 6–7
intensively managed forests, 293, 307, 312
intensively managed landscapes, 293, 309–
10
intermediate clutter foragers, 200–201
intermediate treatments, 181, 183, 191, 201
inventory of populations, 265
island biogeography, 107
Isoptera, 90

LANDIS, 247, 251
landscape: composition, 249; configuration, 249; definition of, 237; ecology,
237–38, 241, 244; models, 238, 247
landscape-level: diversity, 299, 302–3;
management, 248; plans, 238, 296, 242;
studies, 247; variables, 247
large-diameter: snags, 208–9, 212, 223;
stumps, 31; trees, 30, 34, 49, 155, 208–9
large ownership landowners, 311
large-scale disturbance, 184
lasiurines, 62–66, 72, 76, 84, 155, 159, 162–
64, 196, 219
late-rotation forests, 310
leave trees, 215, 218, 307, 310–11
legacy: elements, 198; forest, 299; patches,
311; stands, 312; structures, 213, 222;
trees, 202, 305
Lepidoptera, 6, 88, 90–91, 221; macro-, 93;
micro-, 93
linear landscape elements, 4
litter roosting, 163–65, 168, 207, 215
low-frequency, narrow-bandwidth calls,
199

malaise traps, 93
maneuverability, 29, 66, 85, 87, 95, 216
mark-recapture techniques, 271, 277–78,
280

matrix habitat, 238–40, 250
maximum distance traveled, 102–3, 243
maximum likelihood estimation, 282
meta-analysis, 5
metric: outcome-oriented, 301; process-
oriented, 301
microclimate, 3, 39–40, 73, 156, 169, 214–
16; at roost, 18, 31–33, 48
migration, 40, 63–64, 83, 153–60, 164–65,
167–69, 213, 237; autumn, 167; corridors, 166, 168; diurnal, 157; routes, 156,
159–60, 166, 168; spring, 164–65, 167
mist nets, 95, 103, 243, 272, 285
mitochondrial DNA, 87
mobility: spatial, 237; temporal, 237
moderate-intensity disturbance, 185
molecular genetic analysis, 3, 7, 272
monitoring of populations, 265
morphological: attributes, 84, 86; differences, 86–87; variation, 5
mortality of bats, 167, 169, 207, 215
moth-eating bats, 88
moths, 92–93; Arctiid, 88; Noctuid, 88;
soft-bodied, 85; tympanate, 88

narrow space bats, 87
natal philopatry, 45, 65–66, 77, 166
NatureServe, 297, 299
nectar-feeding bats, 7
neighbor rule: 4-, 241; 8-, 241
Neuroptera, 90
night roost, 39, 90, 129–45, 221, 266; definition of, 129; frequency of night roosting, 141–42; information transfer, 130,
132–33; social cohesion, 133–34
Noctuoidea, 88
nongovernmental organizations, 295
nonindustrial private forests, 294
normothermic body temperature, 134
North American: bats, 9, 19, 26, 37, 39, 61,
64–65, 84, 91, 103, 105, 133, 142, 214,
264; forests, 2, 61, 85, 87–88, 90, 92, 94,
96, 111–12, 210, 224, 293, 296; *Myotis*,
84; research, 1
North American Bat Conservation Partnership, 301; strategic plan, 301

Odanata, 91
old growth: forest, 222–23, 247, 250; redwoods, 107; stands, 209, 218, 223
ontogenetic niche, 88
ontogeny, 84, 88
open space foragers, 88, 200
optimal foraging theory, 83
Orthoptera, 90

passive integrated transponders, 7, 278–
80
patch, 49, 109, 239–42, 244, 246, 250, 309,
312; cut, 217–19, 222, 250; definition of,
238; resource-based, 239–40; retention,
306; size, 106–7, 306; suitability of, 239

pine forests, 69, 74, 180–82, 197, 209, 217–18, 303–4

pitfall traps, 93

population: decline, 263; definition of, 266; metrics, 268, 270, 284–85; viability, 237, 252

predation: by bats, 83, 88, 91–92; on bats, 1, 17–19, 31, 34, 42, 45, 61, 72–73, 76, 131, 137, 145, 168

prey: airborne, 84; availability of, 83, 90, 165, 208, 216–17, 219–20, 227; density, 246; nonairborne, 84; selection, 5

pruning, 197, 200

pseudoreplication, 243

Pyraloidea, 88

quasi-constant frequency components, 87

rabies, 8

radiotelemetry, 3, 7, 19, 34, 36–37, 41, 43, 45, 64, 66, 70, 72, 74, 77, 96–98, 100, 103, 112, 129, 139, 141, 143, 166, 169, 263, 278; aerial, 97, 106; global positioning systems, 97; ground-based, 97, 106; satellite, 97

radiotransmitters, 6–7, 19, 66, 96–97, 101, 139, 243, 263, 280, 284

randomized-block design, 243

recurrent roost switching, 39, 41–45

regeneration treatments, 181, 183, 189, 193, 201; artificial, 189–90; natural, 189–90

real estate investment trusts, 294

release treatments, 190–91; cleaning, 191; liberation cutting, 191

relict trees, 196–99, 202

reproduction cutting harvest, 181, 183, 190, 193, 195, 201

residual: basal area, 195, 197; patches, 218; snags, 308; trees, 185, 192

resting: behavioral, 130; physiological, 130–31

riparian: areas, 76, 86, 109, 208, 215, 220–21, 250, 309; buffer strips, 4, 215; management zones, 249, 298

roosts: day, 129–30, 132–34, 142, 144; destruction of, 61; feeding, 132; foliage, 72; maternity, 129, 132, 246, 279; retention of, 211–12; selection of, 3, 17–49; switching of, 3–4, 8, 34, 39–45, 64–65, 132, 141, 210, 266

roosting area, 66

rotary traps, 93

rotation, 183, 194, 294, 298, 305, 307; age, 193; late, 310; sawtimber, 308; short, 209, 295, 308–9

roundwood production, 295–96

salvage harvest, 180, 212

seed-tree harvest, 180, 183–85, 188, 199

shade-intolerant species, 187, 197

shade-tolerant: midstory, 191; species, 186, 188, 197; trees, 187; tree seedlings, 185–86

shelterwood harvest, 183, 185, 188, 200; preparatory cut, 185,199; removal cut, 185, 199; seed cut, 185, 199

silvicultural, 177–202; constraints, 311; objectives, 179; operations, 298; prescriptions, 107–8, 177–81, 187, 196, 200, 250, 293, 299–300, 302; systems, 179, 193–94, 294, 308; treatments, 94, 180, 195, 303, 307

single-tree selection harvest, 183, 185–88, 193, 199, 201

site preparation, 190, 193, 198, 303

small-diameter snags, 213

snag creation, 145, 198–99, 202, 209, 308

snag retention, 145, 198–99, 202, 215, 302, 305, 309, 311; strategies, 249

social cohesion at maternity roosts, 19, 37

social forestry, 178

soft-bodied insects, 85, 89

solar: exposure, 75; heating, 73; radiation, 28–29, 32, 73, 140, 197, 214–15

spatial: configuration, 242, 247; distribution, 64, 237; resolution, 241

spatiotemporal variability, 265, 268–69, 281

species-area relationships, 240

species occurrence, 281–82

Sphingoidea, 88

stable-isotope analysis, 7

statistical power, 41

sticky traps, 93

stomach contents, 92

streamside management zones, 74, 195, 198, 201, 215, 298, 302, 309, 311

subterranean, 153–54, 156, 162

succession: changes in, 293; early seral, 177, 191; midseral, 191; late seral, 191, 247, 305–6

suction traps, 93

survey: acoustic, 268–70, 273–74, 278, 280; capture, 269–71, 278, 280; cost, 283; landscape, 246; population, 265, 281; protocol, 265, 282, 285; roost, 269–70, 277–78

survivorship, 61, 76, 83, 96, 111, 214, 237, 246, 264, 278–80

sustainable forest: ecosystems, 177; management, 305–6; programs, 297–98; resources, 178; yield, 180

Sustainable Forestry Initiative, 297, 302, 307

swarming, 133

temperate-zone bats, 83, 133, 245

tests of habitat use: chi-square goodness-of-fit, 103; classification-based, 103–4; discrete choice models, 103; distance-based, 103; logistic regression, 103; polytomous logistic regression, 103; rank-order, 103

thermal imaging, 7, 169, 271; infrared, 280

thermal protection, 64

thermoregulatory: benefits, 40; costs, 19, 25, 29, 32–33, 45, 157, 214; strategy, 33, 154–56

thinning harvest, 5, 74, 108–9, 191, 193–95, 197, 199–200, 209, 217–18, 223, 226, 250, 293–94, 309; crown, 192; late-rotation, 185, 193; precommercial, 180, 190; pulpwood, 180

timber investment management organizations, 294

timber management zones, 303–4

torpor, 18, 25, 34, 63–64, 134–35, 154–55, 157, 160, 162–63, 165, 213

tree bats, 153–69

tree-roosting bats, 8, 18–19, 24–25, 32, 36, 39, 45, 77

Trichoptera, 90–91

tympanic organ, 88

uneven-aged: management, 213, 217; methods, 183, 185, 188–89, 201; stands, 194–95, 201; systems, 183, 193–95, 250, 294

upland: forest, 76, 86, 215, 223, 309; habitat, 74, 220–21, 249, 267, 309

use distributions: 95%, 102; 100%, 102; core areas, 102

vespertilionid, 64

Washington Department of Fish and Wildlife's Priority Habitats, 304

water: availability, 208; distance to, 2, 25; quality, 221; sources of, 75–76, 106

whirligig traps, 93

wind turbines, 167, 169

wing: area, 85; length, 85; loading, 63, 66, 85–87, 216; morphology, 85–87, 89; shape, 85; span, 85; tip, 85

woodland edge foragers, 200–201

wood products, 293–94, 297

yellow bats, 63, 70–71, 77